SPACE SCIENCE, EXPLORATION AND POLICIES

NEUTRON STARS

THE ASPECT OF HIGH DENSITY MATTER, EQUATIONS OF STATE AND OBSERVABLES

SPACE SCIENCE, EXPLORATION AND POLICIES

Additional books in this series can be found on Nova's website under the Series tab.

Additional E-books in this series can be found on Nova's website under the E-book tab.

PHYSICS RESEARCH AND TECHNOLOGY

Additional books in this series can be found on Nova's website under the Series tab.

Additional E-books in this series can be found on Nova's website under the E-book tab.

SPACE SCIENCE, EXPLORATION AND POLICIES

NEUTRON STARS

THE ASPECT OF HIGH DENSITY MATTER, EQUATIONS OF STATE AND OBSERVABLES

HIROSHI UECHI
SCHUN T. UECHI
AND
BRIAN D. SEROT
EDITORS

nova publishers
New York

For permission to use material from this book please contact us:
Telephone 631-231-7269; Fax 631-231-8175
Web Site: http://www.novapublishers.com

Additional color graphics may be available in the e-book version of this book.

LIBRARY OF CONGRESS CATALOGING-IN-PUBLICATION DATA

Neutron stars : the aspect of high density matter, equations of state and observables / [edited by] Hiroshi Uechi, Brian D. Serot.
p. cm.
Includes index.
ISBN 978-1-62081-123-8 (hardcover)
1. Neutron stars. I. Uechi, Hiroshi, 1957- II. Serot, Brian David.
QB843.N4N497 2011
523.8'874--dc23

2012004471

Published by Nova Science Publishers, Inc. † New York

CONTENTS

PREFACE

Stars, pulsars, galaxies in our universe exhibit a variety of dynamical phenomena of particle and matter as macroscopic space-time manifestations, and similarly, atoms and molecules, nucleons and many-body system of nucleons, high density hadronic matter, display a diversity of microscopic space-time phenomena. Although academic investigations of both macroscopic and microscopic worlds have developed as distinct fields of physics, it is recognized that they are intimately interrelated in views of frontiers of field theoretical approach to nuclear and high energy physics. The modern views of theoretical physics suggest that our universe is a wonderful laboratory for examining models of nuclear and high energy physics.

The cosmological phenomena are composed of numerous microscopic particles and energies beyond one's daily-life imagination. We are, what is called, space-time travelers on the spaceship *earth*, who observe cosmological phenomena and wander intelligently about macroscopic world. The nuclear particles are invisible, but it is common knowledge that we can extract enormous energy from an atomic nucleus. The sun in our solar system generates energies by nuclear fusions, in the form of heat, light, radiations of elementary and charged particles. Hence, we are observing microscopic dynamics of particles through macroscopic space-time phenomena. In frontiers of physics, concepts of conservation laws, space-time symmetries, gauge symmetries and spontaneous symmetry breakings, thermodynamic relations and equations of state of matter are employed to understand and observe microscopic world. It is exciting to know that our imaginations can go infinite in all directions.

This book illustrates observed values and properties of neutron stars, field theoretical approaches of hadrons to high density matter, such as nuclear and hyperonic matter, signals of dynamical phenomena (meson condensations, nucleon-hyperon, hadron-quark phase transitions), hadronic and hadron-quark neutron stars, radiations from high density matter and supernova phenomena. However, it is not intended to report facts and results at front-line research fields, because numerous conference reports suffice the objectives. The purpose of editing the book is to review related areas of nuclear physics and astrophysics in terms of equations of state of high density matter so as to promote mutual understandings and their consistent progresses. It is focused on theoretical relations between equations of state (EOS) and observables. These topics of high density matter, which extend across many fields of physics, should be frequently reviewed for researchers and people in general.

The equation of state is a relation among energy density, pressure, particle density and temperature. The typical phenomenological equations of state are those for an ideal gas, polytropic equations in equilibrium and radiation pressures in radiation-dominated system. In field theory, the equation of state is derived from the energy-momentum tensor obtained by a corresponding lagrangian with constituent fields of fermions and bosons, and the underlying lagrangian or hamiltonian is supposed to have space-time and gauge symmetry properties connected to interactions of particles. The symmetry properties could be perceived in EOS as self-consistency of many-body interactions, defined by the theory of conserving approximations and density functional theory. Hence, the equation of state enables us, in principle, to understand macroscopic observables from fundamental interactions and symmetry properties of constituent particles.

The modern technologies for astronomical and high energy areas have accumulated observed data for neutron stars, pulsars, and supernova explosions as well as elementary and composite particles. Thus, this would render models of nuclear and high energy physics the more reliable and conclusive, the better precise data become available. The properties of neutron stars, such as masses and radii, and those of nuclear matter (incompressibility and symmetry energy) can be calculated when an equation of state is employed, and therefore, macroscopic phenomena related to compact stars, supernova and high density matter are important for investigating and understanding dynamics of microscopic world. This fact would enlighten one's mind and knowledge of nature and stimulate persisting search for microscopic and macroscopic space-time explorations.

An overview on the relativistic field theory for nuclear and high energy density matter is given in the Introduction, and observables related to neutron stars and high density matter, general relativity for spherically symmetric compact stars, fundamental properties of nuclear matter at saturation density, self-consistency and thermodynamic consistency of many-body theory, and Landau's theory of Fermi liquid are explained in chapter I. The introductory relativistic approach to nuclear physics, equation of state and the role of density functional theory (DFT) in nuclear physics, linear and nonlinear, nonchiral and chiral mean-field models (MFT) of hadrons are explained in chapter II. Some comments on nonrelativistic and relativistic nuclear models, field theories based on hadrons (quantum hadrodynamics, QHD), quarks and gluons (quantum chromodynamics, QCD) are in chapter III, which is necessary to understand limitations of nuclear models and identify true signatures of subhadronic dynamics.

Certain effective models of nuclear physics and hadronic matter, hadron-quark stars (phase transitions of hadrons and hadron-quark) are explained in Chapters IV $\sim$ VI. Noticeable dynamical phenomena inside high density matter such as meson condensations are discussed in Chapter VII. Phase transitions from hadron to quark matter, radiations from magnetars are discussed in chapters VIII and IX, and gravitational collapse of massive stars and supernova phenomena are discussed in chapter X. These topics are also interesting in terms of thermofield dynamics and nonequilibrium nuclear many-body theory.

Graduate students and researchers in other fields could make use of this book for their guidance to the nuclear many-body theory and nuclear astrophysics. Essential theoretical points, such as self-consistency, constraints at nuclear matter saturation, thermodynamic consistency and gauge symmetries, are emphasized; they are fundamental to examine effective models of nuclear physics in the following chapters.

We are very grateful to Professor Brian D. Serot of Indiana University and his collaborators, for their contributions to chapters II and III. *The relativistic nuclear many-body problem* by B. D. Serot and J. D. Walecka, in Advances in Nuclear Physics, Vol. 16 (1986), guided many researchers into the relativistic field approach to nuclear and astrophysics. In this vast intermediate fields of nuclear, high energy and astrophysics, it is helpful to have their expertise to overview relativistic field theories in this rapidly progressing field. We would also like to express our gratitude to Doctors T. Maruyama (Japan Atomic Energy Agency), T. Muto (Chiba Institute of Technology), T. Tatsumi (Kyoto university), S. Yamada (Waseda University), N. Yasutake (National Astronomical Observatory of Japan), for their contributions to the work.

The editors, H. Uechi and S. T. Uechi, are very grateful to Mieko T. Uechi for her help and patient support to publishing the book.

May physics be beautiful as ever like cherry blossoms. Osaka, Japan, in April, 2011.

H. Uechi and S. T. Uechi

In: Neutron Stars
Editors: Hiroshi Uechi et al, pp. 1-37
ISBN: 978-1-62081-123-8

Chapter 1

RELATIONS BETWEEN MACROSCOPIC AND MICROSCOPIC OBSERVABLES

Hiroshi Uechi[1,*] ***and Schun T. Uechi***[2,†]
[1]Department of Distributions and Communication Sciences,
Osaka Gakuin University 2-36-1 Kishibe-minami, Suita, Osaka 564-8511 Japan
[2]Research Center for Nuclear Physics, Osaka University,
Ibaraki, Osaka 567-0047 Japan.

Abstract

In this chapter, we will review and discuss fundamental and theoretical relations between macroscopic and microscopic dynamical quantities and observables. The equations of state, binding energies and their relations to observables as well as self-consistency, requirements of thermodynamic consistency and density functional theory, are discussed. These fundamental topics should be examined in terms of space-time and gauge symmetries, renormalizations in modern views of theoretical physics.

1. Neutron Stars and Nuclear Physics

Since numerous review articles, reports, and books on history and observables for neutron stars are available, this section is not intended to do a comprehensive review of neutron stars, but in terms of nuclear many-body theory, we would like to briefly explain physics relevant to neutron stars and nuclear physics [1–5].

A neutron star is a high density object that is anticipated from the gravitational collapse of a massive star during a supernova explosion. Such stars are mainly composed of neutrons and supported against further collapse because of Pauli exclusion principle exerted by nuclear particles. Hence, the balance between gravitational force and quantum mechanical force exerted by nucleons should be attributed to Fermi energy of nucleons at a given density. In the book, we focus on relations between many-body system of strongly interacting particles (nuclear matter) and high density matter, such as neutron, hyperon and quark matter.

*E-mail address: uechi@ogu.ac.jp
†E-mail address: suechi@rcnp.osaka-u.ac.jp

明月記

八日、乙未、霜凝、天晴、北山雪白、客星事、依不審問泰俊朝臣、返事如此、曉夕東西之條驚而有餘、客星一昨日夜前全現候了、出現以後去二日陰雲不見候、其外者天快晴、連日見候、而此兩日者無引運、天曉見艮方候、曉夕東西出現候之條以外候、
客星出現例、　皇極天皇元年秋七月、甲寅、客星八月、陽成院貞觀十九年正月廿五日、丁酉、戌時、客星在辟、見西方、宇多天皇寛平三年三月廿九日、己卯、亥時、客星在東成星東方、相去一寸所、醍醐天皇延長八年五月以後七月以前、客星入羽林中、一條院寛弘三年四月二日、癸酉、夜以降騎官中有大客星、如熒惑、光明動耀、連夜正見南方、或云、騎陣將軍星變本體增光歟、後冷泉院天喜二年四月中旬以後丑時、客星出觜參度、見東方、孛天關星、大如歳星、二條院永萬二年四月廿二日、乙丑、亥時、客星彗見大微宮事、高倉院治承五年六月廿五日、庚午、戌時、客星見北方、近王良星守傳舍星、午終許心寂房來、事外付減之由加詞、但惡血之充滿非飼蛭時極難治云々、常隨給侍之小婢、依母病危急行南京、老病之最中失手臂之便、

寛喜二年　十一月

Figure 1. This is a record on the astronomical phenomena found in the Japanese classical document: Meigetuki by Fujiwarano Teika, which describes the appearance of supernovas presently known as the Crab Nebura (SN1054) in the year 1054, and the Lupus (SN1006) in the year 1006. The similar record is also found in a Chinese classical document. Transcription: courtesy of Kokusho Kanko-kai ed., Tokyo, Japan, (1969).

The discovery of the neutron in 1932 motivated Landau to predict the existence of neutron stars. In seeking an explanation for the origin of supernova, neutron stars were proposed to be formed in a supernova explosion process. Supernovae are suddenly appearing dying stars, whose luminosity outshines an entire galaxy for days and weeks. W. Baade and Zwicky proposed that the release of the gravitational binding energy of neutron stars powers the supernova and if the central part of a massive star before its collapse contains about 3 $M_\odot$, a neutron star of 2 $M_\odot$ can be formed (the solar mass: $M_\odot \sim 1.989 \times 10^{30}$ kg). A Typical neutron star has a mass between 1.35 to about 2.0 $M_\odot$, with a radius $R \sim 12$ km.

Theoretical calculations of neutron stars were performed by Tolman, Oppenheimer and Volkoff, and in 1967, a radio pulsar was discovered. Radio pulsars are rapidly rotating with periods in the range $0.033\ \mathrm{s} \leq P \leq 4.0\ \mathrm{s}$. They are believed to be powered by rotational energy loss and are rapidly spinning down with period derivative of order $\dot{P} \sim 10^{-12} - 10^{-16}$. A millisecond pulsar has a period between $1.56\ \mathrm{ms} \leq P \leq 100\ \mathrm{ms}$. X-ray pulsars and bursters are believed to be accreting neutron stars from high ($M \gtrsim 10\ M_\odot$) and low mass ($M \lesssim 1.2\ M_\odot$) companions. Nonrotating and non-accreting neutron stars were also studied, whose temperature is expected to be $T \sim 60$ eV and a radius $R \leq 14$ km. A magnetar is a type of neutron star with an extremely powerful magnetic field, and the emission of copious amounts of high energy electromagnetic radiation is considered as X-rays and gamma rays. The radiation can only be observed when the beam of emission is pointing towards the Earth, which is called the lighthouse effect. The strong magnetic

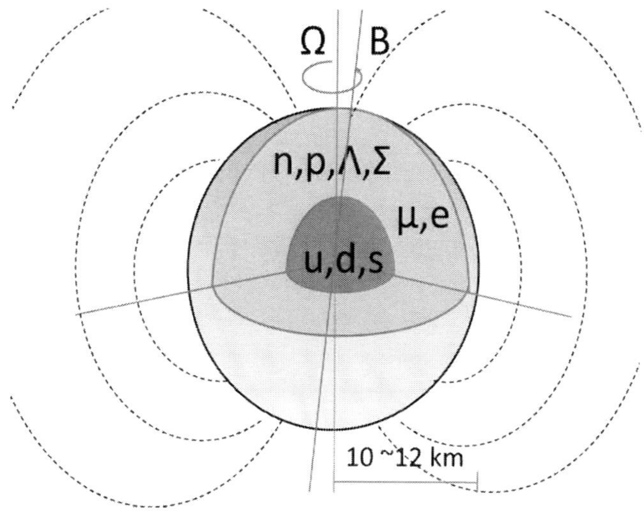

Figure 2. This is a schematic view of a magnetized neutron star with the dipole magnetic axis of the magnetic field $\boldsymbol{B}$ and the rotation axis Ω. The neutron star is assumed to have a hadron-mantle and a quark-core.

field may have significant effects on both the equation of state (EOS) and the structure of neutron stars. The magnitude of the magnetic field strength field, B, is estimated as,

$$B \sim 1.4 \times 10^{18} \Big(\frac{M}{1.4\mathrm{M}_\odot}\Big) \Big(\frac{R}{10\ \mathrm{km}}\Big)^{-2} \ \mathrm{Gauss}, \tag{1}$$

where M and R are mass and radius of a neutron star. The effect of strong magnetic field on EOS of magnetars is one of important problems to be investigated.

The polytropic equation of state is a useful approximation to consider many cosmological phenomena, because at a high enough temperature, any material will be expected to behave like an ideal gas. When the temperature is far enough above the chemical potential or single particle energy of an material, its macroscopic behavior is considered well described by a polytropic EOS, for example,

$$p = \rho(1+Z)k_B T/(Am_p)\ , \qquad C_s^2 = (\partial p/\partial \rho)_s = \gamma p/\rho\ , \tag{2}$$

and

$$\rho\varepsilon = p/(\gamma - 1)\ . \tag{3}$$

where ε is the single particle energy (internal energy per particle or internal energy per molecular weight), and k_B and T are the Boltzmann constant and temperature, respectively. Z is the average ionization number, γ is assumed to be a constant, and A is the molecular weight (the particle mass divided by 1 amu) and m_p is the proton mass (divided by 1 amu). The radiation pressure of blackbody and systems in which radiation is important or dominant is expressed as

$$p_R = \frac{4}{3}\frac{\sigma}{c}T^4\ , \tag{4}$$

where T is the temperature, c is the speed of light, and σ is the Stefan-Boltzmann constant familiar from blackbody radiation. Because this pressure depends on T to the fourth power, the radiation pressure will completely dominate at sufficiently high temperature, while material pressures generally depend on T to the first power [24]. It is important to check whether or not it is the case near and inside neutron stars, black holes, within matter during supernova explosions.

The radius R and mass M of neutron stars are important macroscopic observables. For example, the radiation radius

$$R_{\infty} = R/\sqrt{1 - 2GM/Rc^2}\ , \tag{5}$$

is implied from thermal observations of neutron star's surface. Redshift is defined as

$$z = \left(1 - \frac{2GM}{Rc^2}\right)^{-1/2} - 1\ . \tag{6}$$

Moments of inertia, I $\mathrm{M_{\odot}km^2}$, the Eddington limited flux from burst sources, and the total neutrino emission released in the gravitational collapse to form a neutron star,*etc*, are used to check theoretical calculations and consistency of EOSs [2].

The physics of neutron stars offers an interesting interplay between nuclear physics and astrophysical observables. The relevant densities of neutron stars are far from that of normal nuclear matter density, $\rho_0 \sim 0.148\ \mathrm{fm}^{-3}$. Hence, the determination of EOS for dense matter is essential for calculations of neutron star properties, and nonrelativistic and relativistic many-body approximations, various hadronic models and effective QCD models are employed to construct EOS of neutron stars. As shown briefly, the EOS determines many properties such as the mass, moments of inertia, mass-radius relation, redshifts, cooling rate, and the total neutrino emission in a supernova explosion. The relevant degrees of freedom may not be the same in the crust region where the density is much smaller than the saturation density of nuclear matter, and also, hadronic models may be questionable in the center of the star where density is so high. It is essential to study these questions quantitatively by comparing hadronic and QCD equations of state and exhibit discrepancies of predictions or model-independent predictions in order to understand nuclear and microscopic dynamics.

Within non-relativistic lowest-order Brueckner theory, it suggests that all the new phase-shift equivalent nucleon-nucleon potentials yield essentially similar equations of state up to densities of 3 - $4\rho_0$ for both pure neutron matter and β-stable matter. Other properties such as symmetry energy and proton fractions show a similar quantitative agreement. The inclusion of complicated many-body terms at the level of two-body interactions does not alter the conclusion [1], which implies that all the non-relativistic many-body approaches mentioned just above give similar equations of state around nuclear matter saturation densities ρ_0. However, the relativistic approaches, three-body interactions, onset of hyperons, pion and kaon condensations, superfluidity, and hadron-quark phase transitions are perceived to yield significant corrections to EOS above saturation density. In order to obtain consistent results from different approaches, it is essential to examine conditions and constraints to our models and calculations. Consistency and constraints to theoretical calculations will be discussed and examined in detail in the following chapters and sections.

2. The Relativistic Wave Equation

In this chapter the relativistic wave equation, Klein-Gordon equation (spin-0) and Dirac equation (spin-$\frac{1}{2}$), shall be introduced to confirm signs, notations and definitions. The natural unit, $\hbar = c = 1$, is used.

2.1. The Klein-Gordon Equation

The extension of the nonrelativistic Schrödinger equation to a relativistic equation was attempted by replacing the relativistic energy-momentum relation, $E^2 = \boldsymbol{p}^2 + m^2$, by means of the operator replacements: $E \to i\partial/\partial t$, $\boldsymbol{p} \to -i\boldsymbol{\nabla}$. In addition to the correct relativistic energy-momentum relation, covariance of relativistic wave equation in space-time is required. Energy and momentum are introduced as the four momentum vector, $p^\mu = (E, \boldsymbol{p})$ $(\mu = 0, 1, 2, 3)$, which satisfies the on-shell mass condition and Lorenz covariance defined as

$$p^2 = g_{\mu\nu} p^\mu p^\nu = p_\mu p^\mu = E^2 - \boldsymbol{p}^2 = m^2 \,. \tag{7}$$

The metric tensor , $g_{\mu\nu}$, in Minkowski space is given by

$$g_{\mu\nu} = \begin{pmatrix} 1 & 0 & 0 & 0 \\ 0 & -1 & 0 & 0 \\ 0 & 0 & -1 & 0 \\ 0 & 0 & 0 & -1 \end{pmatrix} . \tag{8}$$

Since the relativistic energy and momentum are related in the form of quadrature, $E^2 = \boldsymbol{p}^2 + m^2$, the square root operator defined from $E = \pm(\boldsymbol{p}^2 + m^2)^{1/2}$ has obvious difficulties in interpretations. Klein and Gordon attempted to build a relativistic quantum mechanics from the squared energy relation by using the operator replacements for E and $\boldsymbol{p}$, and they obtained a relativistic wave equation,

$$-\frac{\partial^2 \psi}{\partial t^2} = (-\boldsymbol{\nabla}^2 + m^2)\psi \,. \tag{9}$$

The Klein-Gordon (KG) equation is known as the relativistic equation for one-component scalar wave function $\phi(\boldsymbol{x}, t)$, expected to be appropriate for the description of spin-0 bosons. However, the KG equation has apparent difficulties despite the fact that the nonrelativistic formalism in the Schrödinger equation was well extended to relativistic one:

1. In constructing a wave equation in close correspondence with the squared energy-momentum relation
$$E^2 = \boldsymbol{p}^2 + m^2 \,, \tag{10}$$
it allows positive and negative-energy solutions, $E = \pm(\boldsymbol{p}^2 + m^2)^{1/2}$.

2. The KG equation has a $\partial^2/\partial t^2$ term: this leads to a continuity equation with a 'probability density' containing $\partial/\partial t$. This means that the probability density is not constrained to be positive, which is inconsistent with the positive-definiteness of a probability density.

Historically, Dirac solved the problem of negative probabilities by requiring that the relativistic equation should be linear in $\partial/\partial t$ and $\boldsymbol{\nabla}$. At the same time, he interpreted vacuum filled with negative energy particles. The absence of a negative-energy particle is equivalent to the presence of a positive-energy antiparticle. The prediction of a positron as an antiparticle for an electron led to the discovery of positron in 1932, which was one of the great achievements of theoretical physics.

2.2. The Dirac Field

Dirac postulated a relativistic and Lorenz covariant equation as follows:

$$\begin{aligned} i\frac{\partial}{\partial t}\psi(\boldsymbol{x},t) &= \left\{-i\left(\alpha_1\frac{\partial}{\partial x_1}+\alpha_2\frac{\partial}{\partial x_2}+\alpha_3\frac{\partial}{\partial x_3}\right)+\beta m\right\}\psi(\boldsymbol{x},t) \\ &= (-i\boldsymbol{\alpha}\cdot\boldsymbol{\nabla}+\beta m)\psi(\boldsymbol{x},t) \equiv \hat{H}\psi(\boldsymbol{x},t)\ , \end{aligned} \tag{11}$$

where α_1, α_2, α_3 and β must be determined so that the operator definition of (11) produces KG equation when the operators on both sides are squared. By squaring the operator on both sides, one obtains

$$\begin{aligned} &\left(i\frac{\partial}{\partial t}\right)^2\psi = (-i\boldsymbol{\alpha}\cdot\boldsymbol{\nabla}+\beta m)(-i\boldsymbol{\alpha}\cdot\boldsymbol{\nabla}+\beta m)\psi \\ &= -\sum_{i=1}^{3}\alpha_i^2\frac{\partial^2}{\partial x_i^2}\psi-\sum_{i,j=1}^{3}(\alpha_i\alpha_j+\alpha_j\alpha_i)\frac{\partial^2}{\partial x_i\partial x_j}\psi-im\sum_{i=1}^{3}(\alpha_i\beta+\beta\alpha_i)\frac{\partial}{\partial x_i}\psi+\beta^2m^2\psi\ . \end{aligned} \tag{12}$$

The equation (12) must be equal to the KG equation,

$$\left(i\frac{\partial}{\partial t}\right)^2\psi = -\sum_{i=1}^{3}\frac{\partial^2}{\partial x_i^2}\psi+m^2\psi\ . \tag{13}$$

The equation (12), and (13) must satisfy the following anticommutation relations in order to eliminate the unwanted terms on the right-hand side of the equation (12):

$$\alpha_i\beta+\beta\alpha_i=0, \qquad i=1,2,3\ . \tag{14}$$

$$\alpha_i\alpha_j+\alpha_j\alpha_i=0, \qquad i,j=1,2,3; \quad i\neq j\ . \tag{15}$$

and

$$\alpha_i^2=\beta^2=\mathbf{1}\ . \tag{16}$$

Thus, it is evident that the $\boldsymbol{\alpha}$ and β cannot be ordinary, classical commuting quantities. The Pauli matrices for nonrelativistic spin $\frac{1}{2}$ particles at that time were known to have the properties of the anticommutation relations (14) $\sim$ (16) [6, 7], and it is natural to interpret $\boldsymbol{\alpha}$ and β as matrices acting on a column vector field ψ.

The Dirac equation is expected to represent a particle with spin-$\frac{1}{2}$, and the smallest possible dimension of the matrices, α's and β, is shown to be 4×4. The conventional choice for the α and β is

$$\alpha_i=\begin{pmatrix}0 & \sigma_i\\ \sigma_i & 0\end{pmatrix}, \qquad \beta=\begin{pmatrix}\mathbf{1} & 0\\ 0 & -\mathbf{1}\end{pmatrix}, \tag{17}$$

where we have written these 4×4 matrices in 2×2 block diagonal form, and Pauli matrices σ_i are the usual 2×2 matrices:

$$\sigma_1 = \begin{pmatrix} 0 & 1 \\ 1 & 0 \end{pmatrix}, \qquad \sigma_2 = \begin{pmatrix} 0 & -i \\ i & 0 \end{pmatrix}, \qquad \sigma_3 = \begin{pmatrix} 1 & 0 \\ 0 & -1 \end{pmatrix}, \tag{18}$$

and $\mathbf{1}$, the 2×2 unit matrix. Since the Dirac Hamiltonian now involves 4×4 matrices, the wavefunction ψ must be interpreted as a four-component column vector.

By analogy with the nonrelativistic free-particle case, we may write a general spin-$\frac{1}{2}$ spinor in term of factorized solutions of Dirac equation in the form: $\psi = \omega e^{-ip_\mu x^\mu}$, where ω is a four-component spinor and $e^{-ip_\mu x^\mu}$ is the plane wave solution. We substitute this into the Dirac equation and obtain

$$i\frac{\partial}{\partial t}\psi = (-i\boldsymbol{\alpha} \cdot \boldsymbol{\nabla} + \beta m)\psi \,. \tag{19}$$

In order to use the 2×2 block form, it is conventional to split up the spinor ω into two component spinors ϕ and χ:

$$\omega = \begin{pmatrix} \phi \\ \chi \end{pmatrix}, \tag{20}$$

and we obtain the matrix equation

$$E\begin{pmatrix} \phi \\ \chi \end{pmatrix} = \begin{pmatrix} m\mathbf{1} & \boldsymbol{\sigma} \cdot \boldsymbol{p} \\ \boldsymbol{\sigma} \cdot \boldsymbol{p} & m\mathbf{1} \end{pmatrix}\begin{pmatrix} \phi \\ \chi \end{pmatrix}, \tag{21}$$

which represents two coupled equations for ϕ and χ:

$$(E - m)\phi = \boldsymbol{\sigma} \cdot \boldsymbol{p}\chi, \qquad (E + m)\chi = \boldsymbol{\sigma} \cdot \boldsymbol{p}\phi \,. \tag{22}$$

The general four-component spinor can be obtained from (22) as

$$\omega = \frac{1}{E + m}\begin{pmatrix} (E + m)\phi \\ \boldsymbol{\sigma} \cdot \boldsymbol{p}\phi \end{pmatrix}, \tag{23}$$

and because $(\boldsymbol{\sigma} \cdot \boldsymbol{p})^2 = \boldsymbol{p}^2\mathbf{1}$, it leads to $(E - m)(E + m)\phi = \boldsymbol{p}^2\phi$, for any ϕ. Hence, we arrive at the same result, $E^2 = \boldsymbol{p}^2 + m^2$, for the KG equation.

The conservation of probability current and Lorentz covariance of the Dirac wave equation are shown as follows. The probability density is given by $\rho = \psi^\dagger\psi$, and expressed as

$$\rho = (\psi_1^*, \psi_2^*, \psi_3^*, \psi_4^*)\begin{pmatrix} \psi_1 \\ \psi_2 \\ \psi_3 \\ \psi_4 \end{pmatrix} = \sum_{n=1}^{4} |\psi_n|^2 > 0 \,. \tag{24}$$

It shows that all components of ρ have independent probability densities which are explicitly positive-definite. By calculating $\psi^\dagger \times \{i\frac{\partial}{\partial t}\psi - (-i\boldsymbol{\alpha} \cdot \boldsymbol{\nabla} + \beta m)\psi\} - \{i\frac{\partial}{\partial t}\psi - (-i\boldsymbol{\alpha} \cdot \boldsymbol{\nabla} + \beta m)\}^\dagger \times \psi$ and noting the hermiticity of Pauli matrices, $\sigma_i^\dagger = \sigma_i$, the conservation law of the probability density is written as

$$\frac{\partial \rho}{\partial t} + \boldsymbol{\nabla} \cdot \boldsymbol{j} = 0 \,, \tag{25}$$

where $\boldsymbol{j}(x) = \psi^\dagger(x)\boldsymbol{j}\psi(x) = (\psi^\dagger\alpha_1\psi, \psi^\dagger\alpha_2\psi, \psi^\dagger\alpha_3\psi)$.

Next, the Lorentz covariance of Dirac equation is explained by introducing γ-matrix notations. The γ-matrices are defined by,

$$\gamma^0 = \beta, \qquad \boldsymbol{\gamma} = \beta\boldsymbol{\alpha} = (\gamma^1, \gamma^2, \gamma^3) \,, \tag{26}$$

and $\gamma^\mu(\mu = 0, 1, 2, 3)$ matrices satisfy the anticommutation relation,

$$\{\gamma^\mu, \gamma^\nu\} \equiv \gamma^\mu\gamma^\nu + \gamma^\nu\gamma^\mu = 2g^{\mu\nu} \,. \tag{27}$$

The Dirac equation is written in a simple form:

$$(i\gamma_\mu\partial^\mu - m)\psi(x) = 0 \,. \tag{28}$$

Denoting an arbitrary space-time Lorentz transformation by some 4×4 matrix Λ,

$$x^\mu \to x'^\mu = \Lambda^\mu{}_\nu x^\nu \,, \tag{29}$$

the transformation of the Lorentz scalar field $\phi(x)$ is written as

$$\phi(x) \to \phi'(x') = \phi(x) = \phi(\Lambda^{-1}x') \,. \tag{30}$$

This is the requirement of Lorentz scalar quantity, which means that the transformed field evaluated at the boosted point gives the same value as the original field evaluated at the point before boosting. The transformation of $\partial_\mu\phi(x)$ is expressed as

$$(\partial_\mu\phi(x)) \to \partial_\mu\left(\phi(\Lambda^{-1}x)\right) = (\Lambda^{-1})^\nu{}_\mu(\partial_\nu\phi)(\Lambda^{-1}x) \,. \tag{31}$$

By employing the 4×4 representation of the boost and rotation generators of Lorentz algebra defined by

$$S^{\mu\nu} = \frac{i}{4}[\gamma^\mu, \gamma^\nu] \,, \tag{32}$$

the infinitesimal form of Lorentz transformation Λ, is defined as

$$\Lambda_{\frac{1}{2}}^{-1}\gamma^\mu\Lambda_{\frac{1}{2}} = \Lambda^\mu{}_\nu\gamma^\mu \,. \tag{33}$$

The $\Lambda_{\frac{1}{2}}$ is the infinitesimal spinor representation of the Lorenz transformation, which is given by

$$\Lambda_{\frac{1}{2}} \equiv \exp\left(-\frac{i}{2}\omega_{\mu\nu}S^{\mu\nu}\right) \,, \tag{34}$$

where $\omega_{\mu\nu}$ is an antisymmetric tensor which gives the infinitesimal angles.

It is now ready to show that the Dirac equation is Lorentz invariant. Let us write down the Lorentz-transformed version of the left-hand side and calculate:

$$\begin{aligned}
(i\gamma_\mu\partial^\mu - m)\psi(x) &\to \left(i\gamma^\mu(\Lambda^{-1})^\mu{}_\nu\partial_\nu - m\right)\Lambda_{\frac{1}{2}}\psi(\Lambda^{-1}x') \\
&= \Lambda_{\frac{1}{2}}\Lambda_{\frac{1}{2}}^{-1}\left(i\gamma^\mu(\Lambda^{-1})^\mu{}_\nu\partial_\nu - m\right)\Lambda_{\frac{1}{2}}\psi(\Lambda^{-1}x') \\
&= \Lambda_{\frac{1}{2}}\left(i\Lambda_{\frac{1}{2}}^{-1}\gamma^\mu\Lambda_{\frac{1}{2}}(\Lambda^{-1})^\mu{}_\nu\partial_\nu - m\right)\Lambda_{\frac{1}{2}}\psi(\Lambda^{-1}x') \\
&= \Lambda_{\frac{1}{2}}\left(i\Lambda^\mu{}_\sigma\gamma^\sigma(\Lambda^{-1})^\nu{}_\mu\partial_\nu - m\right)\psi(\Lambda^{-1}x') \\
&= \Lambda_{\frac{1}{2}}(i\gamma^\mu\partial_\mu - m)\psi(\Lambda^{-1}x') = 0 \,.
\end{aligned} \tag{35}$$

The equation (28) and above results show the Lorentz-invariance of Dirac equation [8]. Although the Dirac approach demanded the vacuum filled with negative-energy particles, it displayed remarkable properties of physics. Later, Feynman introduced a consistent particle-antiparticle interpretation to the positive and negative energies. Feynman's interpretation has the advantage that the correct Feynman rules can be obtained for fermions and bosons by means of a single interpretation of the negative-energy solutions, which works for both the Dirac and Klein-Gordon equations. There is no need to introduce the idea of the vacuum as filled negative-energy states. Feynman's prescription is stated as follows:

negative-energy particle solutions propagating backward in time
$\equiv$ *positive-energy antiparticle solutions propagating forward in time .*

The relativistic field theory was thus rigorously established and evolved as the fundamental approach to modern physics.

3. General Relativity

In this section, we will introduce basic relations for space-time geometry to produce Bianchi equation, Einstein equation and Tolman-Oppenheimer-Volkoff (TOV) equation, for the calculation of compact stars, such as white dwarf stars, neutron stars and black holes.

3.1. Curved Space and Parallel Displacement

The space-time points are denoted as $t = x^0$, $x = x^1$,$y = x^2$, $z = x^3$. The metric tensor, $g_{\mu\nu}$, in Minkowski space is given by (8). We consider a curved space, called Riemann space, in which a small region of it is approximately flat. The invariant distance ds between a point x^μ and a neighboring point is given by

$$ds^2 = g_{\mu\nu} dx^\mu dx^\nu \ , \tag{36}$$

where ds is real for a timelike interval $(g_{\mu\nu} dx^\mu dx^\nu > 0)$ and imaginary for a spacelike $(g_{\mu\nu} dx^\mu dx^\nu < 0)$ interval. The metric $g_{\mu\nu}$ determines both the coordinate system and the curvature of the space.

Let us consider equations for the parallel displacement of a vector supposing our four-dimensional physical space to be immersed in a flat space of a higher number of N-dimensions, and discuss rectilinear coordinates $z^n (N = 1, 2, \cdots, N)$ in the N-dimensional space, whose distance ds is given by

$$ds^2 = h_{nm} dz^n dz^m \ , \tag{37}$$

where n and m are summed for $n, m = 1, 2, \cdots, N$. The h_{nm} may be constants different from $g_{\mu\nu}$. We may use them to lower suffixes in the N-dimensional space such that,

$$dz_n = h_{nm} dz^m \ . \tag{38}$$

A four dimensional physical space is assumed in a small, flat space in the N-dimensional space. Each point x^μ in the physical surface determines a definite point $y^n(x^\mu)$

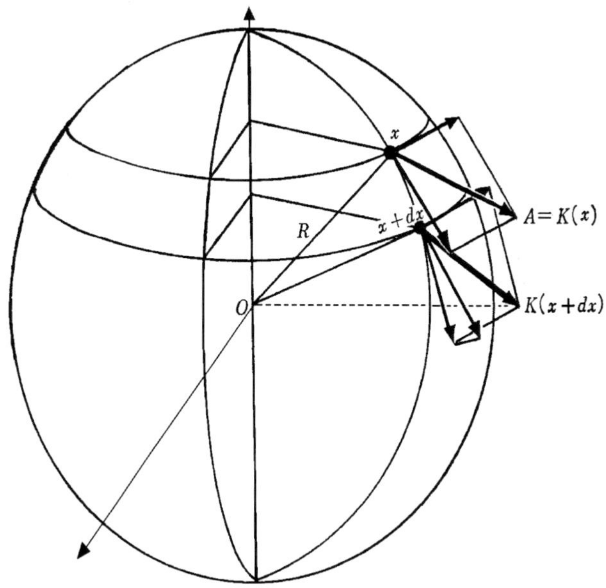

Figure 3. The schematic view of the parallel displacement of a vector on the sphere.

in the N-dimensional space. By differentiating the $y(x)$ with respect to the parameters x^μ, we get

$$\frac{\partial y^n(x)}{\partial x^\mu} = y^n_{\ ,\mu} \ . \tag{39}$$

For the distance of two neighboring points on the surface separated by δx^μ, we have

$$\delta y^n = y^n_{\ ,\nu} \delta x^\nu \ . \tag{40}$$

The squared distance ds^2 is, from (39) and (40),

$$\delta s^2 = h_{nm} \delta y^n \delta y^m = h_{nm} y^n_{\ ,\mu} y^m_{\ ,\nu} \delta x^\mu \delta x^\nu = \delta y^n \delta y^m = h_{nm} y^n_{\ ,\mu} y^m_{\ ,\nu} \delta x^\mu \delta x^\nu \ . \tag{41}$$

We also identify $\delta s^2 = g_{\mu\nu} \delta x^\mu \delta x^\nu$, and hence, the metric $g_{\mu\nu}$ is given by

$$g_{\mu\nu} = y^n_{\ ,\mu} y_{n,\nu} \ . \tag{42}$$

Since a parallel displacement of the vector A^μ, keeping it parallel to itself to a neighboring point $x + dx$ on the surface will no longer lie on the surface at the new point due to the curvature of the surface, we project it on to the surface in order to get a definite vector lying on the surface. The projection process consists in splitting the vector into two parts, a tangential part and a normal part,

$$A^n = A^n_{\rm tan} + A^n_{\rm nor} \ . \tag{43}$$

The normal part $A^n_{\rm nor}$ will be neglected because of the flatness in the small displacement of the geometry, and if K^μ denotes the components of A^n referred to the x coordinate system on the surface, we have

$$A^n_{\text{tan}} = K^\mu y^n_{,\mu}(x + dx) , \tag{44}$$

with the coefficients $y^n_{,\mu}$ taken at the new point $x + dx$.

By multiplying (43) with $y_{n,\nu}(x + dx)$ and dropping A^n_{nor} term, we obtain

$$\begin{aligned} A^n y_{n,\nu}(x + dx) &= K^\mu y^n_{,\mu}(x + dx) y_{n,\nu}(x + dx) \\ &= K^\mu g_{\mu\nu}(x + dx) , \end{aligned} \tag{45}$$

from (42). To the first order in dx, $K_\nu(x + dx)$ is written as

$$\begin{aligned} K_\nu(x + dx) &= A^n[y_{n,\nu}(x) + y_{n,\nu,\sigma} dx^\sigma] \\ &= A^\mu y^n_{,\mu}[y_{n,\nu} + y_{n,\mu,\sigma} dx^\sigma] \\ &= A_\nu + A^\mu y^n_{,\mu} y_{n,\nu,\sigma} dx^\sigma . \end{aligned} \tag{46}$$

This is the result of the parallel displacement of A_ν to the point $x + dx$, and we may define

$$K_\nu - A_\nu = dA_\nu , \tag{47}$$

so that dA_ν denotes the change in A_ν under parallel displacement. Therefore, the infinitesimal displacement of a vector A_ν is given by

$$dA_\nu = A^\mu y^n_{,\mu} y_{n,\nu,\sigma} dx^\sigma . \tag{48}$$

3.2. Christoffel Symbols

By differentiating (42), one obtains

$$g_{\mu\nu,\sigma} = y^n_{,\mu\sigma} y_{n,\nu} + y^n_{,\mu} y_{n,\nu\sigma} = y_{n,\mu\sigma} y^n_{,\nu} + y_{n,\mu} y^n_{,\nu\sigma} , \tag{49}$$

where the suffix n is moved freely up and down on account of the constancy of the h_{mn}. Interchange of μ and σ in (49) leads to

$$g_{\sigma\nu,\mu} = y_{n,\sigma\mu} y^n_{,\nu} + y_{n,\nu\mu} y^n_{,\sigma} , \tag{50}$$

and again, interchanging ν and σ in (49), one obtains

$$g_{\mu\sigma,\nu} = y_{n,\mu\nu} y^n_{,\sigma} + y_{n,\sigma\nu} y^n_{,\mu} . \tag{51}$$

Performing the calculation (49)+(51)-(50) and dividing by 2, the following result

$$\frac{1}{2}(g_{\mu\nu,\sigma} - g_{\mu\sigma,\nu} - g_{\nu\sigma,\mu}) = y_{n,\nu\sigma} y^n_{,\mu} , \tag{52}$$

is obtained. The LHS is denoted as

$$\Gamma_{\mu\nu\sigma} = \frac{1}{2}(g_{\mu\nu,\sigma} - g_{\mu\sigma,\nu} - g_{\nu\sigma,\mu}) , \tag{53}$$

which is called a Christoffel symbol of the first kind. It is symmetrical for the interchange of the last two suffixes, and it is a nontensor.

A simple consequence of (53) is

$$\Gamma_{\mu\nu\sigma} + \Gamma_{\nu\mu\sigma} = g_{\mu\nu,\sigma} \ , \tag{54}$$

and we see that (48) can be written as

$$dA_\nu = A^\mu \Gamma_{\mu\nu\sigma} dx^\sigma \ . \tag{55}$$

It is useful to define raising and lowering the first suffix of the Christoffel symbol so as to form

$$\Gamma^\mu_{\nu\sigma} = g^{\mu\lambda}\Gamma_{\lambda\nu\sigma} \ , \tag{56}$$

which is called a Christoffel symbol of the second kind. It is symmetrical for the interchange of its two lower suffixes. The formula (55) may be rewritten as

$$dA_\nu = \Gamma^\mu_{\nu\sigma} A_\mu dx^\sigma \ , \tag{57}$$

which is the formula for contravariant vector components. In order to obtain a formula referring to covariant vector components, we use the fact that the constancy of the length of any vector is unchanged by parallel displacement. The constancy of the scalar product $A_\mu B^\mu$ of any vectors A and B means that $d(A_\nu B^\nu) = 0$, which is used to calculate

$$A_\nu dB^\nu = -B^\nu dA_\nu = -B^\nu \Gamma^\mu_{\nu\sigma} A_\mu dx^\sigma = -B^\mu \Gamma^\nu_{\mu\sigma} A_\nu dx^\sigma \ . \tag{58}$$

Because this must hold for any A_ν, we get

$$dB^\nu = -\Gamma^\nu_{\mu\sigma} B^\mu dx^\sigma \ . \tag{59}$$

This is the standard formula for parallel displacement referring to covariant vector components.

3.3. Covariant Differentiation

Since the direct differentiation of a vector transformation law, $A_{\mu'} = A_\rho x^\rho_{\ \mu'}$, shows that

$$A_{\mu',\nu'} = (A_\rho x^\rho_{\ ,\mu'})_{,\nu'} = A_{\rho,\sigma} x^\sigma_{\ ,\nu'} x^\rho_{\ ,\mu'} + A_\rho x^\rho_{\ ,\nu'\mu'} \ , \tag{60}$$

which is a nontensor, we modify the process of differentiation so as to get a covariant tensor. Let us take the vector A_μ at the point x and shift it to $x + dx$ by parallel displacement. It is, to the first order,

$$A_\mu(x + dx) - [A_\mu(x) + \Gamma^\alpha_{\mu\nu} A_\alpha dx^\nu] = (A_{\mu,\nu} - \Gamma^\alpha_{\mu\nu} A_\alpha) dx^\nu \ . \tag{61}$$

This quantity is a vector for any vector dx^ν, and hence, the coefficient, $A_{\mu,\nu} - \Gamma^\alpha_{\mu\nu} A_\alpha$, is a tensor. It is denoted as the covariant derivative of A_μ,

$$A_{\mu:\nu} = A_{\mu,\nu} - \Gamma^\alpha_{\mu\nu} A_\alpha \ . \tag{62}$$

The column-symbol at the lower suffix will always denote a covariant derivative, and the comma denotes an ordinary derivative.

The covariant derivative of a tensor is defined from the outer product $A_\mu B_\nu$,

$$(A_\mu B_\nu)_{:\sigma} = A_{\mu:\sigma} B_\nu + A_\mu B_{\nu:\sigma} \,, \tag{63}$$

which is written as,

$$\begin{aligned}(A_\mu B_\nu)_{:\sigma} &= (A_{\mu,\sigma} - \Gamma^\alpha_{\mu\sigma} A_\alpha) B_\nu + A_\mu (B_{\nu,\sigma} - \Gamma^\alpha_{\nu\sigma} B_\alpha) \\ &= (A_\mu B_\nu)_{,\sigma} - \Gamma^\alpha_{\mu\sigma} A_\alpha B_\nu - \Gamma^\alpha_{\nu\sigma} A_\mu B_\alpha \,.\end{aligned} \tag{64}$$

It is used to define the covariant derivative of a tensor quantity,

$$T_{\mu\nu:\sigma} = T_{\mu\nu,\sigma} - \Gamma^\alpha_{\mu\sigma} T_{\alpha\nu} - \Gamma^\alpha_{\nu\sigma} T_{\mu\alpha} \,. \tag{65}$$

The rule can be extended to the covariant derivative of a tensor $Y_{\nu\mu\cdots}$ with any number of subscripts

$$Y_{\mu\nu\cdots:\sigma} = Y_{\mu\nu\cdots,\sigma} - (\text{a } \Gamma \text{ term for each suffix}) \,. \tag{66}$$

The case of a scalar is included in the general formula (66) as

$$Y_{:\sigma} = Y_{,\sigma} \,. \tag{67}$$

When the covariant derivative is applied on $g_{\mu\nu}$, it gives

$$\begin{aligned}g_{\mu\nu:\sigma} &= g_{\mu\nu,\sigma} - \Gamma^\alpha_{\mu\sigma} g_{\alpha\nu} - \Gamma^\alpha_{\nu\sigma} g_{\mu\alpha} \\ &= g_{\mu\nu,\sigma} - \Gamma_{\nu\mu\sigma} - \Gamma_{\mu\nu\sigma} = 0 \,,\end{aligned} \tag{68}$$

which shows that $g_{\mu\nu}$ is constant under the covariant differentiation.

The covariant derivative is assumed to hold also for the derivative of scalar product of two vectors, such that

$$(A^\mu B_\mu)_{:\sigma} = A^\mu{}_{:\sigma} B_\mu + A^\mu B_{\mu:\sigma} \,. \tag{69}$$

According to (62) and (67), one obtains

$$(A^\mu B_\mu)_{,\sigma} = A^\mu{}_{:\sigma} B_\mu + A^\mu (B_{\mu,\sigma} - \Gamma^\alpha_{\mu\sigma} B_\alpha) \,, \tag{70}$$

and

$$A^\mu{}_{:\sigma} B_\mu = A^\mu{}_{:\sigma} B_\mu - A^\alpha \Gamma^\mu_{\alpha\sigma} B_\mu \,, \tag{71}$$

which holds for any B_μ, resulting in

$$A^\mu{}_{:\sigma} = A^\mu{}_{,\sigma} + \Gamma^\mu_{\alpha\sigma} A^\alpha \,. \tag{72}$$

This is the basic formula for the covariant derivative of a contravariant vector.

3.4. The Curvature Tensor

Let us consider two successive covariant differentiations on a scalar field S and a vector field A_ν. From the formula (62), we have

$$S_{:\mu:\nu} = S_{:\mu,\nu} - \Gamma^\alpha_{\mu\nu} S_{:\alpha} = S_{,\mu\nu} - \Gamma^\alpha_{\mu\nu} S_{,\alpha} \,. \tag{73}$$

This is symmetrical for the exchange of μ and ν. From the formula (62) and (65) for $A_{\nu:\rho}$ and $T_{\nu\rho}$, the two-covariant differentiation on A_ν produces

$$\begin{aligned} A_{\nu:\rho:\sigma} &= A_{\nu:\rho,\sigma} - \Gamma^\alpha_{\nu\sigma} A_{\alpha:\rho} - \Gamma^\alpha_{\rho\sigma} A_{\nu:\alpha} \\ &= (A_{\nu,\rho} - \Gamma^\alpha_{\nu\rho} A_\alpha)_{,\sigma} - \Gamma^\alpha_{\nu\sigma}(A_{\alpha,\rho} - \Gamma^\beta_{\alpha\rho} A_\beta) - \Gamma^\alpha_{\rho\sigma}(A_{\nu,\alpha} - \Gamma^\beta_{\nu\alpha} A_\beta) \\ &= A_{\nu,\rho,\sigma} - \Gamma^\alpha_{\nu\rho} A_{\alpha,\sigma} - \Gamma^\alpha_{\nu\sigma} A_{\alpha,\rho} - \Gamma^\alpha_{\rho\sigma} A_{\nu,\alpha} - A_\beta(\Gamma^\beta_{\nu\rho,\sigma} - \Gamma^\alpha_{\nu\sigma}\Gamma^\beta_{\alpha\rho} - \Gamma^\alpha_{\rho\sigma}\Gamma^\beta_{\nu\alpha}) \,. \end{aligned} \tag{74}$$

The interchange of ρ and σ of (74) and subtraction gives,

$$A_{\nu:\rho:\sigma} - A_{\nu:\sigma:\rho} = A_\beta R^\beta_{\nu\rho\sigma} \,. \tag{75}$$

The tensor $R^\beta_{\nu\rho\sigma}$ is defined as

$$R^\beta_{\nu\rho\sigma} = \Gamma^\beta_{\nu\sigma,\rho} - \Gamma^\beta_{\nu\rho,\sigma} + \Gamma^\alpha_{\nu\sigma}\Gamma^\beta_{\alpha\rho} - \Gamma^\alpha_{\nu\rho}\Gamma^\beta_{\alpha\sigma} \,, \tag{76}$$

which is called the Riemann-Christoffel tensor or the curvature tensor. The Riemann-Christoffel tensor has symmetry properties from (76), such as

$$R^\beta_{\nu\rho\sigma} = -R^\beta_{\nu\sigma\rho} \,, \tag{77}$$

and

$$R^\beta_{\nu\rho\sigma} + R^\beta_{\rho\sigma\nu} + R^\beta_{\sigma\nu\rho} = 0 \,. \tag{78}$$

From (56), the superscript μ in $R^\mu_{\nu\rho\sigma}$ can be changed to a subscript and we obtain,

$$R_{\mu\nu\rho\sigma} = g_{\mu\beta} R^\beta_{\nu\rho\sigma} = g_{\mu\beta}\Gamma^\beta_{\nu\sigma,\beta} + \Gamma^\alpha_{\nu\sigma}\Gamma_{\mu\alpha\rho} - <\rho\sigma> \,, \tag{79}$$

where the symbol $<\rho\sigma>$ is used to denote the preceding terms with ρ and σ interchanged. Thus we have

$$\begin{aligned} R_{\mu\nu\sigma\rho} &= \Gamma_{\mu\nu\sigma,\rho} - g_{\mu\beta,\rho}\Gamma^\beta_{\nu\sigma} + \Gamma_{\mu\beta\rho}\Gamma^\beta_{\nu\sigma} - <\rho\sigma> \\ &= \Gamma_{\mu\nu\sigma,\rho} - \Gamma_{\beta\mu\rho}\Gamma^\beta_{\nu\sigma} - <\rho\sigma> \\ &= \frac{1}{2}(g_{\mu\sigma,\nu\rho} - g_{\nu\sigma,\mu\rho} - g_{\mu\rho,\nu\sigma} + g_{\nu\rho,\mu\sigma}) + \Gamma_{\beta\mu\sigma}\Gamma^\beta_{\nu\rho} - \Gamma_{\beta\mu\rho}\Gamma^\beta_{\nu\sigma} \,, \end{aligned} \tag{80}$$

and additional symmetry properties

$$R_{\mu\nu\rho\sigma} = -R_{\nu\mu\rho\sigma} \,, \tag{81}$$

and

$$R_{\mu\nu\rho\sigma} = R_{\rho\sigma\mu\nu} = R_{\sigma\rho\nu\mu} \,. \tag{82}$$

All these symmetries indicate that only 20 components are independent in the 256 components of $R_{\mu\nu\rho\sigma}$.

3.5. The Bianchi Equation

The second covariant derivative of a tensor produces the fundamental equation known as Bianchi equation for space-time geometry. The covariant derivative of the outer product of two vectors $A_\mu B_\tau$ gives,

$$\begin{aligned}(A_\mu B_\tau)_{:\rho:\sigma} &= (A_{\mu:\rho}B_\tau + A_\mu B_{\tau:\rho})_{:\sigma} \\ &= A_{\mu:\rho:\sigma}B_\tau + A_{\mu:\rho}B_{\tau:\sigma} + A_{\mu:\sigma}B_{\tau:\rho} + A_\mu B_{\tau:\rho:\sigma} \,.\end{aligned} \tag{83}$$

The interchange of ρ and σ of (83) and subtraction, gives from (74)

$$(A_\mu B_\tau)_{:\rho:\sigma} - (A_\mu B_\tau)_{:\sigma:\rho} = A_\alpha R^\alpha_{\mu\rho\sigma}B_\tau + A_\mu R^\alpha_{\tau\rho\sigma}B_\alpha \,. \tag{84}$$

A general tensor $T_{\mu\tau}$ is expressible as a sum of terms like $A_\mu B_\tau$, so it must satisfy

$$T_{\mu\tau:\rho:\sigma} - T_{\mu\tau:\sigma:\rho} = T_{\alpha\tau}R^\alpha_{\mu\rho\sigma} + T_{\mu\alpha}R^\alpha_{\tau\rho\sigma} \,. \tag{85}$$

Taking $T_{\mu\tau}$ as the covariant derivative of a vector $A_{\mu:\tau}$, one obtains

$$A_{\mu:\tau:\rho:\sigma} - A_{\mu:\tau:\sigma:\rho} = A_{\alpha:\tau}R^\alpha_{\mu\rho\sigma} + A_{\mu:\alpha}R^\alpha_{\tau\rho\sigma} \,. \tag{86}$$

Making cyclic permutations of τ, ρ, σ and adding the three generated equations, one obtains

$$\begin{aligned}A_{\mu:\rho:\sigma:\rho} - A_{\mu:\sigma:\rho:\tau} + (\text{cyc perm}) &= (A_\alpha R^\alpha_{\mu\rho\sigma})_{:\tau} + (\text{cyc perm}) \\ &= A_{\alpha:t}R^\alpha_{\mu\rho\sigma} + A_\alpha R^\alpha_{\mu\rho\sigma:\tau} + (\text{cyc perm}) \,.\end{aligned} \tag{87}$$

The terms $A_{\alpha:\tau}R^\alpha_{\mu\rho\sigma} + (\text{cyc perm})$ in the RHS are canceled by (78), and we are left with

$$A_{\alpha:\tau}R^\alpha_{\mu\rho\sigma:\tau} + (\text{cyc perm}) = 0 \,. \tag{88}$$

The factor $A_{\alpha:\tau}$ occurs throughout this equation and can be factored out. Hence, we have

$$R^\alpha_{\mu\rho\sigma:\tau} + R^\alpha_{\mu\sigma\tau:\rho} + R^\alpha_{\mu\tau\rho:\sigma} = 0 \,. \tag{89}$$

This is the covariant differential equation which the curvature tensor must satisfy as well as all the symmetry relations in Section 3.1. This is known as the Bianchi equation to determine the fundamental equation for a space-time geometry.

3.6. Ricci Tensor

Let us contract two subscripts of $R_{\mu\nu\rho\sigma}$ as

$$R^\mu_{\nu\rho\mu} = R_{\nu\rho} \,, \tag{90}$$

which is called Ricci tensor. By multiplying (82) by $g^{\mu\sigma}$ we obtain

$$R_{\nu\rho} = R_{\rho\nu} \,, \tag{91}$$

which shows that the Ricci tensor is symmetrical. The Ricci tensor can be contracted as

$$g^{\nu\rho}R_{\nu\rho} = R^\nu_\nu = R \,. \tag{92}$$

This R is a scalar and called the scalar curvature or total curvature, and it is defined in such a way that it is positive on the surface of a sphere in three dimensions.

Let us contract two of the five suffixes of the Bianchi equation (89) setting $\tau = \alpha$ and multiplying by $g^{\mu\rho}$. The result is

$$g^{\mu\rho}(R^{\alpha}_{\mu\rho\sigma:\alpha} + R^{\alpha}_{\mu\sigma\alpha:\rho} + R^{\alpha}_{\mu\alpha\rho:\sigma}) = 0 \,, \tag{93}$$

and

$$(g^{\mu\rho}R^{\alpha}_{\mu\rho\sigma})_{:\alpha} + (g^{\mu\rho}R^{\alpha}_{\mu\sigma\alpha})_{:\rho} + (g^{\mu\rho}R^{\alpha}_{\mu\alpha\rho})_{:\sigma} = 0 \,. \tag{94}$$

Because

$$g^{\mu\rho}R^{\alpha}_{\mu\rho\sigma} = g^{\mu\rho}g^{\alpha\beta}R_{\beta\mu\rho\sigma} = g^{\mu\rho\rho}g^{\alpha\beta}R_{\mu\beta\sigma\rho} = g^{\alpha\beta}R_{\beta\sigma} = R^{\alpha}_{\sigma} \,, \tag{95}$$

the equation (94) becomes

$$R^{\alpha}_{\sigma:\alpha} + (g^{\mu\rho}R_{\mu\sigma})_{:\rho} - R_{:\sigma} = 0 \,. \tag{96}$$

Using equations (91) and (77), one obtains

$$2R^{\alpha}_{\sigma:\alpha} - R_{:\sigma} = 0 \,, \tag{97}$$

which is the Bianchi relation for the Ricci tensor. By raising the suffix σ, we get

$$(R^{\sigma\alpha} - \frac{1}{2}g^{\sigma\alpha}R)_{:\alpha} = 0 \,. \tag{98}$$

From (76), the explicit expression for the Ricci tensor is

$$R_{\mu\nu} = \Gamma^{\alpha}_{\mu\alpha,\nu} - \Gamma^{\alpha}_{\mu\nu,\alpha} - \Gamma^{\alpha}_{\mu\nu}\Gamma^{\beta}_{\alpha\beta} + \Gamma^{\alpha}_{\mu\beta}\Gamma^{\beta}_{\nu\alpha} \,. \tag{99}$$

The three terms other than the first term $\Gamma^{\alpha}_{\mu\alpha,\nu}$ evidently are symmetrical. The first term is also symmetrical. By differentiating the determinant g and multiplying with the cofactor $gg^{\lambda\mu}$ ($g^{\mu\nu}g_{\nu\alpha} = g^{\mu}_{\alpha}$) we have $g_{,\nu} = gg^{\lambda\mu}g_{\lambda\mu,\nu}$ and so

$$\begin{aligned}\Gamma^{\mu}_{\nu\mu} &= g^{\lambda\mu}\Gamma_{\lambda\nu\mu} = \frac{1}{2}g^{\lambda\mu}(g_{\lambda\nu,\mu} + g_{\lambda\mu,\nu} - g_{\mu\nu,\lambda}) \\ &= \frac{1}{2}g^{\lambda\mu}g_{\lambda\mu,\nu} = \frac{1}{2}g^{-1}g_{,\nu} = (\log g)_{,\nu} \,,\end{aligned} \tag{100}$$

which shows that the first term of (99) is symmetrical.

3.7. Einstein's Equation and Schwarzschild Radius

The law of gravitation assumed by Einstein is that the Ricci tensor constitutes

$$R_{\mu\nu} = 0 \,, \tag{101}$$

in an empty or flat space where no matter and physical physical fields exist except the gravitational field. The gravitational field is assumed not to disturb the space-time geometry and emptiness, but other fields alter the space-time geometry.

The geodesics are the straight lines in a flat space and so particles move along straight lines. The space where is not flat, Einstein's law (101) puts restrictions on the curvature. Looking upon the $g_{\mu\nu}$ as potentials, the metric tensor $g_{\mu\nu}$ is determined from (101). The Ricci tensor equations are similar to the usual field equations of physics in the sense that they are of the second order derivatives, because the second order derivatives appear in (99), as the Christoffel symbols involve the first order derivatives. They are nonlinear field equations, and space-time geometry is determined by matter and fields in a too much complicated way to get accurate solutions.

There is one special case that can be solved approximately; namely, the static, spherically symmetric field produced by a spherically symmetric body at the rest frame. The static condition means that the components of $g_{\mu\nu}$ are independent of the time t and also $g_{0m} = 0$. The spatial coordinates may be taken to be spherical polar coordinates $x^1 = r$, $x^2 = \theta$, $x^3 = \phi$ and the most general form for ds^2 compatible with spherical symmetry is

$$ds^2 = Udt^2 - Vdr^2 - Wr^2(d\theta^2 + \sin^2 d\phi^2)\ , \tag{102}$$

where U, V and W are any functions of r and this freedom is used to simplify things as much as possible. The most convenient arrangement is to take $W = 1$, and the expression for ds^2 may then be written as

$$ds^2 = e^{2\nu}dt^2 - e^{2\lambda}dr^2 - r^2d\theta^2 - r^2\sin^2\theta d\phi^2\ , \tag{103}$$

where ν and λ are functions of r only. The components of $g_{\mu\nu}$ must be chosen to satisfy the Einstein's assumption (101). We can read off the values of $g_{\mu\nu}$ from (103), namely,

$$g_{00} = e^{2\nu}, \qquad g_{11} = -e^{2\lambda}, \qquad g_{22} = -r^2, \qquad g_{33} = -r^2\sin^2\theta\ , \tag{104}$$

and $g_{\mu\nu} = 0$ for $\mu \neq \nu$. The covariant components $g_{\mu\nu}$ are found from $g_{\mu\nu}g^{\nu\rho} = \delta^\rho_\mu$

$$g^{00} = e^{-2\nu}, \quad g^{11} = -e^{-2\lambda}, \quad g^{22} = -r^{-2}, \quad g^{33} = -r^{-2}\sin^{-2}\theta\ , \tag{105}$$

and $g^{\mu\nu} = 0$ for $\mu \neq \nu$.

It is now able to calculate all the Christoffel symbols $\Gamma^\sigma_{\mu\nu}$. Many of them vanish, and the components that do not vanish are, denoting differentiations with respect to r,

$$\begin{aligned} \Gamma^1_{00} &= \nu' e^{2\nu-2\lambda}, & \Gamma^0_{10} &= \nu'\ , \\ \Gamma^1_{11} &= \lambda', & \Gamma^2_{12} &= \Gamma^3_{13} = r^{-1}\ , \\ \Gamma^1_{22} &= -re^{-2\lambda}, & \Gamma^3_{23} &= \cot\theta\ , \\ \Gamma^1_{33} &= -r\sin^2\theta e^{-2\lambda}, & \Gamma^2_{33} &= -\sin\theta\cos\theta\ . \end{aligned} \tag{106}$$

These expressions are to be substituted in the Ricci tensor (99). The results are coupled nonlinear differential equations for $\lambda(r)$ and $\nu(r)$ determined by

$$\begin{aligned} R_{00} &= \left(-\nu'' + \lambda'\nu' - \nu'^2 - \frac{2\nu'}{r}\right)e^{2\nu-2\lambda}, & R_{11} &= \nu'' - \lambda'\nu' + \nu'^2 - \frac{2\lambda'}{r}, \\ R_{22} &= (1 + r\nu' - r\lambda')e^{-2\lambda} - 1, & R_{33} &= R_{22}\sin^2\theta\ , \end{aligned} \tag{107}$$

and the other components of $R_{\mu\nu}$ vanish due to Einstein's law of gravitation (101). $R_{00} = 0$ and $R_{11} = 0$ lead to $\lambda' + \nu' = 0$. In the spherically symmetric system of fields and matter the space must be approximately flat for large values of r so that both λ and ν tend to zero as $r \to \infty$. It generates that $\lambda + \nu = 0$. R_{22} gives $(1 + 2r\nu')e^{2\nu} = 1$, or $(re^{2\nu})' = 1$, resulting in $re^{2\nu} = r - 2m$, where $-2m$ is a constant of integration chosen to be consistent with Newtonian approximation in the gravitational unit ($\hbar = c = G = 1$). One should note that the constant of integration should be chosen as $-2Gm$, compared to the hydrostatic equation for the balance of forces between pressure and gravity in the Newtonian limit [4,5]. Thus, we get

$$g_{00} = 1 - \frac{2m}{r} . \tag{108}$$

The constant of integration m that has appeared in (107) is interpreted as the mass of the central body producing the gravitational field. The complete solution is

$$ds^2 = \left(1 - \frac{2m}{r}\right) dt^2 - \left(1 - \frac{2m}{r}\right)^{-1} dr^2 - r^2 d\theta^2 - r^2 \sin^2\theta d\phi^2 , \tag{109}$$

which is known as the Schwarzschild solution . It holds outside the surface of the body, where there is no matter.

The Einstein equation (101) in the absence of matter leads to

$$R^{\mu\nu} - \frac{1}{2} g^{\mu\nu} R = 0 , \tag{110}$$

or $g^{\mu\nu} \times (101)$ leads to $R = 0$. In the presence of matter these equations must be modified as

$$R^{\mu\nu} - \frac{1}{2} g^{\mu\nu} R = Y^{\mu\nu} , \tag{111}$$

where $Y^{\mu\nu}$ is symmetric two-index tensor indicating the presence of matter. The Bianchi relation (98) requires that the covariant derivative of the symmetric tensor $Y^{\mu\nu}$ vanish:

$$Y^{\mu\nu}_{\ :\nu} = 0 . \tag{112}$$

Any tensor $Y^{\mu\nu}$ produced by matter must satisfy this condition to be consistent with the requirement of the general relativity. For example $Y^{\mu\nu}$ is taken as the energy-momentum tensor, $Y^{\mu\nu} = kT^{\mu\nu}$ (k is a constant). The coefficient of tensor, k, is traditionally defined as -8π and the equation (111) is rewritten as

$$R^{\mu\nu} - \frac{1}{2} g^{\mu\nu} R = -8\pi T^{\mu\nu} . \tag{113}$$

We shall assume that the tensor $T^{\mu\nu}$ with the coefficient be interpreted as the density and flux of energy and momentum. In flat space, the equation (112) would become

$$T^{\mu\nu}_{\ ,\nu} = 0 , \tag{114}$$

and the world then will be consistent with the conservation of energy and momentum. In curved space the conservation of energy and momentum is considered approximate. The error is to be ascribed to the gravitational field working on the matter because it carries out some energy and momentum.

3.8. The Tolman-Oppenheimer-Volkoff (TOV) Equation

For the region outside a star, the Ricci tensor assumed to be zero; however, this is not the case with the interior of the star. We need the Einstein equation in the presence of matter (111). From (105) and (107) the scalar curvature becomes

$$\begin{aligned} R = g^{\mu\nu} R_{\mu\nu} &= e^{-2\nu} R_{00} - e^{-2\lambda} R_{11} - \frac{2}{r^2} R_{22} \\ &= e^{-\lambda} \left(-2\nu'' + 2\lambda'\nu' - 2\nu'^2 - \frac{2}{r^2} + \frac{4\lambda'}{r} - \frac{4\nu'}{r} \right) + \frac{2}{r^2} \,. \end{aligned} \tag{115}$$

It is convenient to work with mixed tensors, because $g_{\mu\nu} g^{\nu\sigma} = g^{\sigma}_{\mu}$ simplifies calculations of Einstein equation. The components of the Einstein equation are obtaind from (111),

$$\begin{aligned} r^2 Y_0{}^0 &= e^{-2\lambda}(1 - 2r\lambda') - 1 = -\frac{d}{dr}\{r(1 - e^{-2\lambda})\} \,, \\ r^2 Y_1{}^1 &= e^{-2\lambda}(1 + 2r\nu') - 1 \,, \\ Y_2{}^2 &= (\nu'' + \nu'^2 - \lambda'\nu' + \frac{\nu' - \lambda'}{r}) \,, \qquad Y_3{}^3 = Y_2{}^2 \,. \end{aligned} \tag{116}$$

The energy-momentum tensor in the present metric ($Y^{\mu\nu} = kT^{\mu\nu}$) is equal to

$$T_0{}^0 = \mathcal{E} \,, \quad T_i{}^i = -p \quad (i \text{ is not summed}) \,. \tag{117}$$

The term $r^2 Y_0{}^0$ gives

$$r^2 Y_0{}^0 = -\frac{d}{dr}\{r(1 - e^{-2\lambda})\} = kr^2 T_0{}^0 = kr^2 \mathcal{E}(r) \,. \tag{118}$$

This can be integrated immediately to yield

$$e^{-2\lambda(r)} = 1 + \frac{k}{r} \int_0^r x^2 dx \mathcal{E}(x) \,. \tag{119}$$

Let us define

$$M(r) \equiv 4\pi \int_0^r x^2 dx \mathcal{E}(x) \,, \tag{120}$$

which is the mass of a sphere with the radius r, and we denote R as the radius of the star. The edge of the star is defined by the $p(R) = 0$ because zero pressure can support no material against the gravitational attraction. The constant k should be chosen as $k = -8\pi$ or $k = -8\pi G$. It is ready to derive the differential equations for the structure of a static, spherically symmetric relativistic star. The Schwarzschild solution requires

$$e^{2\lambda} = \left(1 - \frac{2M(r)}{r}\right)^{-1} \,, \tag{121}$$

$$r^2 T_1{}^1 = e^{-2\lambda}(1 + 2r\nu') - 1 = 8\pi r^2 p(r) \,. \tag{122}$$

For convenience, we introduce the prime symbol for differentiation by r, and derivatives of (121) and (122) by r give

$$\lambda' = (4\pi\mathcal{E}(r) - M(r)) \frac{e^{2\lambda}}{r^2} \,, \qquad \nu' = \left(4\pi r^3 p(r) + M(r)\right) \frac{e^{2\lambda}}{r^2} \,. \tag{123}$$

Noticing that

$$\nu'^2 = \left(4\pi r^3 p(r) + M(r)\right)^2 \frac{e^{4\lambda}}{r^4}\,, \tag{124}$$

$$\lambda'\nu' = \left(4\pi r^3 p(r) + M(r)\right)\left(4\pi r^3 \mathcal{E}(r) - M(r)\right) \frac{e^{4\lambda}}{r^4}\,, \tag{125}$$

$$\frac{\nu' - \lambda'}{r} = \left\{4\pi r^3 (p(r) - \mathcal{E}(r)) + 2M(r)\right\} \frac{e^{2\lambda}}{r^3}\,, \tag{126}$$

$$\begin{aligned}\nu'' &= \left(4\pi r^3 p(r) + M(r)\right)\left(4\pi r^3 \mathcal{E}(r) - M(r)\right) \frac{2e^{4\lambda}}{r^4} \\ &\quad + \left\{4\pi r^5 p'(r) + 4\pi r^4 \left(p(r) + \mathcal{E}(r)\right) - 2rM(r)\right\} \frac{e^{2\lambda}}{r^4}\,,\end{aligned} \tag{127}$$

we obtain from (116),

$$\begin{aligned}T_2^{\ 2} &= e^{-2\lambda}\left(\nu'' + \nu'^2 - \lambda'\nu' + \frac{\nu' - \lambda'}{r}\right) \\ &= e^{-2\lambda}\frac{e^{4\lambda}}{r^4}\big[\left(4\pi r^3 p(r) + M(r)\right)\left\{4\pi r^3 p(r) + M(r) + 8\pi r^3 \mathcal{E} - 2M(r)\right\} \\ &\quad - \left(4\pi r^3 p(r) + M(r)\right)\left(4\pi r^3 \mathcal{E}(r) - M(r)\right)\big] \\ &\quad + e^{-2\lambda}\frac{e^{2\lambda}}{r^4}\left\{4\pi r^5 p'(r) + 4\pi r^4 \left(p(r) + \mathcal{E}(r)\right) - 2rM(r)\right. \\ &\qquad\qquad \left. + 4\pi r^4 \left(p(r) - \mathcal{E}(r)\right) + 2rM(r)\right\} \\ &= \frac{e^{2\lambda}}{r^4}\left(4\pi r^3 p(r) + M(r)\right)\left\{4\pi r^3 \left(p(r) - \mathcal{E}(r)\right) + 8\pi r^3 \mathcal{E}(r)\right\} \\ &\qquad\qquad + 4\pi\left(rp'(r) + 2p(r)\right) \\ &\equiv 8\pi p(r)\,.\end{aligned} \tag{128}$$

By arranging the last two lines, one obtains

$$\begin{aligned}p'(r) &= -\frac{e^{2\lambda}}{r^2}\left(4\pi r^3 p(r) + M(r)\right)\left(p(r) + \mathcal{E}(r)\right) \\ &= -\frac{\left(4\pi r^3 p(r) + M(r)\right)\left(p(r) + \mathcal{E}(r)\right)}{r^2\left(1 - \frac{2M(r)}{r}\right)}\,.\end{aligned} \tag{129}$$

This is the TOV equation,

$$\frac{dp}{dr} = -\frac{G\left[p(r) + \mathcal{E}(r)\right]\left[M(r) + 4\pi r^3 p(r)\right]}{r\left[r - 2GM(r)\right]}\,, \tag{130}$$

where G is inserted for a mnemonic purpose.

4. Introduction to Nuclear Many-Body Problem

4.1. Weizsäcker Mass Formula

We have discussed macroscopic dynamics of space-time and matter. From this section , microscopic dynamics and the characteristic properties of nucleus are briefly reviewed.

The radius of stable nucleus is proportional to the nuclear mass number $A = Z + N$; Z and N express the number of proton and neutron respectively. The radius of nucleus is described as

$$R_0 = r_0 A^{1/3}, \quad r_0 \sim 1.2\,\text{fm} \quad (1\,\text{fm} = 10^{-13}\,\text{cm})\,, \tag{131}$$

The binding energy of nucleus is roughly understood by $B(Z,N) = -\{ZM_p + NM_n - M(Z,N)\}c^2$, where $M(Z,N)$ is the total mass of nucleus, $M_p = 938\text{MeV}$ and $M_n = 939.6\text{MeV}$ are proton and neutron masses, respectively. However, the origin of binding energy is well expressed by the semi-empirical mass formula :

$$B(Z,N) = -a_1 A + a_2 A^{2/3} + a_3 Z^2 A^{-1/3} + a_4 \frac{(N-Z)^2}{A} + \delta(A)\,, \tag{132}$$

which is based on the concept of an incompressible, uniformly-charged liquid-drop model, that constitutes a liquid sphere composed of A baryons with the charge eZ. This formula is called Weizsäcker semi-empirical mass formula known to describe nuclear binding energies well throughout the periodic table.

The first term of (132) describes the volume energy, which corresponds to the binding energy of nuclear matter at saturation density ($k_F = 1.30\text{fm}^{-1}$, $|a_1| = |B/A - M| \sim 15.75\text{MeV}$). The second term is the contribution from finite nuclear surface energy and the third term expresses the coulomb force which works between protons. The fourth term expresses the tendency of symmetrization with respect to neutrons and protons, known as symmetry energy. The fifth term expresses the pairing energy, which comes form the fact that the binding energy depends on the combination of even or odd number of Z and N. These values are given by

$$\begin{aligned} a_1 &= 15.75\text{ MeV}\,, & a_2 &= 17.2\text{ MeV}\,, \\ a_3 &= 0.06\text{ MeV}\,, & a_4 &= 30.0\text{ MeV}\,. \end{aligned} \tag{133}$$

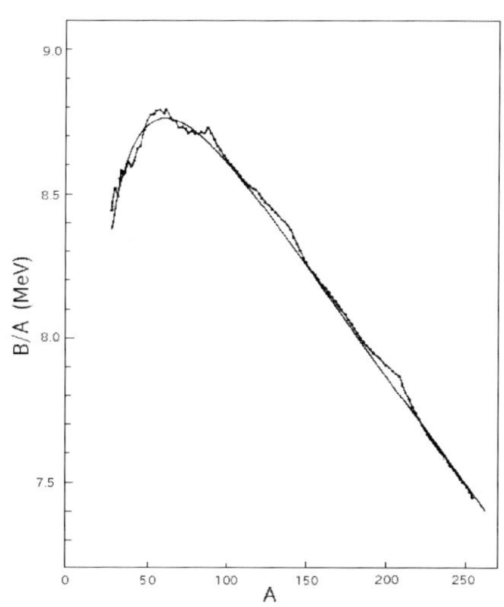

Figure 4. A schematic figure of the experimental binding energies and semi-empirical mass formula. The dots denote experimental binding energies [A. Bohr and B. R. Mottelson, 1969]. The smooth curve represents the semi-empirical mass formula (132).

The pairing energy is

$$\delta(A) = \begin{cases} \frac{12}{\sqrt{A}}\ \mathrm{MeV}, & (Z = \text{odd number}, \mathrm{N} = \text{odd number}) \\ 0, & (A = Z + N = \text{odd}) \\ -\frac{12}{\sqrt{A}}\ \mathrm{MeV}, & (Z = \text{even}, \mathrm{N} = \text{even})\ . \end{cases} \tag{134}$$

These are the essential values of nuclear system [14].

4.2. The Rectangle Problem of Binding Energy

The experimental values of binding energy (133) are essential to examine theoretical consequences of nuclear models, because those values are manifestations of hadronic interactions in nucleus. Nuclear models may be constructed with hadronic (mesons and nucleons) and quark (gluons and quarks) degree of freedom. If a nuclear model could not satisfy the values of (133) within certain experimental errors, the model will be rejected. A famous example is the rectangle problem of binding energy of nuclear matter on the value $a_1 = 15.75\mathrm{MeV}$.

If coulomb forces were negligible and the number of nucleus was so large that the nucleus was the size of neutron stars ($\mathrm{R} = 10 \sim 12\mathrm{km}$) and in addition $N = Z$, then, Weizsäcker mass formula becomes simple with binding energy proportional to the number of nucleons, A. The hypothetical system with equal number of protons and neutrons is called nuclear matter. The calculation of binding energy of nuclear matter is evidently the calculation of a_1. The values a_2 and a_3 will be neglected in nuclear matter, and the resulting empirical saturation point has an energy/particle between -15 MeV and -17 MeV and a Fermi momentum k_F in the range $1.29 \sim 1.44\mathrm{fm}^{-1}$. The saturation point must agree with the empirical one if the potential is to account for the properties of nuclei.

B. D. Day checked nuclear models if they can describe the density and binding energy at the saturation point with several potentials that fits nucleon-nucleon scattering data and the properties of the deuteron [10]. The nucleons obey the nonrelativistic Schrödinger equation with a Hamiltonian consisting of kinetic energies and two-body potentials,

$$H = \sum_{i=1}^{A} T_i + \sum_{i<j}^{A} V_{ij}\ , \tag{135}$$

where A is the number of nucleons. The saturation point is calculated by extrapolation from finite nuclei, using either the liquid-drop model and semi-empirical mass formula of Hartree-Fock calculations with phenomenological forces fitted to finite nuclei.

The result is shown in Fig. 5, and the resulting saturation points of potentials for a wide variety of potentials lie on a narrow band in the energy-density plane that does not include the empirical saturation point. In spite of a lot of effort and two-body Brueckner-Bethe calculations, the reproduction of empirical saturation point shown in Fig. 5 was difficult to satisfy [10]. The relativistic effects, many-body forces, and three body forces may repair the rectangle discrepancy.

The first requirement for nuclear potential is that it correctly reproduces the saturation point of nuclear matter. Only if this calculation can be done accurately it makes sense to compare empirical properties of nuclear matter and hence to accept or reject the potential. The reproduction of binding energy extremely restricts the model of nuclear matter.

Moreover, at the saturation point, pressure is zero. This leads us to Hugenholtz-Van Hove theorem which is discussed in detail in the next section. The rectangle problem or Hugenholtz-Van Hove theorem is deeply related with self-consistency of nuclear models. Any models of hadrons, effective QCD, Lattice QCD which describe nuclear physics must satisfy these conditions.

5. Hugenholtz-Van Hove Theorem at Nuclear Matter Saturation

The derivation of a well-defined force between nucleons is a complicated problem, because the strong short-range repulsion makes it difficult to decide upon an appropriate treatment of a finite system of many particles. A number of simplifications are possible in a theory of nuclear matter, which is essentially an infinite nucleus with a given ratio of neutrons and protons. The mean-field approach is a prominent example, which assumes that at high enough density, fluctuations in the meson fields could be ignored and they could be replaced by their classical expectation values or mean fields [11]. J. D. Walecka assumed that these conditions be valid at ordinary nuclear density and in many-body system of hadrons. The mean-field theory is relativistic, renormalizable quantum field theory known as Quantum Hadrodynamics (QHD). QHD calculations have exhibited numerous successes but also apparent failures which motivated extensions and evolutions of its own theoretical scheme [12].

The QHD was applied to many-body system of symmetric nuclear matter. Symmetric

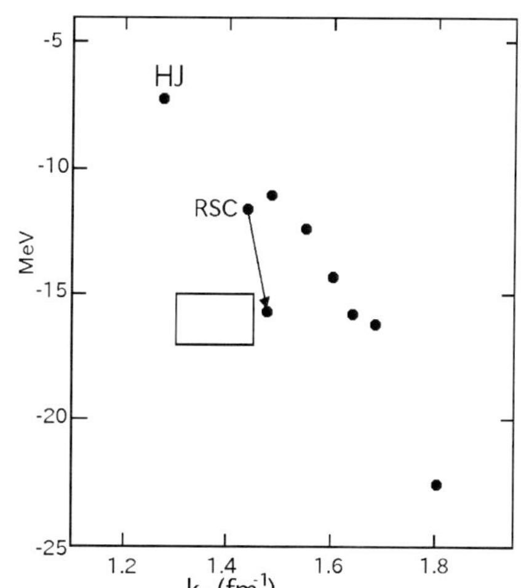

Figure 5. A schematic figure of saturation points given by two-body potentials. Saturation points calculated with the Brueckner-Bethe method, including the two-hole-line term with the conventional spectrum. The arrow shows the shift in the saturation point for the Reid potential when three-hole-line terms are included. The empirical saturation point lies in the open rectangle [10]. The capital symbols describe potentials (RSC: Reid potential, HJ: Hamada and Johnston, etc.).

nuclear matter is self-bound and the binding energy of nuclear matter becomes the minimum at saturation density (see, Fig. 6) because of the balance between attractive and repulsive forces, which shows that the pressure at saturation vanishes ($p = 0$). The self-bound property manifests a realization of fundamental self-consistency of many-body nucleon system, resulting in a theorem on the single particle energy of nucleons, known as the Hugenholtz-Van Hove (HV) theorem . This is important to check internal consistency of an approximation, which requires equality between Fermi energy and the average energy at saturation density. If a proper dynamical definition of the single particle energy is formulated in terms of the propagation of an additional particle (or a hole) of given momentum through the medium, the single particle energy must satisfy the relation:

$$E(k_F) = \frac{\mathcal{E}}{\rho_B} + \rho_B \frac{d}{d\rho_B}\left(\frac{\mathcal{E}}{\rho_B}\right), \tag{136}$$

where $\mathcal{E}$ is the energy density, ρ_B the baryon density and $E(k_F)$ the single particle energy on the Fermi surface. The eq. (136) is equivalent to a thermodynamic relation: $\mathcal{E} + p = \rho_B \mu$, where the chemical potential $\mu = E(k_F)$. However, energy density and pressure in the eq. (136) must be given by lagrangian or hamiltonian through the formalism of quantum statistical mechanics.

The HV theorem can be interpreted in terms of a transformation of a general interacting Fermi system into a quasiparticle system where the sharp Fermi surface is maintained exactly as in the noninteracting Fermi system. The quasiparticle system is defined properly on condition that the chemical potential, μ, of a quasiparticle system is equal to $E(k_F)$. The relation can be used to test the internal consistency of a theory for nuclear matter and enables us to examine theoretical consistency of the single particle energy defined by any given approximation. A famous example is the work by Brueckner *et.al.* [13]. Hugenholtz and Van Hove [11] published their general proof of a theorem for a system of an interacting

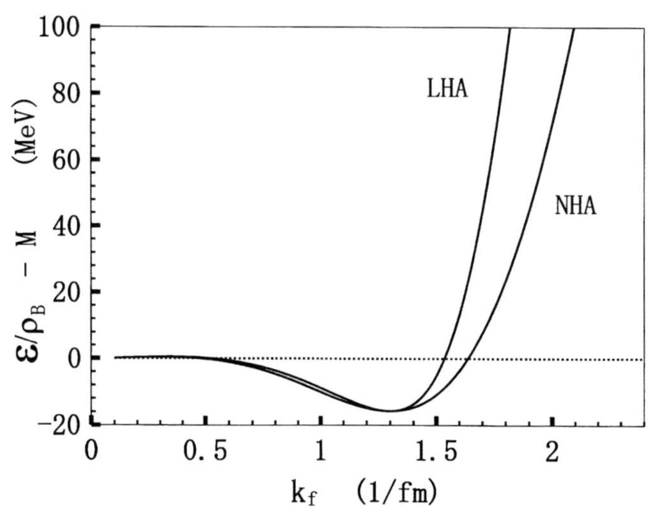

Figure 6. Binding energies of symmetric nuclear matter, $\mathcal{E}/\rho_B - M$ ($M = 939$ MeV), given by LHA (linear Hartree approximation) and NHA (nonlinear Hartree approximation).

Fermi gas, and they pointed out the inconsistency of Brueckner's theory of nuclear matter. At nuclear matter saturation, the second term of (136) vanishes; $E(k_F)$ must then be equal to the average energy per particle $\mathcal{E}/\rho_B$. The original Brueckner's theory of nuclear matter exhibited a large discrepancy between the values of $E(k_F)$ and $\mathcal{E}/\rho_B$. In subsequent papers, Brueckner reconsidered the problem in the framework of his theory and improved his self-consistent calculations to maintain the requirement of HV theorem numerically to a good approximation.

In the nonrelativistic many-body theory, energy density and pressure are given by model hamiltonian and pressure tensor. Similarly, in the relativistic field theory, they are expressed by the ground state expectation value of the energy-momentum tensor $\langle \hat{T}^{\mu\nu} \rangle$ ($\mu, \nu = 0, 1, 2, 3$). The operator is defined with an employed lagrangian $\mathcal{L}$ as:

$$\hat{T}^{\mu\nu} = -g^{\mu\nu}\mathcal{L} + \frac{\partial q_i}{\partial x_\nu} \frac{\partial \mathcal{L}}{\partial(\partial q_i/\partial x^\mu)} \ . \tag{137}$$

where the repeated index i is summed over all generalized coordinate or fields q_i. In the ground state of nuclear matter at $T = 0$, energy density is defined as,

$$\mathcal{E} = \langle \hat{T}^{00} \rangle \ , \tag{138}$$

and pressure is defined as,

$$P_{hy} = \frac{1}{3}\langle \hat{T}^{ii} \rangle \ , \tag{139}$$

where i is summed ($i = 1, 2, 3$), and the pressure calculated from the energy-momentum tensor is denoted as the hydrodynamic pressure, P_{hy}. A couple of independent physical pressures of a normal Fermi system can be introduced: thermodynamic pressure,

$$P_T = \rho_B^2 \frac{\partial(\langle \hat{T}^{00} \rangle/\rho_B)}{\partial \rho_B} \ , \tag{140}$$

and pressure consistent with the HV theorem as,

$$P_{HV} = E(k_F)\rho_B - \mathcal{E} \ , \tag{141}$$

where $E(k_F)$ is the self-consistent single particle energy on the Fermi surface, and ρ_B is the baryon density at $T = 0$,

$$\rho_B = \zeta k_F^3/6\pi^2 \ , \tag{142}$$

where ζ is the spin-isospin degeneracy; $\zeta = 4$ for nuclear matter and $\zeta = 2$ for neutron matter [14]. P_{HV} shows explicitly the requirement of the HV theorem at saturation of nuclear matter. Because pressure is zero at saturation, the single particle energy is equal to the average energy per particle: $E(k_F) = \mathcal{E}/\rho_B$. This requirement is used to check self-consistency to approximations for nuclear matter.

Hence, the HV theorem in nuclear matter is written as,

$$\mathcal{E} + \rho_B^2 \frac{\partial}{\partial \rho_B}\left(\frac{\mathcal{E}}{\rho_B}\right) = E(k_F)\rho_B \ , \tag{143}$$

and equality of thermodynamical and hydrodynamical pressures, $P_{hy} = P_T$, is known as the virial theorem . Self-consistency is therefore summarized as:

$$\rho_B^2 \frac{\partial(\langle \hat{T}^{00} \rangle / \rho_B)}{\partial \rho_B} = \frac{1}{3} \langle \hat{T}^{ii} \rangle \,, \tag{144}$$

with

$$\frac{\partial(\langle \hat{T}^{00} \rangle)}{\partial \rho_B} = E(k_F) \,. \tag{145}$$

The properties of nuclear matter at saturation (saturation density and binding energy) are specified by two conditions. The saturation density is determined by the condition that the pressures (139)-(141) are zero at saturation. The binding energy at saturation is then examined by the HV theorem, $\mathcal{E}/\rho_B = E(k_F)$, or by the eq. (145). The experimental data used for the binding energy at saturation is about $\mathcal{E}/\rho_B - M_n = -15.75$ MeV, ($M_n = 939$ MeV), and the density is $\rho_B = 0.148$ fm^{-3} ($k_F = 1.30$ fm^{-1}). The diagonal components of the energy-momentum tensor of an ideal perfect fluid are related to each other through eq. (144), and macroscopic quantities are related to the single particle energy by eq. (145). In the relativistic Hartree approximation of Walecka model, it is readily shown that the single particle energy satisfies eq. (144) and (145) exactly. The analytical proof of self-consistency in the static limit of Dirac-Hartree-Fock approximation is discussed [15]; however, it is difficult to construct self-consistent approximations. The problem of self-consistency evolved and produced fundamental many-body theories, such as the concept of *thermodynamic consistency* , the theory of conserving approximations and the density functional theory [16–21].

Since the condition of thermodynamic consistency is connected to the determination of self-consistent single particle energy (chemical potential), it is also related to a fundamental problem, ‘sign problem ’, in the lattice QCD . The extension of lattice QCD analyses to (real) non-zero chemical potentials is one of fundamental problems by the fact that in this case the fermion determinant in the QCD partition function becomes complex. As a consequence, the standard statistical weight for the sampling is no longer positive definite which causes difficulty in the convergence of calculations [22]. Although several methods to circumvent the problem have been developed, the saturation mechanism is not yet solved sufficiently in lattice QCD as shown in the theory of Kadanoff-Baym, or the theory of density functional theory in hadronic models. It is certainly one of fundamental problems for the lattice QCD approach to maintain the requirement: $\mathcal{E}/\rho_B = E(k_F)$ and the pressure $p = 0$ at nuclear matter saturation density.

6. Landau Theory of Fermi Liquid

Landau’s Fermi liquid theory [23] provides us with a consistent way of parameterizing properties of Fermi systems and makes experimentally measured physical quantities relate directly to each other using a small number of parameters. The theory is the application of the concept of elementary excitations to normal Fermi systems [24], and it is also applied to finite Fermi systems [25]. The states of a Fermi liquid are assumed to be characterized by the quasiparticle distribution function $n(k_i)$ and the energy density $\mathcal{E}(n_i)$, which is a functional of the distribution function $n(k_i)$. If we add a quasiparticle $n(k_i)$ to an unoccupied quasiparticle state, then the total energy of the system will increase by an amount $E(k_i)$, which is called the quasiparticle energy and is also a functional of the distribution function.

Assuming the smooth dependence of the distribution function on the momentum $n(k_i)$ and the energy density on the distribution function n_i, Landau's quasiparticle energy is defined by the first derivative of the energy density, $\mathcal{E}$, with respect to the occupation number of quasiparticles, n_i, as:

$$\delta\mathcal{E} = \sum_i E(k_i)\delta n_i \ . \tag{146}$$

The quasiparticle interaction is defined by the second derivative of $\mathcal{E}$ with respect to n_i or by the first derivative of $E(k_i)$ with n_j,

$$\delta E(k_i) = \sum_j f_{ij}\delta n_j \ . \tag{147}$$

The equations (146) and (147) are Landau's fundamental assumption to relate the total energy of the system to the single particle energy and the single particle energy to the quasiparticle interaction. Note that one-to-one correspondence between noninteracting and interacting energy states is assumed as it is also needed for the proof of the HV theorem. Landau's parameters are defined from quasiparticle interaction, f_{ij}, considering equilibrium distribution of quasiparticles in the absence of macroscopic current, $\mathbf{j}_B = 0$, with momenta $\mathbf{k}_i$ and $\mathbf{k}_j$ both at the Fermi surface. The eq. (146) is compatible with the first law of thermodynamics with a constant valume at $T = 0$ when Landau's quasiparticle energy is evaluated on the Fermi surface; the equation requires that the chemical potential of the system be equal to the quasiparticle Fermi energy: $\mu = E(k_F)$. Therefore, it is important to check whether the quasiparticle Fermi energy is equal to the chemical potential of the system in order to define Landau's quasiparticle interactions consistently. They can be checked by comparing the chemical potential given by the thermodynamic relation, $\mu = (\mathcal{E} + p)/\rho_B$, and the self-consistent single particle energy on the Fermi surface $E(k_F)$.

The Legendre expansion of f_{ij} with $\mathbf{k}_i$ and $\mathbf{k}_j$ both taken at the Fermi surface:

$$f_l \equiv \frac{2l+1}{4\pi}\int d\Omega_{ij} P_l(\cos\theta_{ij}) f_{ij}|_{|k_i|=|k_j|=k_F} \ , \tag{148}$$

and the relativistic density of states on the Fermi surface is defined as,

$$N_F = \sum_i \left(\frac{\partial n_i}{\partial E_i}\right)_{E_i=\mu} = \frac{\zeta k_F}{2\pi^2}\left(\frac{\partial k_i}{\partial E_i}\right)_{k_i=k_F} \ , \tag{149}$$

and then the dimensionless Landau parameters are defined: $F_l = N_F f_l$ $(l = 0, 1, 2, \cdots)$. The degeneracy factor ζ is $(\zeta = 4)$ for nuclear matter and $(\zeta = 2)$ for neutron matter.

Example: Fermi-liquid properties in a relativistic σ-ω mean-field theory
Let us consider the relativistic σ-ω mean-field theory originally proposed by Walecka:

$$\mathcal{L} = \bar{\psi}\left[\gamma_\mu(i\partial^\mu - g_\omega\omega^\mu) - (M - g_\sigma\phi)\right]\psi - \frac{1}{2}m_\sigma^2\phi^2 + \frac{1}{2}m_\omega^2\omega_\mu^2 \ , \tag{150}$$

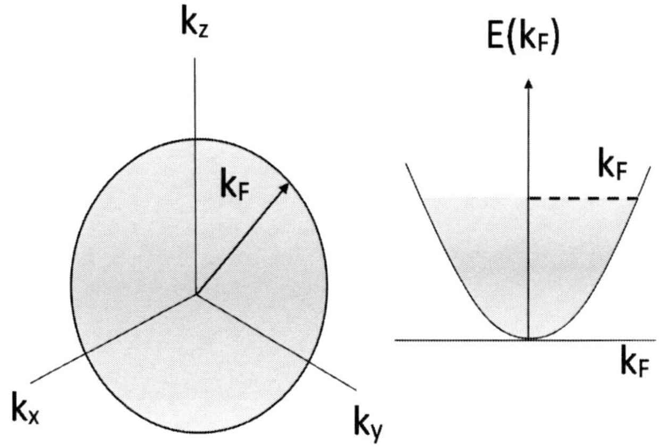

Figure 7. The schematic view of the Fermi sphere and $E(k_F)$ vs. k_F in the rest frame of nuclear matter, $\mathbf{j}_B \equiv 0$. The single particle energy is $E(k_F) = \frac{g_\omega^2}{m_\omega^2}\rho_B + \sqrt{k_F^2 + M^{*2}}$ with $E(k_F \to 0) = M$.

where the vector and scalar coupling constants are denoted by g_ω and g_σ. The meson fields are replaced by expectation values as $\phi = \langle \phi_0 \rangle$ and $\omega_\mu = (V_0, \mathbf{V})$ in the mean-field approximation. The general solution to the Dirac field equation is discussed in detail in chap. 3 of Serot and Walecka [14], and using the results, equations of motion for mesons are derived and denoted as

$$\begin{aligned} \phi_0 &= \frac{g_\sigma}{m_\sigma^2}\langle \bar{\psi}\psi \rangle = \frac{g_\sigma}{m_\sigma^2}\sum_i n_i \frac{M^*}{(\boldsymbol{\kappa}_i^2 + M^{*2})^{1/2}} \equiv \frac{g_\sigma}{m_\sigma^2}\rho_s\,, \\ \langle \omega^\mu \rangle &= (\omega_0, \boldsymbol{\omega}) = \frac{g_\omega}{m_\omega^2}\langle \bar{\psi}\gamma^\mu\psi \rangle = \frac{g_\omega}{m_\omega^2}(\rho_B, \mathbf{j}_B)\,. \end{aligned} \tag{151}$$

(1) Based on the above consequences, show the energy density derived from the σ-ω mean-field lagrangian using (137) and (138):

$$\mathcal{E} = \frac{1}{2}\frac{g_\omega^2}{m_\omega^2}\rho_B^2 + \frac{1}{2}\frac{g_\omega^2}{m_\omega^2}\mathbf{j}_B^2 + \frac{1}{2}\frac{g_\sigma^2}{m_\sigma^2}(M^* - M)^2 + \sum_i n_i\sqrt{\left(\mathbf{k}_i - \frac{g_\omega^2}{m_\omega^2}\mathbf{j}_B\right)^2 + M^{*2}}\,, \tag{152}$$

where n_i denotes the occupation number of the quasinucleon and $\rho_B = \sum_i n_i$.

The effective nucleon mass M^* and baryon current density $\mathbf{j}_B$ are determined by minimizing $\mathcal{E}$ with respect to M^* and $\mathbf{j}_B$ respectively, and this procedure gives self-consistent equations for M^* and $\mathbf{j}_B$:

$$\begin{aligned} M^* &= M - \frac{g_\sigma^2}{m_\sigma^2}\sum_i n_i \frac{M^*}{(\boldsymbol{\kappa}_i^2 + M^{*2})^{1/2}}\,, \\ \mathbf{j}_B &= \sum_i n_i \frac{\boldsymbol{\kappa}_i^2}{(\boldsymbol{\kappa}_i^2 + M^{*2})^{1/2}}\,, \end{aligned} \tag{153}$$

where $\boldsymbol{\kappa}_i = \mathbf{k}_i - (g_\omega^2/m_\omega^2)\mathbf{j}_B$.

(2) Derive (153) by minimizing the energy density (152), and find the self-consistent single particle energy from (146), which is given by:

$$\epsilon_i = \frac{g_\omega^2}{m_\omega^2}\rho_B + \sqrt{\boldsymbol{\kappa}^2 + M^{*2}}\,. \tag{154}$$

The quasiparticle interaction f_{ij} is similarly derived from (147) (see, ref. [26]) and Landau parameters can be obtained. The dimensionless Landau parameters F_l (mostly, F_0 and F_1) are closely related to observables as explained in the following section.

7. Properties of Nuclear Matter

Thermodynamic properties such as incompressibility K, first sound velocity and symmetry energy a_4 are expressed by Landau parameters in a simple form. The incompressibility is defined by the second derivative of energy density $\mathcal{E}$ with respect to baryon density ρ_B in the absence of baryon current,

$$K = 9\rho_B \frac{\partial^2 \mathcal{E}}{\partial \rho_B^2}\,. \tag{155}$$

The first derivative of $\mathcal{E}$ is equal to the single particle energy on the Fermi surface: $E(k_F) = \partial\mathcal{E}/\partial\rho_B = \mu$ (one should note that this relation must be proved in an employed calculation). The incompressibility is derived from the derivative of $E(k)$ as

$$\begin{aligned} K = 9\rho_B \frac{\partial \mu}{\partial \rho_B} &= 9\rho_B \left[\frac{2\pi^2}{\zeta k_F^2}\left(\frac{\partial E(k)}{\partial k}\right)_{k=k_F} + f_0\right], \\ &= 9\rho_B(1+F_0)/N_F\,, \end{aligned} \tag{156}$$

where the density of states is introduced:$N_F = \zeta k_F^2/2\pi^2(\partial k/\partial E(k))_{k=k_F}$. The incompressibilities for symmetric nuclear matter and neutron matter calculated from linear σ-ω Hartree and Dirac-Hartree-Fock approximations are shown in Fig. 8.

The hydrodynamic first sound velocity C_1 in the relativistic case is given by

$$C_1^2 = \left(\frac{\partial p}{\partial \mathcal{E}}\right), \tag{157}$$

where p is the pressure of the system. Using the relation $p = -\mathcal{E} + \rho_B\mu$ and $\partial p/\partial \rho_B = \rho_B \partial\mu/\partial\rho_B = \frac{1}{9}K$, one obtains

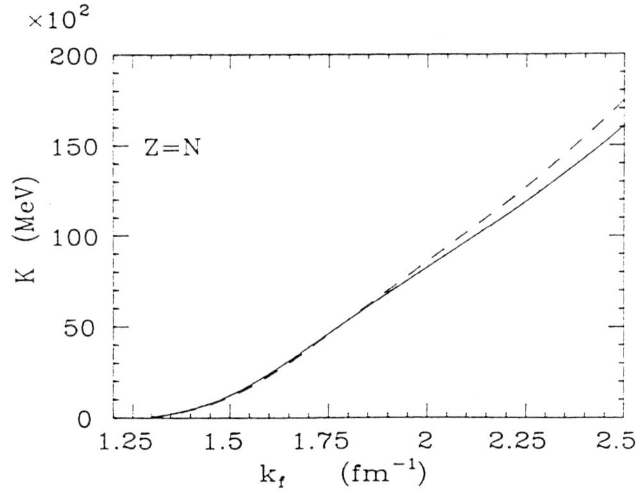

Figure 8. Incompressibilities for symmetric nuclear matter. The solid line is the result of Dirac-Hartree-Fock (DHF) and the dashed line is that of linear σ-ω Hartree approximation (LHA).

$$C_1^2 = \frac{\partial p}{\partial \rho_B}\frac{\partial \rho_B}{\partial \mathcal{E}} = \frac{K}{9\mu} \,. \tag{158}$$

The first sound velocities are shown in Fig. 9. The mean-value of the nucleon velocity in the medium is about $0.3c$ (c is the velocity of light) at normal nuclear density, and C_1 is smaller than the velocity of light at all densities, which is consistent with causality.

The symmetry energy a_4 is calculated from

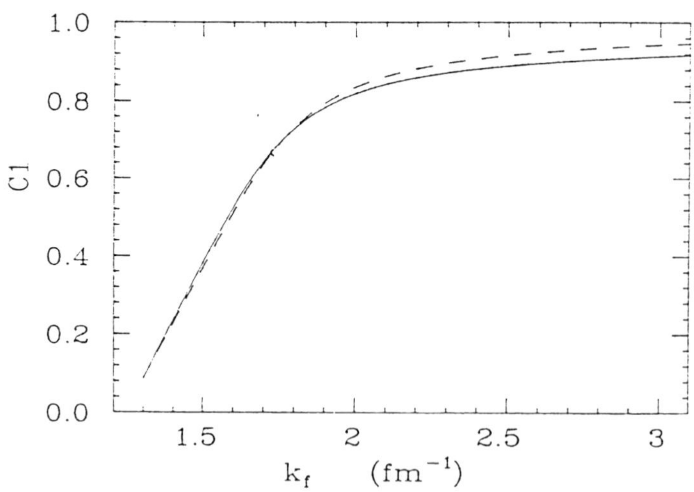

Figure 9. The first sound velocities for symmetric nuclear matter.

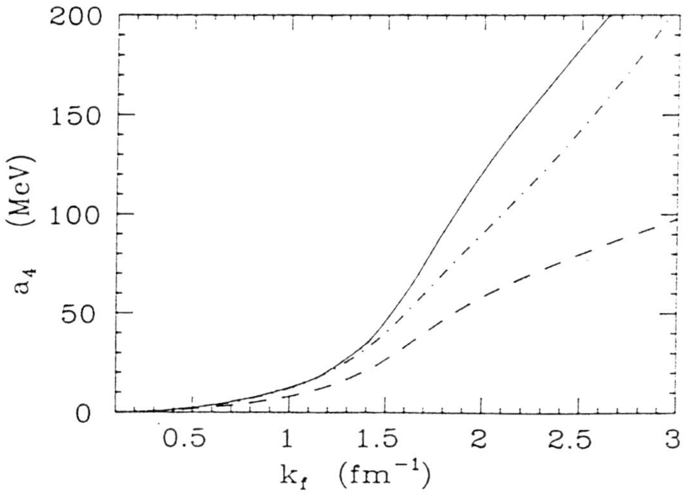

Figure 10. Symmetry energies. The static DHF (solid line), MFT-I (dashed-line) and MFT-II [27] (dot-dashed line).

$$a_4 = \frac{1}{2}\rho_B \left[\left(\frac{\partial^2 \mathcal{E}}{\partial \rho_3^2} \right)_{\rho_B} \right]_{\rho_3 = 0} = \frac{1}{6} \frac{k_F}{[\partial k / \partial E(k)]_{k=k_F}} (1 + F_0') \, , \tag{159}$$

where ρ_3 is the difference between the proton and neutron density: $\rho_3 = (k_{F_p}^3 - k_{F_n}^3)/3\pi^2$, and F_0' is the dimensionless parameter for the isovector quasiparticle interaction. The symmetry energy is shown in Fig. 10. The dimensionless relativistic Landau parameters for symmetric nuclear matter are shown in Fig. 11.

The definition of 'the effective mass of nucleon M^*' and 'the density of states' in non-

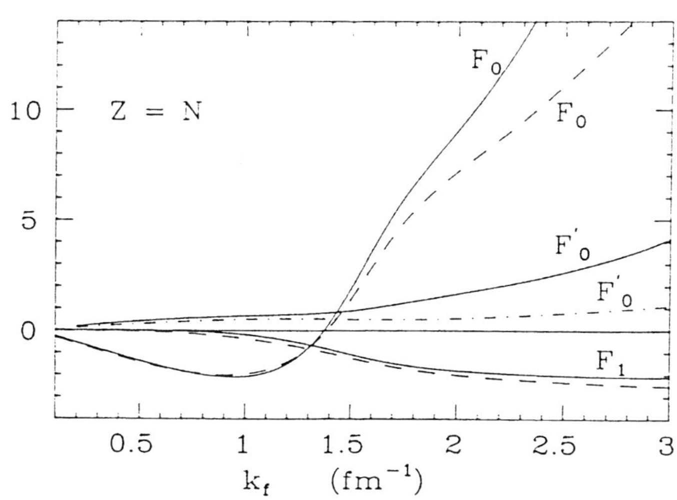

Figure 11. Dimensionless relativistic Landau parameters for symmetric nuclear matter.

relativistic and relativistic formalism should be commented in order to avoid any confusion on these quantities. In the nonrelativistic calculations, the density of states is used to define the effective mass; however, in the relativistic formalism, the effective mass is defined as a scalar quantity in Lorentz transformation properties. Hence, the physical meaning of nonrelativistic and relativistic effective masses is essentially different. Let us consider the relativistic σ-ω mean-field theory, and self-energies in the Hartree approximation are density-dependent but readily obtained. In this case, the density of states on the Fermi surface ρ_B/N_F can be calculated and one obtains,

$$\frac{\rho_B}{N_F} = \frac{k_F}{3}\Big(\frac{\partial E(k)}{\partial k}\Big)_{k=k_F} = \begin{cases} \frac{1}{3}\frac{k_F^2}{E^*(k_F)} & \text{relativistic;} \\ \frac{1}{3}\frac{k_F^2}{M^*(k_F)} & \text{nonrelativistic,} \end{cases} \tag{160}$$

where $E^*(k_F) = (k_F^2 + M^{*2})^{1/2}$. The nonrelativistic result is listed for comparison. It is evident from (160) that the nonrelativistic effective mass M^* should be compared with the relativistic kinetic energy, $E^*(k_F) = (k_F^2 + M^{*2})^{1/2}$. The relativistic effective mass might be small, but it would be close to the value of nonrelativistic effective mass: $E^*(k_F)/M \simeq M^*/M$ at saturation density of nuclear matter. The direct comparison between nonrelativistic and relativistic effective masses should be carefully checked by comparing the density of states. The Landau parameter F_1 is derived from,

$$\Big(\frac{\partial k}{\partial E(k)}\Big)_{k=k_F} = \frac{\mu}{k_F}(1 + \frac{1}{3}F_1) \tag{161}$$

where $\mu = E(k_F)$. This is known as the Landau's effective mass relation [28].

The Landau theory provides us with a consistent macroscopic approach to an underlying dynamical system by defining a single particle energy in terms of total energy of the system.

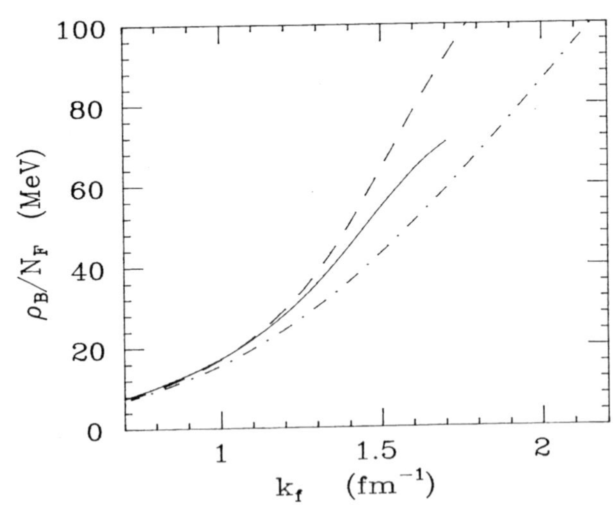

Figure 12. The densities of states at the Fermi surface.

Landau's theory of Fermi liquid parametrizes physical quantities using a small number of parameters defined by microscopic dynamics. The concept of quasiparticles in the strongly interacting nuclear many-body system should be defined consistently by the relation among Fermi energy, density and total energy constrained by symmetry of constituent particles. It is important to check conditions of self-consistency to extract consistent conclusions from employed approximations.

8. The Density Functional Theory

The density functional theory (DFT) is based on the rigorous phenomenology of thermodynamics, and the motivation of DFT is to define a generating functional in order to produce a single particle energy as a functional of a particle density, $n(r)$. The condition of the functional is the self-consistency defined by the functional derivative of the grand potential with respect to $n(r)$: $\delta\Omega/\delta n(r)$. The grand potential Ω is expressed in terms of the Helmholtz free energy F:

$$F(N,T,V) = \Omega(N,T,V) + \mu N \,, \tag{162}$$

and the chemical potential μ is defined self-consistently with $\partial\Omega/\partial N = \partial F/\partial N - \mu = 0$ at thermodynamic equilibrium. The density distribution of particles, $n(r)$, is defined in an external potential, $v(r)$, as $n(r) = \delta\Omega_v/\delta v(r)$, and the Hohenberg-Kohn free energy :

$$F_{HK}(n(r)) = \Omega_v(n(r)) - \int drn(r)v(r) \tag{163}$$

is defined as a functional Legendre transform of the grand potential $\Omega_v(n(r))$. If the free energy is minimized with respect to $n(r)$ at thermodynamic equilibrium: $\delta\Omega_v/\delta n(r) = 0$, the relation

$$\frac{\delta F_{HK}}{\delta n(r)} = -v(r) \,, \tag{164}$$

is obtained. The definition (163) shows that the Hohenberg-Kohn theorem restores the properties of thermodynamic equilibrium with the external, single particle potential defined by the eq. (164), and when $v(r) = 0$ and temperature $T = 0$, the ground state is derived as the functional derivative of Ω with respect to the density ρ:

$$\frac{\delta F_{HK}}{\delta\rho} = \frac{\delta\Omega}{\delta\rho} = 0 \,. \tag{165}$$

The essential point of the DFT approach depends on the definition and self-consistency of the functional F_{HK}. The theory of conserving approximations reveals the structure and condition for the self-consistency.

9. The Theory of Conserving Approximations

We will consider the relativistic mean-field theory of infinite nuclear matter. The Green's function, $G(k)$, is defined in the rest frame of nuclear matter by the Schwinger-Dyson equation, and it is assumed that the baryon Green's function has the similar structure as the one of noninteracting Green's function, which assumes the existence of the (on-shell) single quasiparticle energy, $E(k)$, and one to one correspondence of the single particle energy states. The assumptions are maintained in the self-consistent condition required by the theory of conserving approximations.

In the homogeneous infinite matter, the matrix structure of the baryon self-energy will be reduced generally to the following Lorentz-scalar form:

$$\Sigma(k) = \Sigma^s(k) - \gamma_\mu \Sigma^\mu(k) \,, \\ = \Sigma^s(|\mathbf{k}|, k^0) - \gamma^0 \Sigma^0(|\mathbf{k}|, k^0) + \gamma \cdot \mathbf{k} \Sigma^v(|\mathbf{k}|, k^0) \,, \tag{166}$$

and so, we have the three independent self-energy functions, $\Sigma^s(k)$, $\Sigma^0(k)$ and $\Sigma^v(k)$. The self-consistent dynamical variables (all are functions of $|\mathbf{k}|, k^0$ and k_F) are defined with Dirac equation of motion as:

$$M^*(k) \equiv M + \Sigma^s(k) \,, \qquad \mathbf{k}^*(k) \equiv \mathbf{k}(1 + \Sigma^v(k)) \,, \\ k^{*0} \equiv E^*(k) \equiv (\mathbf{k}^{*2}(k) + M^*(k)^2)^{1/2} \,, \\ k^{*\mu} \equiv k^\mu + \Sigma^\mu(k) = (k^0 + \Sigma^0(k), \mathbf{k}^*(k)) \,, \qquad k^0 = E(k) \,. \tag{167}$$

The Green's function, $G(k)$, is defined in the rest frame of nuclear matter in terms of the particle-antiparticle propagator, $G_F(k)$, and the hole propagator inside the Fermi sea, $G_D(k)$, as follows:

$$G(k) = G_F(k) + G_D(k) \,, \tag{168}$$

$$G_F(k) = (\gamma^\mu k^*_\mu + M^*(k)) \frac{1}{k^{*2}_\mu - M^*(k)^2 + i\epsilon} \,, \tag{169}$$

$$G_D(k) = (\gamma^\mu k^*_\mu + M^*(k)) \frac{i\pi}{E^*(k)} \delta(k^0 - E(k)) \theta(k_F - |\mathbf{k}|) \,. \tag{170}$$

Since the evaluation of observables in the infinite nuclear matter is given by the expectation value of a bilinear product of fermion operators [14, 29] in the baryon ground state $|\Psi\rangle$, the energy density, pressure and other thermodynamic quantities are all expressed by the self-consistent Green's function.

The self-consistent relations between macroscopic conservation laws and energy-momentum conservations at vertices of Feynman diagrams are discussed as properties of conserving approximations [16]. The fundamental requirements of conserving approximations state that:

(A) The self-consistent one-particle Green's function $G(k)$ must satisfy the equation of motion defined by a basic Hamiltonian, or a Lagrangian.

(B) A given approximate two-particle Green's function $G_2(k_1, k_2)$ expressed in terms of $G(k)$ must satisfy the symmetry condition: $G_2(k_1, k_2) = G_2(k_2, k_1)$.
The self-consistent Green's function with a given class of Feynman diagrams should be constructed to maintain the fundamental requirement (A) and (B).

The fundamental requirements are discussed as the properties of Φ-derivable approximation [16], and in an infinite nuclear matter, the thermodynamic potential, $\omega(T, \mu)$, at finite temperature T is given by:

$$\omega(T, \mu) = -k_B T \int \frac{d^3k}{(2\pi)^3} \mathrm{Tr}\Big[\ln(-G^{-1})\Big] - k_B T \int \frac{d^3k}{(2\pi)^3} \mathrm{Tr}\Big[\Sigma G\Big] + k_B T \phi \,, \tag{171}$$

where k_B is the Boltzmann constant and μ is a chemical potential. The thermodynamic potential ω is stationary with respect to variations of the Green's function: $\delta\omega/\delta G(k) = 0$, since the self-energy is given by $\Sigma(k) = \delta\phi/\delta G(k)$. The variations of thermodynamic potential and the energy density, $\mathcal{E}$, are related with $\delta\omega|_{T,\mu} = \delta\mathcal{E}|_{\rho_B,\sigma}$ (σ is the entropy density), and consequently, the self-consistent condition follows [31]:

$$\delta\mathcal{E}/\delta\Sigma(k)|_{\rho_B,\sigma} = 0\ . \tag{172}$$

This is one of the fundamental conditions for the theory of conserving approximations .

The self-consistent condition (172) should be understood for the variation of the three independent self-energies, $\Sigma^s(k)$, $\Sigma^0(k)$ and $\Sigma^v(k)$, respectively, and the dynamical variables, $M^*(k)$, $k^*(k)$ and $E(k)$, are given by the self-energies as (167). The functional derivative of energy density, $\delta\mathcal{E}/\delta n_i$, is given by

$$\frac{\delta\mathcal{E}}{\delta n_i} = E(k_i)+\sum_l n_l\left[\frac{\delta M^*(k_l)}{\delta n_i}\frac{\delta\mathcal{E}}{\delta M^*(k_l)}+\frac{\delta\mathbf{k}^*(k_l)}{\delta n_i}\cdot\frac{\delta\mathcal{E}}{\delta\mathbf{k}^*(k_l)}+\frac{\delta\Sigma^0(k_l)}{\delta n_i}\frac{\delta\mathcal{E}}{\delta\Sigma^0(k_l)}\right]. \tag{173}$$

The functional derivative with respect to self-energies, or dynamical variables $M^*(k)$, $k^*(k)$ and $\Sigma^0(k)$, on the right-hand side will vanish by the eq. (172). When the functional differential equations (172) are solved, the chemical potential is equal to the single particle energy: $\mu = E(k_F)$ at the Fermi surface, and $\delta\mathcal{E}/\delta n_i = E(k_i) = E^*(k_i) - \Sigma^0(k_i)$, is maintained exactly [31].

The self-energies derived from (172), $\delta\mathcal{E}/\delta\Sigma(k) = 0$, will be denoted as $\Sigma_L(k)$, and the HV theorem and Landau's requirement of quasiparticle energy are maintained with $\Sigma_L(k)$. The self-energies constructed by the standard Feynman-diagram method will be denoted as $\Sigma_F(k)$. It should be noted that the self-energy $\Sigma_F(k)$ constructed directly by Feynman rules are not always equivalent to $\Sigma_L(k)$. The self-consistent condition may be difficult to maintain, since the self-consistency depends on the functional form and accuracy to an approximate Green's function. The eq. (172) generates complicated functional differential equations for self-energies $\Sigma_L(k)$. The solutions are difficult to obtain directly, however, the Feynman self-energy, $\Sigma_F(k)$, is very useful to find the exact self-energy solution to the functional differential equation.

The condition of conserving approximation will produce the self-consistent free energy $F_{HK}(n(k))$ discussed in DFT. For example, the Hohenberg-Kohn functional, $F_{HK}(n(k))$, is equivalent to the energy density, $\mathcal{E}$, for fixed σ and ρ_B. The deficiency of the self-consistency of the kinetic energy contribution pointed out in the ref. [30] is unambiguously solved in the theory of conserving approximations, as the construction of self-energy functions by the eq. (172). The self-consistent method is applied to Hartree-Fock approximation in nuclear matter [21]. It is possible to analyze the characteristic scheme and consistency of one's approximate calculations by comparing self-energies, $\Sigma_F(k)$ and $\Sigma_L(k)$.

References

[1] H. Heiselberg and M. Hjorth-Jensen, *Phys. Rep.*, V328, (2000) 237.

[2] J. M. Lattimer and M. Prakash, *Phys. Rep.*, 442, (2007) 109.

[3] R. P. Drake, *High-Energy-Density Physics*, Springer-verlag, (2006).

[4] R. Kippenhahn and A. Weigert, *Stellar Structure and Evolution*, Springer-verlag, (1994).

[5] N. Straumann, *General Relativity With Applications to Astrophysics*, Springer-verlag, (2004).

[6] I. J. R. Aitchison, *An Informal Introduction to Gauge Field Theories*, Cambridge University Press, 1982.

[7] C. Itzykson and J. Zuber, *Quantum Field Theory*, McGraw-Hill, 1986.

[8] M. E. Peskin and D. V. Schroeder, *An Introduction to Quantum Field Theory*, Westview Press, 1995.

[9] A. Bohr and B. R. Mottelson, *Nuclear Structure: Single-Particle Motion/Nuclear Deformations*, World Scientific Pub Co Inc, 1969.

[10] B. D. Day, *Reviews of Modern Phys.* 50 (1978) 495.

[11] N. M. Hugenholtz and L. Van Hove, *Physica,* 24 (1958) 363.

[12] B. D. Serot, *Rep. Prog. Phys.* 55 (1992) 1855.

[13] K. A. Brueckner and C. A. Levinson, *Phys. Rev.* 97 (1955) 1344; K. A. Brueckner, *Phys. Rev.* 110 (1958) 597; K. A. Brueckner and D. T. Goldman, Phys. Rev. 117 (1960) 207; K. A. Brueckner, J. L. Gammel and J. T. Kubis, *Phys. Rev.* 118 (1960) 1438.

[14] B. D. Serot and J. D. Walecka, *Advances in Nuclear Physics*, edited by J. W. Negele and E. Vogt (Plenum, New York, 1986), Vol. 16.

[15] H. Uechi, *Nucl. Phys.* A501 (1989) 813; H. Uechi, *Phys. Rev.* C41 (1990) 744.

[16] G. Baym and L. P. Kadanoff, *Phys. Rev.* 124 (1961) 287; G. Baym, *Phys. Rev.* 127 (1962) 1391.

[17] Y. Takada, *Phys. Rev.* B52 (1995) 12708.

[18] *Progress in Nonequilibrium Green's Functions, Progress in Nonequilibrium Green's Functions II*, edited by M. Bonitz, R. Nareyka and D. Semkat, *World Scientific*, 2000.

[19] W. Kohn and L. J. Sham, *Phys. Rev.* 140 (1965) A1133.

[20] W. Kohn, *Rev. Mod. Phys.*, 71 (1999) 1253.

[21] H. Uechi, *Prog. Theor. Phys.* 111 (2004) 525.

[22] M. Buballa, *Phys. Rep.* 407 (2005) 205.

[23] L. D. Landau, *Sov. Phys. JETP* 3 (1956) 920; 5 (1957) 101.

[24] G. Baym and C. Pethick, in The physics of liquid and solid helium, Part II, ed. K. H. Bennemann and J. B. Ketterson (John Wiley, New York,1978).

[25] A. Migdal, *Theory of finite fermi systems and applications to atomic nulcei*, J. Wiley, New York, 1967.

[26] T. Matsui, *Nucl. Phys.* A370, (1981) 365; H. Uechi, *Nucl. Phys.* A501, (1989) 813.

[27] B. D. Serot, *Phys. Lett.* 86 B (1979) 146 [Erratum: *Phys. Lett.* 87B (1979) 403].

[28] G. Baym and S. A. Chin, *Nucl. Phys.* A262 (1976) 527.

[29] A. L. Fetter and J. D. Walecka, *Quantum Theory of Many-Particle Systems*, McGraw-Hill, Inc., New York, 1971.

[30] R. M. Dreizler and E. K. U. Gross, *Density Functional Theory*, Springer-Verlag, Berlin Heidelberg, 1990.

[31] H. Uechi, *Nucl. Phys.* A696, (2001) 511.

In: Neutron Stars
Editors: Hiroshi Uechi et al, pp. 39-72
ISBN: 978-1-62081-123-8

Chapter 2

THE HIGH-DENSITY NUCLEAR EQUATION OF STATE IN RELATIVISTIC MEAN-FIELD THEORY

Brian D. Serot*
Department of Physics and Center for Exploration of Energy and Matter
Indiana University, Bloomington, IN 47405.

Abstract

The properties of high-density nuclear and neutron matter are studied using a relativistic mean-field approximation to the nuclear matter energy functional. Based on ideas of effective field theory, nonlinear interactions between the fields are introduced to parameterize the density dependence of the energy functional. Various types of nonlinearities involving scalar-isoscalar (σ), vector-isoscalar (ω), and vector-isovector (ρ) fields are studied. After calibrating the model parameters at equilibrium nuclear matter density, the model and parameter dependence of the resulting equation of state is examined in the neutron-rich and high-density regime. It is possible to build different models that reproduce the same observed static properties at normal nuclear densities, but which yield maximum neutron star masses that differ by more than one solar mass. Implications for the existence of kaon condensates or quark cores in neutron stars are discussed. Recent developments involving isoscalar-isovector interactions and the inclusion of dynamical nuclear properties into the calibration procedure are shown to lead to important constraints on the high-density, neutron-rich equation of state.

1. Introduction

Numerous calculations have established that relativistic mean-field models provide a realistic description of the bulk properties of finite nuclei and nuclear matter [1–8]. In addition to this successful low-energy phenomenology, these models are often extrapolated into regimes of high density and temperature to extract the nuclear equation of state (EOS), which is the basic ingredient in many astrophysical applications and in microscopic models of energetic nucleus–nucleus collisions.

Based on the original version of Walecka [9] and its extensions [3,4,10–14], relativistic mean-field models generally involve the interaction of Dirac nucleons with neutral scalar

*E-mail address: serot@indiana.edu

and vector mesons and with isovector ρ mesons. One of the key observations in their success is that to provide sufficient flexibility, nonlinear self-interactions for the scalar meson must be included [2, 10, 15–20]. Since these models were proposed to be renormalizable, the scalar self-interactions are limited to a quartic polynomial, and scalar–vector and vector–vector interactions are not allowed [21]. One of the motivations for renormalizability, as discussed in Walecka's seminal paper, is that once the model parameters are calibrated to observed nuclear properties, one can extrapolate into regimes of high density or temperature without the appearance of new, unknown parameters.

An alternative approach is inspired by *effective* field theories, such as chiral perturbation theory [22–27], which successfully describes the low-energy phenomenology of hadronic Goldstone bosons [28–31]. Although a lagrangian usually serves as the starting point, the meson and baryon fields are no longer considered elementary, and the constraint of renormalizability is dropped. This has several important consequences. First, there is no reason to restrict meson self-interactions to a simple quartic polynomial in the scalar field; on the contrary, one should include all interaction terms that are consistent with the underlying symmetries of QCD. Second, since there are an infinite number of coupling constants, one must find suitable expansion parameters for the systems under consideration, and one must develop a systematic truncation scheme for the effective theory to have any predictive power. Third, extrapolation of calculated results into new regimes of the physical parameters becomes problematic, because the truncation scheme may break down, and predictions can become sensitive to unknown parameters.

Within the framework of effective field theory, mean-field models of nuclear structure and the EOS must be interpreted in a new context. One important observation is that near normal nuclear density, the mean scalar and vector fields (or nucleon self-energies), which we denote as Φ and W, are large on nuclear energy scales but are small compared to the nucleon mass M and vary slowly in finite nuclei. This implies that the ratios Φ/M and W/M and the gradients $|\nabla\Phi|/M^2$ and $|\nabla W|/M^2$ are useful expansion parameters. The assumption of "naturalness " in effective field theory is also important. Naturalness implies that the coefficients of the various terms in the lagrangian, when expressed in appropriate dimensionless form, should all be of order unity. When combined with meaningful expansion parameters, this means that one can anticipate the approximate magnitude of mean-field contributions to the energy (at least up to moderate nuclear densities) and thereby motivate a suitable truncation scheme; if the coefficients are natural, the omitted terms will be numerically unimportant.[1] Naturalness also implies that one should include *all* possible terms (that is, those allowed by the symmetries) through a given order of truncation; it is *unnatural* for some coefficients to vanish without a relevant symmetry argument.

¿From this point of view, it is difficult to justify nuclear mean-field models that include only scalar self-interactions [10, 15, 16, 19], and consequently, generalizations that also include quartic self-interactions for the neutral vector meson have been discussed [3,4,12,13,18,33–36]. Moreover, an analysis involving all meson self-interactions through fourth order in the isoscalar scalar and vector fields has been performed [3, 13]. These extensions give rise to additional model parameters (coupling constants) that must be constrained by calibrating to observed nuclear properties. For the truncation at fourth order to

[1]It has also been shown that the naturalness assumption is consistent with dimensional counting in chiral perturbation theory [3,32].

be sensible, the parameters so obtained should exhibit naturalness.

Although it is possible to discuss effective hadronic field theory from the point of view of a lagrangian, as above, the expansion in powers of the mean fields is a low-density expansion, and it is hard to justify the neglect of many-body corrections, which are known to be relevant in nuclear structure and in the EOS. Alternatively, one effective action [13]. In such a formulation of the relativistic nuclear many-body problem, the central object is an energy functional of scalar and vector densities (or more generally, vector four-currents) [37–40]. Extremization of the functional gives rise to Dirac equations for occupied orbitals with *local* scalar and vector potentials, not only in the Hartree approximation, but in the general case as well. Rather than work solely with the densities, one can introduce auxiliary variables corresponding to the local potentials, so that the functional depends also on mean meson fields. The resulting field equations have the same form as in a Dirac–Hartree calculation [3], but correlation effects can be included, *if* the proper energy functional can be found. This procedure is analogous to the well-known Kohn–Sham [41, 42] approach in density-functional theory, with the local meson fields playing the role of (relativistic) Kohn–Sham potentials ; by introducing nonlinear couplings between these fields, one can implicitly include additional density dependence in the potentials. Thus the nonlinear meson interaction terms simulate more complicated physics, such as one- and two-pion exchange, or vacuum-loop corrections, which might be calculated directly in a more microscopic many-body approach [8,34,35,39,43,44]. The fields (and their gradients) again serve as useful expansion parameters at moderate density, so the nonlinear interaction terms can be truncated, leaving a finite number of unknown couplings.

Rather than focus on the calculation of the nonlinear couplings from an underlying effective lagrangian, we wish to concentrate instead on how well the energy functional can be calibrated by fitting the couplings to observed nuclear properties, and on the limitations on the extrapolation of the resulting EOS into the high-density regime. In general, even with a significant truncation, the number of unknown couplings exceeds the number of normalization conditions, which we initially take to be five properties of infinite nuclear matter: the equilibrium density and binding energy (ρ_0, $-e_0$), the nucleon effective (or Dirac) mass at equilibrium (M_0^*), the compression modulus (K_0), and the bulk symmetry energy (a_4). (Experience has shown that an accurate reproduction of these five properties leads to realistic results when the calculations are extended to finite nuclei [3, 13, 15, 16, 19, 20, 45–47].) Thus families of models can be generated which describe exactly the same nuclear matter properties at equilibrium [33]. We then investigate the differences in the high-density EOS predicted by models within a given family. This question is important for astrophysical applications, particularly in the study of neutron stars; it will be difficult to deduce the existence of "exotic" neutron-star structure (for example, hyperonic matter, kaon condensates, or quark cores) unless the EOS of the more mundane components (neutrons, protons, and electrons) is well constrained.

If our truncation of the energy functional is motivated by low-density behavior, why should we have any confidence at all in a high-density extrapolation? This is indeed the crucial question, and we are not attempting here to justify such an extrapolation; we are merely recognizing that this procedure is often used in neutron-star calculations, even recent ones, without any mention of the implicit assumptions about the absence of additional contributions at high density [48–61]. We therefore feel it is necessary to investigate quantitatively the uncertainties in the extrapolated equation of state.

We begin the theoretical analysis with a model that contains meson–meson and meson self-interactions described by an *arbitrary* finite polynomial in the fields. We find that both the asymptotic (high-density) limit of the EOS and the approach to this limit (*i.e.*, the "stiffness") are model dependent. In particular, one can construct models with the *same* equilibrium nuclear matter properties that yield high-density equations of state ranging all the way from the causal limit ($p = \mathcal{E}$) to one that resembles a free relativistic gas ($p = \mathcal{E}/3$) [18, 33]. (Here p is the pressure and $\mathcal{E}$ is the energy density.)

As an explicit example, we consider a model that includes self-interactions for the isoscalar scalar and vector mesons and for the ρ meson up to fourth order in the fields [36]. To our knowledge, this was the first time nonlinear terms in the ρ meson mean field were included. (Note that the ρ field enters here as an *effective* field whose purpose is to parameterize the isospin dependence; thus, fundamental questions about causal propagation [62, 63] and spin mixing [64] are not relevant. In particular, all of the meson propagators have *spacelike* four-momenta in our many-body calculations.) To provide a quantitative measure of the variations in the EOS, we compute neutron star masses, which turn out to be sensitive to the model types and to changes in the parameterizations, *even for models that reproduce the same equilibrium nuclear matter properties*. In some cases, variations in the calculated maximum mass are more than one solar mass.

These initial results lead us to two basic conclusions. First, earlier methods for calibration of the EOS at normal density are *not* sufficient to provide a satisfactory extrapolation into the density regime relevant for neutron stars. The basic problem rests with the quartic neutral vector meson (ω) interaction, which produces major modifications in the high-density behavior. Improved calibration of this term using additional nuclear input is necessary, as we will show in more modern treatments. Second, we find that if the quartic ω term can be accurately determined, additional higher-order isoscalar terms (and the quartic ρ term) produce relatively minor changes in the neutron star mass. This occurs because the W^4 term softens the high-density EOS so completely that additional isoscalar interactions have little effect, at least in the density regime relevant for neutron stars.

These conclusions have important implications, because the contributions to the EOS from the neutrons and protons are commonly believed to be the best understood part of the physical input necessary to describe a dense stellar object. Without well-constrained results from this part of the EOS, it is impossible to determine the importance of additional dynamics, for example, the transition from nuclear matter to quark matter in the interiors of neutron stars [49, 56, 57, 60, 61, 65, 66], or the role of strangeness in the form of hyperons [53, 58, 67–70] or a kaon condensate [55, 71–74]. As an example of how uncertainties in the basic nuclear EOS can influence these interesting effects, we study the transition from hadronic matter to quark matter using a simple model [1, 66] and show the possible variations in the results.

This chapter is an update of an earlier article on relativistic mean-field theory and the high-density nuclear equation of state [36]. As such, it is intended to review some earlier calculations and then briefly present more recent improvements and advances. The structure of this chapter is as follows: In Sec. 2, we present the general model, which involves an arbitrary number of nonlinear meson interactions, and derive the EOS. Based on this general model, Section 3 is devoted to the high-density limit of the EOS. In Sec. 4, we apply our model to neutron stars. For the quantitative analysis, we initially include self-interactions up to fourth order in the ρ and ω fields and then investigate the consequences of sixth-order and eighth-order ω self-interactions. We also briefly discuss the parameter and model dependence of the transition to quark matter in the central region of the star. In Section 5, we consider newer developments in the calibration of these models, which lead to more stringent constraints on neutron-rich matter above equilibrium nuclear density. Section 6 contains a short summary and our conclusions.

2. The Nuclear Equation of State

We describe the nuclear equation of state using a relativistic approach involving valence Dirac nucleons and effective mesonic degrees of freedom, which are taken to be neutral scalar and vector fields, plus the isovector ρ meson field. Rather than focus directly on a lagrangian, we consider instead as a starting point an *effective action*

$$\Gamma = \Gamma[\phi, V_\mu, \mathbf{b}_\mu] \ , \tag{1}$$

which is a functional of the meson fields denoted by ϕ, V_μ, and $\mathbf{b}_\mu$ for the scalar, vector-isoscalar, and vector-isovector field, respectively. In principle, this functional can be calculated in a many-body approach based on a Lagrangian for the nucleon–nucleon interaction, or its general form might be obtained from an underlying theory. Here we will be satisfied to parameterize the effective action, calibrate it as accurately as we can to observed nuclear properties, and then examine the predicted high-density equation of state. The effective action is related to the thermodynamic potential Ω [2, 75] by

$$i\beta\Omega = \Gamma[\phi, V_\mu, \mathbf{b}_\mu] \ , \tag{2}$$

where the fields are determined by the general thermodynamic principle that they should make Ω stationary:

$$\frac{\partial\Gamma}{\partial\phi} = \frac{\partial\Gamma}{\partial V_\mu} = \frac{\partial\Gamma}{\partial\mathbf{b}_\mu} = 0 \ . \tag{3}$$

A basic property of the functional is that it reflects the underlying symmetries [76]. Thus, if we assume that the system possesses two conserved charges, namely, baryon number B and the third component of total isospin I_3, this gives rise to two chemical potentials μ and ν:

$$\Omega \equiv \Omega(\beta, \mu, \nu) \ , \tag{4}$$

with

$$B = \int d^3x\, \rho = -\frac{\partial\Omega}{\partial\mu} \ , \tag{5}$$

$$I_3 = \frac{1}{2}\int d^3x\, \rho_3 = -\frac{\partial\Omega}{\partial\nu} \ , \tag{6}$$

β the inverse temperature, and $\rho_3 \equiv \rho_{\rm p} - \rho_{\rm n}$. Note that the fields can be held fixed in evaluating the partial derivatives in Eqs. (5) and (6) by virtue of the extremization conditions (3).

In the field theoretical context [77], one can show that the effective action can be expanded as a power series in gradients of the fields. Thus, after taking the zero-temperature limit, one usually writes

$$\Gamma[\phi, V_\mu, \mathbf{b}_\mu] = \int d^4x \Big[-\mathcal{V}_{\text{eff}}(\phi, V_\mu, \mathbf{b}_\mu) + \mathcal{Z}(\phi, V_\mu, \mathbf{b}_\mu, \partial_\nu\phi, \partial_\nu V_\mu, \partial_\nu \mathbf{b}_\mu) \Big] , \qquad (7)$$

where the second term vanishes in a uniform system, and the dependence on the chemical potentials has been suppressed. For the effective potential, we make the following *Ansatz*:

$$\begin{aligned} \mathcal{V}_{\text{eff}}(\phi, V_\mu, \mathbf{b}_\mu; \mu, \nu) &= \frac{1}{2} m_{\text{s}}^2 \phi^2 - \frac{1}{2} m_{\text{v}}^2 V_\mu V^\mu - \frac{1}{2} m_\rho^2 \mathbf{b}_\mu \cdot \mathbf{b}^\mu \\ &\quad + i \operatorname{tr} \ln Z_\psi(\mu, \nu) - i \operatorname{tr} \ln Z_\psi(0, 0) \\ &\quad + \Delta\mathcal{V}(\phi, V_\mu V^\mu, \mathbf{b}_\nu \cdot \mathbf{b}^\nu; \mu, \nu) , \end{aligned} \qquad (8)$$

with a nonlinear potential[2]

$$\Delta\mathcal{V}(\phi, V_\mu V^\mu, \mathbf{b}_\nu \cdot \mathbf{b}^\nu; \mu, \nu) \equiv -\sum_{i,j,k} a_{ijk}(\mu, \nu)\, \phi^i \,(V_\mu V^\mu)^j \,(\mathbf{b}_\nu \cdot \mathbf{b}^\nu)^k \qquad (9)$$

that contains at least three powers of the fields: $i + 2j + 2k \geq 3$. The fermionic contributions are represented by a one-body term $\operatorname{tr} \ln Z_\psi$ (with the appropriate zero-density subtraction [78]) and by terms where the fermions have been "integrated out", which results in a (generally nonanalytic) dependence of the mesonic coefficients a_{ijk} on the chemical potentials μ and ν. There will also be contributions to the a_{ijk} that are independent of the chemical potentials; these arise from integrating out heavy degrees of freedom and vacuum loops. Thus $\mathcal{V}_{\text{eff}}$ contains explicit contributions only from valence nucleons and classical meson fields.

The fermionic part $\operatorname{tr} \ln Z_\psi$ is obtained by evaluating the trace of the kernel

$$K(\mu, \nu) = (i\partial^\mu - g_{\text{v}} V^\mu - \frac{1}{2} g_\rho \boldsymbol{\tau} \cdot \mathbf{b}^\mu)\gamma_\mu + \mu\gamma^0 + \frac{1}{2}\nu\tau_3\gamma^0 - (M - g_{\text{s}}\phi) \qquad (10)$$

using Dirac wave functions calculated in the presence of static background fields. The subtraction removes contributions from negative-energy states, which are already included implicitly in the nonlinear parameters a_{ijk}. (See Ref. [78] for an analogous calculation.) The values of the fields are determined by extremization at the given values of μ and ν, which are determined by inverting Eqs. (5) and (6).

The potential of Eq. (8) represents an effective field theory for the interacting nucleons. Although the exact form of the effective potential is not known, we introduce the meson mean fields as relativistic Kohn–Sham potentials [41,42] and consider the valence nucleons moving in the resulting local fields. The nonlinear interactions of the fields generate implicit density dependence above and beyond that arising from the couplings in Eq. (10). Thus the series in Eq. (9) can be interpreted as a Taylor series parametrization of the unknown part

[2]The reader will notice that we have omitted "off-diagonal" terms like $\sum_{i,j,k} \tilde{a}_{ijk}(\mu, \nu)\, \phi^i \,(V_\mu V_\nu)^j \,(\mathbf{b}^\nu \cdot \mathbf{b}^\mu)^k$ in our potential. Since we will work in the rest frame of infinite nuclear matter, only the zeroth component of the neutral vector mean fields are nonzero. In this case, the "off-diagonal" terms reduce to the terms we have retained.

of the effective potential, which includes the effects of nucleon exchange and correlations, as well as contributions from other mesons and the quantum vacuum. (See the discussion in Refs. [13, 78].)

Although the couplings a_{ijk} generally depend on the chemical potentials, experience with calculations for finite nuclei and nuclear matter, together with explicit computations of exchange and correlation corrections [79, 80], implies that mean fields and *constant* couplings a_{ijk} provide an adequate (albeit approximate) parametrization of these many-body effects. Thus we consider the a_{ijk} as constants in the sequel and leave the study of their dependence on μ and ν as a topic for future investigation. Moreover, at low densities and temperatures, the mean meson fields are small compared to the nucleon mass, and so provide useful expansion parameters [13]. Thus, in practice, the series in Eq. (9) can be truncated at some reasonable order, and the relevant question in this paper is how far one can extrapolate the truncated potential into the high-density regime.

In principle, the unknown coefficients (coupling constants) can be constrained by imposing chiral symmetry and other symmetries of the underlying QCD, such as broken scale invariance. (Lorentz covariance and isospin symmetry are already incorporated explicitly.) As has been discussed previously, however [13, 20], if one assumes a *nonlinear* realization of the chiral symmetry for the pions and nucleons [81], the meson self-interaction terms are essentially unconstrained.[3] We therefore take the couplings as free model parameters in our approach. Obviously, an infinite number of normalization conditions is generally needed to fix their values. Since this is not feasible in practice, we terminate the summation at the finite values $i_{\max}$, $j_{\max}$, and $k_{\max}$.

According to Eq. (3), the thermodynamic potential Ω must be stationary with respect to changes in the fields for fixed values of the proton and neutron chemical potentials

$$\mu_{\rm p} \equiv \mu + \frac{\nu}{2} = (k_{\rm Fp}^2 + M^{*2})^{1/2} + W + \frac{1}{2}R\,, \tag{11}$$

$$\mu_{\rm n} \equiv \mu - \frac{\nu}{2} = (k_{\rm Fn}^2 + M^{*2})^{1/2} + W - \frac{1}{2}R\,. \tag{12}$$

Here, following Bodmer [33], we define the scaled meson fields $\Phi \equiv g_{\rm s}\phi$, $W \equiv g_{\rm v}V_0$, and $R \equiv g_\rho b_0$, with b_0 the timelike, neutral part of the ρ meson field; the effective nucleon mass is $M^* \equiv M - \Phi$. (We work in the rest frame of the infinite matter, where the spatial parts of the vector fields vanish.) The Fermi momenta for protons ($k_{\rm Fp}$) and neutrons ($k_{\rm Fn}$) are related to the conserved baryon density

$$\rho = \frac{1}{3\pi^2}\left(k_{\rm Fp}^3 + k_{\rm Fn}^3\right) \tag{13}$$

and isovector density

$$\rho_3 = \frac{1}{3\pi^2}\left(k_{\rm Fp}^3 - k_{\rm Fn}^3\right). \tag{14}$$

Using Eqs. (2), (5), and (6), together with relations (11) and (12) for the chemical potentials, it is straightforward to eliminate the chemical potentials in favor of the densities

[3] Broken scale invariance leads to restrictions on the purely scalar interactions, as shown in Refs. [82] and [78], but we will not consider these limitations here. As discussed in the next section, the details of the scalar dynamics do not play a major role in our analysis.

and to compute the pressure p and the energy density $\mathcal{E} = -p + \mu\rho + \frac{1}{2}\,\nu\rho_3$:

$$
\begin{aligned}
p \;=\;& \frac{1}{3\pi^2}\int_0^{k_{\mathrm{Fp}}} dk \frac{k^4}{(k^2+M^{*2})^{1/2}} + \frac{1}{3\pi^2}\int_0^{k_{\mathrm{Fn}}} dk \frac{k^4}{(k^2+M^{*2})^{1/2}} \\
&+ \frac{1}{2c_{\mathrm{v}}^2}\,W^2 + \frac{1}{2c_\rho^2}\,R^2 - \frac{1}{2c_{\mathrm{s}}^2}\,\Phi^2 + \sum_{i,j,k} \bar{a}_{ijk}\,\Phi^i\,W^{2j}\,R^{2k}\,, \qquad (15)
\end{aligned}
$$

$$
\begin{aligned}
\mathcal{E} \;=\;& \frac{1}{\pi^2}\int_0^{k_{\mathrm{Fp}}} dk\,k^2(k^2+M^{*2})^{1/2} + \frac{1}{\pi^2}\int_0^{k_{\mathrm{Fn}}} dk\,k^2(k^2+M^{*2})^{1/2} \\
&+ W\rho + \frac{1}{2}\,R\rho_3 - \frac{1}{2c_{\mathrm{v}}^2}\,W^2 - \frac{1}{2c_\rho^2}\,R^2 + \frac{1}{2c_{\mathrm{s}}^2}\,\Phi^2 - \sum_{i,j,k} \bar{a}_{ijk}\,\Phi^i\,W^{2j}\,R^{2k}\,. \qquad (16)
\end{aligned}
$$

Here the ratios $c_i^2 = g_i^2/m_i^2$ and $\bar{a}_{ijk} \equiv a_{ijk}/(g_s^i g_{\mathrm{v}}^{2j} g_\rho^{2k})$ have been introduced for convenience.

As noted earlier, the pressure in Eq. (15) and the energy density in Eq. (16) also contain vacuum contributions arising from the partition function of the nucleons. However, these vacuum terms can be absorbed in the definition of the nonlinear couplings in Eq. (9) [78, 83, 84], and thus we include explicitly only the contributions from valence nucleons.

At zero temperature, the stationarity conditions of Eq. (3) with fixed chemical potentials are equivalent to an extremization of the energy at fixed baryon and isovector density. This leads to the self-consistency equations

$$
\begin{aligned}
\frac{1}{c_{\mathrm{s}}^2}\,\Phi - \sum_{i,j,k} i\,\bar{a}_{ijk}\,\Phi^{i-1}\,W^{2j}\,R^{2k} &= \rho_{\mathrm{s}}\,, \qquad (17)\\
\frac{1}{c_{\mathrm{v}}^2}\,W + \sum_{i,j,k} 2j\,\bar{a}_{ijk}\,\Phi^{i}\,W^{2j-1}\,R^{2k} &= \rho\,, \qquad (18)\\
\frac{1}{c_\rho^2}\,R + \sum_{i,j,k} 2k\,\bar{a}_{ijk}\,\Phi^{i}\,W^{2j}\,R^{2k-1} &= \frac{1}{2}\,\rho_3\,, \qquad (19)
\end{aligned}
$$

where the scalar density is given by

$$
\rho_{\mathrm{s}} = \frac{M^*}{\pi^2}\int_0^{k_{\mathrm{Fp}}} dk \frac{k^2}{(k^2+M^{*2})^{1/2}} + \frac{M^*}{\pi^2}\int_0^{k_{\mathrm{Fn}}} dk \frac{k^2}{(k^2+M^{*2})^{1/2}}\,. \qquad (20)
$$

Because of the factor of $1/2$ on the right-hand side of Eq. (19), it follows that the relevant expansion parameters for the energy density are Φ/M, W/M, and $2R/M$. Moreover, by dividing $\mathcal{E}$ by M^4 and expressing the result in terms of these expansion parameters, one can identify the scaled couplings that should all be of roughly the same size if they are "natural", namely,

$$
\frac{1}{2c_{\mathrm{s}}^2 M^2}\,,\quad \frac{1}{2c_{\mathrm{v}}^2 M^2}\,,\quad \frac{1}{8c_\rho^2 M^2}\,,\quad \text{and} \quad \frac{\bar{a}_{ijk}M^{i+2j+2k-4}}{2^{2k}}\,.
$$

3. The High-Density Limit

The coupling constants c_{v}^2, c_{s}^2, c_{ρ}^2, and $\bar{a}_{ijk}$ in Eqs. (15) and (16) enter as unknown model parameters. According to the generally accepted procedure, these parameters will be chosen to reproduce the properties of nuclear matter near equilibrium. The basic ingredient in many astrophysical problems, *e.g.*, neutron-star calculations, is the EOS

$$p = p(\mathcal{E}) \,, \tag{21}$$

which is then extrapolated into the neutron-rich and high-density regime. Anticipating the results of the next section, one can expect that different parameter sets that lead to identical equilibrium properties produce qualitatively similar equations of state at low densities. The relevant question is whether this qualitatively similar behavior persists at high densities, particularly in the regime important for neutron stars. As a first step in this direction, we investigate the high-density limit of the EOS generated by the model introduced in the previous section.

To make the discussion more transparent, we focus here on pure neutron matter ($k_{\mathrm{Fp}} = 0, k_{\mathrm{Fn}} \equiv k_{\mathrm{F}}$), although nuclear matter in β-decay equilibrium with a finite proton to neutron ratio is necessary to achieve accurate results for maximum neutron star masses. We will return to this issue in the next section.

It is clear that a sufficiently large number of couplings introduces a high degree of flexibility. Due to the nonlinearity of the problem, not all families of parameter sets lead to physically acceptable results, which provides one way to restrict the parameter space. Classes of models can be ruled out if basic physical requirements are violated. For example, one certainly requires that the pressure p be a smooth function of the energy density $\mathcal{E}$. Moreover, it is necessary that the speed of (first) sound c_1 respect causality and also be real, to ensure stability. That is,

$$0 \leq c_1^2 = \frac{\partial p}{\partial \mathcal{E}} \leq 1 \,.$$

In addition to these general principles, we require a positive and bounded value of the nucleon effective mass, *i.e.*,

$$0 \leq M^* \leq M \,. \tag{22}$$

This is motivated by the expectation that physically reasonable models will demonstrate some degree of chiral-symmetry restoration at finite density, leading to a reduction in the nucleon mass, with the most extreme situation corresponding to total restoration of the symmetry. The consequences of Eq. (22) will become more transparent in the following.

By specializing the formalism of the preceding section to pure neutron matter and by using the self-consistency equations (17)–(19), the pressure and energy density can be expressed as

$$\begin{aligned} p &\equiv p_0(k_{\rm F}, M^*) + \Delta p(\Phi, W, R) \\ &= \frac{1}{3\pi^2}\int_0^{k_{\rm F}} dk \frac{k^4}{(k^2+M^{*2})^{1/2}} \\ &\quad + \frac{1}{2c_{\rm v}^2} W^2 + \frac{1}{2c_\rho^2} R^2 - \frac{1}{2c_{\rm s}^2}\Phi^2 + \sum_{i,j,k} \bar{a}_{ijk}\, \Phi^i\, W^{2j}\, R^{2k} \,, \end{aligned} \tag{23}$$

$$\begin{aligned} \mathcal{E} &\equiv \mathcal{E}_0(k_{\rm F}, M^*) + \Delta\mathcal{E}(\Phi, W, R) \\ &= \frac{1}{\pi^2}\int_0^{k_{\rm F}} dk\, k^2 (k^2+M^{*2})^{1/2} \\ &\quad + \frac{1}{2c_{\rm v}^2} W^2 + \frac{1}{2c_\rho^2} R^2 + \frac{1}{2c_{\rm s}^2}\Phi^2 + \sum_{i,j,k} (2j+2k-1)\, \bar{a}_{ijk}\, \Phi^i\, W^{2j}\, R^{2k} \,, \end{aligned} \tag{24}$$

where p_0 and $\mathcal{E}_0$ denote the results for a relativistic, noninteracting gas of spin-1/2 baryons with mass M^*. Note that these expressions include, as a special case, models where only the $\bar{a}_{0jk}$ are nonzero, so that there are no scalar–vector couplings. Moreover, the isovector density in Eq. (19) is replaced by

$$\rho_3 = -\rho = -\frac{1}{3\pi^2}\, k_{\rm F}^3 \,. \tag{25}$$

At high densities, the left-hand sides of the self-consistency equations (18) and (19) must grow linearly in ρ, and thus we start with the *Ansatz*

$$\lim_{\rho\to\infty} W = w_0 \rho^\alpha \quad , \quad \lim_{\rho\to\infty} R = r_0 \rho^\beta \,, \tag{26}$$

where $0 < \alpha, \beta \leq 1$, $w_0 > 0$, and $r_0 < 0$. Since we assume that the effective mass is bounded, we can replace the scalar field by the limit

$$\Phi_\infty \equiv \lim_{\rho\to\infty} (M - M^*) \leq M \,. \tag{27}$$

To fulfill the resulting self-consistency equations:

$$\frac{1}{c_{\rm v}^2} W + \sum_{i,j,k} 2j\, \bar{a}_{ijk}\, \Phi_\infty^i\, W^{2j-1}\, R^{2k} \;=\; \rho \,, \tag{28}$$

$$\frac{1}{c_\rho^2} R + \sum_{i,j,k} 2k\, \bar{a}_{ijk}\, \Phi_\infty^i\, W^{2j}\, R^{2k-1} \;=\; \frac{1}{2}\rho_3 = -\frac{1}{2}\rho \,, \tag{29}$$

there must be integers (j_m, k_m) and (j'_m, k'_m) with

$$(2j_m - 1)\alpha + 2k_m\beta = 1 \quad > \quad (2j-1)\alpha + 2k\beta \quad \text{for all} \quad j \neq j_m \,,\, k \neq k_m \,, \tag{30}$$

$$2j'_m\alpha + (2k'_m - 1)\beta = 1 \quad > \quad 2j\alpha + (2k-1)\beta \quad \text{for all} \quad j \neq j'_m \,,\, k \neq k'_m \,. \tag{31}$$

(This assumes that only one term in each sum produces the leading asymptotic behavior; if this actually happens for more than one term in a sum, the conclusions below are unchanged.) Using

$$\lim_{\rho\to\infty} p_0 = \frac{(3\pi^2)^{1/3}}{4}\rho^{4/3} + O(\rho^{2/3}) \quad , \quad \lim_{\rho\to\infty} \mathcal{E}_0 = \frac{3(3\pi^2)^{1/3}}{4}\rho^{4/3} + O(\rho^{2/3}) \,, \tag{32}$$

the leading contributions to the pressure and the energy density are found to be

$$\lim_{\rho\to\infty} p = \frac{(3\pi^2)^{1/3}}{4}\rho^{4/3} + \frac{1}{2c_{\mathrm{v}}^2}w_0^2\rho^{2\alpha} + \frac{1}{2c_\rho^2}r_0^2\rho^{2\beta}$$
$$+ w_0^{2j_m} r_0^{2k_m}\rho^{1+\alpha}\sum_i \bar{a}_{ij_m k_m}\Phi_\infty^i + w_0^{2j'_m} r_0^{2k'_m}\rho^{1+\beta}\sum_i \bar{a}_{ij'_m k'_m}\Phi_\infty^i \, , \quad (33)$$

$$\lim_{\rho\to\infty} \mathcal{E} = \frac{3(3\pi^2)^{1/3}}{4}\rho^{4/3} + \frac{1}{2c_{\mathrm{v}}^2}w_0^2\rho^{2\alpha} + \frac{1}{2c_\rho^2}r_0^2\rho^{2\beta}$$
$$+ w_0^{2j_m} r_0^{2k_m}\rho^{1+\alpha}(2j_m + 2k_m - 1)\sum_i \bar{a}_{ij_m k_m}\Phi_\infty^i$$
$$+ w_0^{2j'_m} r_0^{2k'_m}\rho^{1+\beta}(2j'_m + 2k'_m - 1)\sum_i \bar{a}_{ij'_m k'_m}\Phi_\infty^i \, . \quad (34)$$

To this point, the discussion is rather general. To make the conclusions more concrete, we discuss two distinct situations separately:

1. No coupling between W and R.

   In this special case, the asymptotic behavior of the fields is governed by their highest powers in the potential (9). ¿From Eqs. (30) and (31), we obtain

   $$\alpha = \frac{1}{2j_{\max} - 1} \quad , \quad \beta = \frac{1}{2k_{\max} - 1} \, .$$

   For $j_{\max} = 1\, , k_{\max} \geq 1$ or $j_{\max} \geq 1\, , k_{\max} = 1$, the quadratic terms dominate the right-hand sides of Eqs. (33) and (34), so that

   $$\lim_{\rho\to\infty} \frac{1}{c_{\mathrm{v}}^2} W^2 \propto \rho^2 \quad \text{or} \quad \lim_{\rho\to\infty} \frac{1}{c_\rho^2} R^2 \propto \rho^2 \, .$$

   This case includes the original version of the Walecka model [9] and generates the limiting behavior

   $$\lim_{\rho\to\infty} p = \mathcal{E} \, . \quad (35)$$

   The sums in Δp and $\Delta\mathcal{E}$ contribute to the leading order only if $j_{\max} = 2\, , k_{\max} \geq 2$ or $j_{\max} \geq 2\, , k_{\max} = 2$. In this case, the quadratic terms can be neglected, and Δp and $\Delta\mathcal{E}$ are of the same order as the contributions from the ideal-Fermi-gas terms. However, the factors in Eq. (34) conspire such that

   $$\lim_{\rho\to\infty} \Delta p = \frac{1}{3}\Delta\mathcal{E} \, , \quad (36)$$

   and the functional form of the limiting EOS resembles that of an ideal Fermi gas:

   $$\lim_{\rho\to\infty} p = \frac{1}{3}\mathcal{E} \, . \quad (37)$$

   In the remaining cases ($j_{\max} > 2, k_{\max} > 2$), the dominant contributions arise solely from p_0 and $\mathcal{E}_0$, which also leads to Eq. (37). Note here the importance of Eq. (22), which implies that in the high-density limit, M^* becomes negligible, at least to leading order.

2. At least one coupling between W and R.

¿From Eqs (30) and (31), it follows directly that

$$\alpha \leq \frac{1}{3} \quad \text{and} \quad \beta \leq \frac{1}{3} \, .$$

The quadratic contributions in the fields are negligible, and Δp and $\Delta \mathcal{E}$ contribute to the leading term only if $\alpha = \beta = 1/3$, where again Eq. (36) holds. In any event, this leads to the limit of Eq. (37).

To summarize, we conclude that the high-density limit of the EOS is strongly influenced by nonlinear meson–meson interactions, which agrees with the conclusion of Bodmer and Price [18,33]. The limit in Eq. (35) obtained in the original version of the Walecka model [9] is *a special case;* in the more general situation, the nuclear matter EOS approaches that of an ideal Fermi gas, given by Eq. (37). We will show in the next section that these two limits can be achieved using different models with parameter sets that reproduce the same equilibrium properties of nuclear matter.

4. Consequences for Neutron Stars

The high-density limit of the EOS and the way in which the matter approaches the asymptotic regime have important consequences in neutron star calculations. The masses and radii of stars are sensitive to the stiffness of the EOS, thus providing a quantitative measure for studying the impact of the nonlinear interaction terms in Eq. (9).

To be specific, it is necessary to choose an explicit potential, and we begin with the form

$$\sum_{i,j,k} a_{ijk}\, \phi^i \, (V_\mu V^\mu)^j \, (\mathbf{b}_\mu \cdot \mathbf{b}^\mu)^k \approx -\frac{\kappa}{3!}\, \phi^3 - \frac{\lambda}{4!}\, \phi^4 + \frac{\zeta}{4!}\, g_{\mathrm{v}}^4 (V_\mu V^\mu)^2 + \frac{\xi}{4!}\, g_\rho^4 (\mathbf{b}_\mu \cdot \mathbf{b}^\mu)^2 \, , \quad (38)$$

which includes a subset of the meson self-interactions up to fourth order in the fields. As discussed in the Introduction, setting some of the allowed cubic and quartic couplings to zero is "unnatural", but as we will discover, the model defined by Eq. (38) is already general enough to produce significant differences in predicted neutron star masses, and restoring the omitted couplings will lead to even more variation in the results [85–88], which we will discuss in the next section. Moreover, the present model can be related to the most common models discussed in the literature, and it generalizes them to include a nonlinear isovector interaction. The motivation for adding the quartic rho-meson term is that one expects this coupling to be essentially unconstrained by normal nuclear observables, where the neutron–proton asymmetry is low, but it may have significant impact on the neutron-rich matter in neutron stars. As noted earlier, since the meson fields are effective (Kohn–Sham) potentials, we are not concerned here with their elementary excitations, and considerations of causality [62,63] are unimportant.

In nuclear matter calculations, this model contains seven free parameters. The polynomial in Eq. (38) contains four couplings that we may write as $\bar{\kappa} \equiv \kappa/g_{\mathrm{s}}^3$, $\bar{\lambda} \equiv \lambda/g_{\mathrm{s}}^4$, ζ, and ξ; in addition, values for the three ratios $c_i^2 = g_i^2/m_i^2$ ($i =$ s, v, ρ) are needed. Five of the seven parameters can be chosen to reproduce the equilibrium properties of symmetric nuclear matter, which we take as the equilibrium density and binding energy (ρ_0, $-e_0$), the nucleon effective (or Dirac) mass at equilibrium (M_0^*), the compression modulus (K_0), and the bulk symmetry energy (a_4). The first three of these are tightly constrained [5, 20], whereas the latter two are not. In principle, the sensitivity of the high-density EOS to reasonable variations in K_0 and a_4 could be examined, but for simplicity, we keep their values fixed in our initial calculations and discuss more recent input to these parameters later. The "standard" set of equilibrium properties used here are listed in Table 1; these are motivated by successful descriptions of bulk and single-particle nuclear properties [5, 13, 20, 78]. The nucleon mass is fixed at its empirical value ($M = 939\,\mathrm{MeV}$).

Table 1. Equilibrium Properties of Nuclear Matter

$(k_{\mathrm{F}})_0$	ρ_0	M_0^*/M	e_0	K_0	a_4
1.30 fm^{-1}	0.1484 fm^{-3}	0.60	$-15.75\,\mathrm{MeV}$	$250\,\mathrm{MeV}$	$35\,\mathrm{MeV}$

Our primary goal is to study the influence of the nonlinear vector-meson interactions on neutron star masses. Since there are more free couplings than normalization conditions, we proceed as follows: We choose values for the couplings ζ and ξ and determine the remaining couplings by requiring that they reproduce the desired equilibrium properties. This is achieved by solving a set of transcendental equations that relate the parameters directly to the nuclear matter properties [13, 33]. Although we have no specific guidance on the allowed values of ζ and ξ, we rely on the assumption of naturalness, and based on the discussion at the end of Sec. 2, we observe that the following parameter combinations should all be of roughly equal size:

$$\frac{1}{2c_{\mathrm{s}}^2 M^2}, \quad \frac{1}{2c_{\mathrm{v}}^2 M^2}, \quad \frac{1}{8c_{\rho}^2 M^2}, \quad \frac{\bar{\kappa}}{6M}, \quad \frac{\bar{\lambda}}{24}, \quad \frac{\zeta}{24}, \quad \text{and} \quad \frac{\xi}{384}.$$

Typical values for the first three parameters are between 0.001 and 0.002, so that the natural values of ζ and ξ are roughly limited to $0 \le \zeta \lesssim 0.06$ and $0 \le \xi \lesssim 1.0$. (To avoid abnormal solutions of the vector field equations, *i.e.*, those with finite mean fields at zero density, ζ and ξ must be positive.[4]) We will include results for vanishing ζ and ξ, which are in a strict sense unnatural, in order to make contact with earlier calculations. For a more thorough discussion of naturalness and its implications, see Ref. [13, 83, 89, 90].

[4]Note that positive ζ and ξ imply that the resulting nonlinear interactions are *attractive*. This constraint on the highest-order vector interactions appears to be general and explains why our earlier analysis finds that nonlinear interactions soften the equation of state.

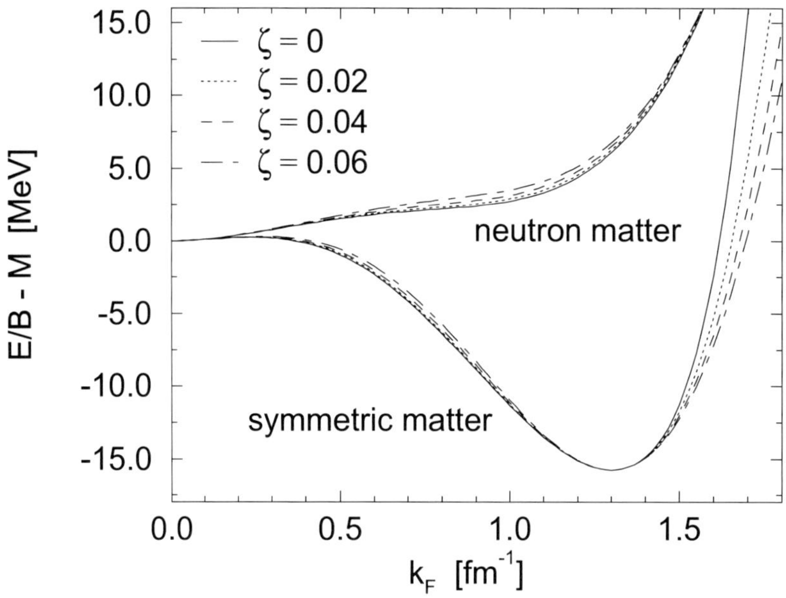

Figure 1. Binding energy of symmetric and neutron matter.

Using the notation of Sec. 2, the self-consistency equations (17)–(19) can be written as

$$\frac{1}{c_{\mathrm{s}}^2}\,\Phi + \frac{\bar{\kappa}}{2}\Phi^2 + \frac{\bar{\lambda}}{6}\Phi^3 = \rho_{\mathrm{s}}\;, \tag{39}$$

$$W\left(\frac{1}{c_{\mathrm{v}}^2} + \frac{\zeta}{6}W^2\right) = \rho\;, \tag{40}$$

$$R\left(\frac{1}{c_{\rho}^2} + \frac{\xi}{6}R^2\right) = \frac{1}{2}\rho_3 = -\frac{1}{2}\rho\;. \tag{41}$$

The expressions for the pressure and the energy density follow as

$$\begin{aligned} p &= \frac{1}{3\pi^2}\int_0^{k_{\mathrm{F}}} dk\frac{k^4}{(k^2+M^{*2})^{1/2}} \\ &\quad + \frac{1}{2c_{\mathrm{v}}^2}\,W^2 + \frac{\zeta}{24}\,W^4 + \frac{1}{2c_{\rho}^2}\,R^2 + \frac{\xi}{24}\,R^4 - \frac{1}{2c_{\mathrm{s}}^2}\,\Phi^2 - \frac{\bar{\kappa}}{6}\Phi^3 - \frac{\bar{\lambda}}{24}\Phi^4\;, \end{aligned} \tag{42}$$

$$\begin{aligned} \mathcal{E} &= \frac{1}{\pi^2}\int_0^{k_{\mathrm{F}}} dk\,k^2(k^2+M^{*2})^{1/2} \\ &\quad + \frac{1}{2c_{\mathrm{v}}^2}\,W^2 + \frac{\zeta}{8}\,W^4 + \frac{1}{2c_{\rho}^2}\,R^2 + \frac{\xi}{8}\,R^4 + \frac{1}{2c_{\mathrm{s}}^2}\,\Phi^2 + \frac{\bar{\kappa}}{6}\Phi^3 + \frac{\bar{\lambda}}{24}\Phi^4\;. \end{aligned} \tag{43}$$

We begin our discussion with the model introduced by Bodmer and Price [18], which corresponds to $\xi = 0$. According to the discussion in the preceding section, this model has the interesting feature that the high-density EOS of pure neutron matter approaches

$p = \mathcal{E}$, while in symmetric matter, where the mean-field of the ρ meson vanishes, the EOS approaches the massless Fermi gas limit, given by Eq. (37). In Fig. 1, we show the binding energy curves for symmetric and pure neutron matter for different values of the nonlinear coupling ζ. *We emphasize that all parameterizations reproduce the same equilibrium properties listed in Table 1.*

At low densities, all the curves approach a common limit, because the nonlinear terms do not contribute at leading order in a low-density expansion. At higher densities, the softening of the EOS as ζ increases is clearly visible, at least for symmetric matter. The softening in neutron matter is more apparent in Fig. 2. In the regime of intermediate density, $200 \lesssim \mathcal{E} \lesssim 1000\,\mathrm{MeV/fm^3}$, the EOS becomes softer with increasing values of ζ [33, 88]. To study the approach to the asymptotic limit, we examine the nonleading terms in the high-density expansion:

$$\lim_{\mathcal{E}\to\infty} p = \mathcal{E} - \frac{1}{6\pi^2}\left(\frac{72\pi^4}{4c_{\rm v}^2 + c_\rho^2}\right)^{2/3} \mathcal{E}^{2/3} + O(\mathcal{E}^{1/3}) \quad \text{for} \quad \zeta = 0\,, \tag{44}$$

$$\lim_{\mathcal{E}\to\infty} p = \mathcal{E} - \frac{1}{6\pi^2}\left(\frac{72\pi^4}{c_\rho^2}\right)^{2/3} \left[1+\left(\frac{2}{\pi^2\zeta}\right)^{1/3}\right] \mathcal{E}^{2/3} + O(\mathcal{E}^{1/3}) \quad \text{for} \quad \zeta \neq 0 \tag{45}$$

which reveals two important features. First, the $\zeta \neq 0$ results are *nonanalytic* in ζ. Thus one cannot reproduce Eq. (44) by taking the $\zeta \to 0$ limit of Eq. (45); at least as far as ζ is concerned, the high-density expansion is essentially a strong coupling expansion. Second, we observe that the coefficient of the nonleading term is smaller for $\zeta = 0$ for two reasons: the appearance of the isoscalar coupling $c_{\rm v}^2$ in the denominator and the absence of the multiplicative factor containing ζ. (Note that c_ρ^2 is independent of ζ.) These two features produce a coefficient that is roughly an order of magnitude smaller for $\zeta = 0$ than for $\zeta \neq 0$, which explains the relatively slow approach to the asymptotic limit in the latter case, as is evident from Fig. 2.

The consequences for neutron stars can be studied in Fig. 3, where the star masses are shown as a function of the central mass density ρ_c. For a given value of ρ_c, the mass is obtained by integrating the Tolman–Oppenheimer–Volkoff (TOV) equation, which follows from hydrostatic equilibrium in strong gravitational fields and the Einstein equations [91].

As expected from Fig. 2, the maximum mass decreases with increasing ζ. This decrease is substantial: from $M_{\rm max} = 2.9\,M_\odot$ for $\zeta = 0$ to $M_{\rm max} = 2.1\,M_\odot$ for $\zeta = 0.06$, which is roughly 30%. The shifts in the maximum mass are most dramatic for small couplings; for larger couplings, the softening effects begin to saturate.

To understand this result, it is useful to identify the regime of energy density that is most important in determining the mass of the star. This regime can be deduced from Fig. 4, where we show the radial mass density distributions for several neutron stars, as well as the corresponding energy densities. Observe that most of the mass is generated at radii between 6 and 12 km, which corresponds to energy densities of several hundred MeV/fm^3. As can be seen from Fig. 2, this includes the regime where the EOS is sensitive to ζ, because this is where the W^4 contribution to the EOS begins to become important. Note, however, that the contribution of the W^4 term in this regime is still smaller than that of the W^2 term; the quartic term does not begin to dominate until the energy density reaches several thousand MeV/fm^3, as indicated by the coalescence of the dashed and dotted curves in Fig. 2.

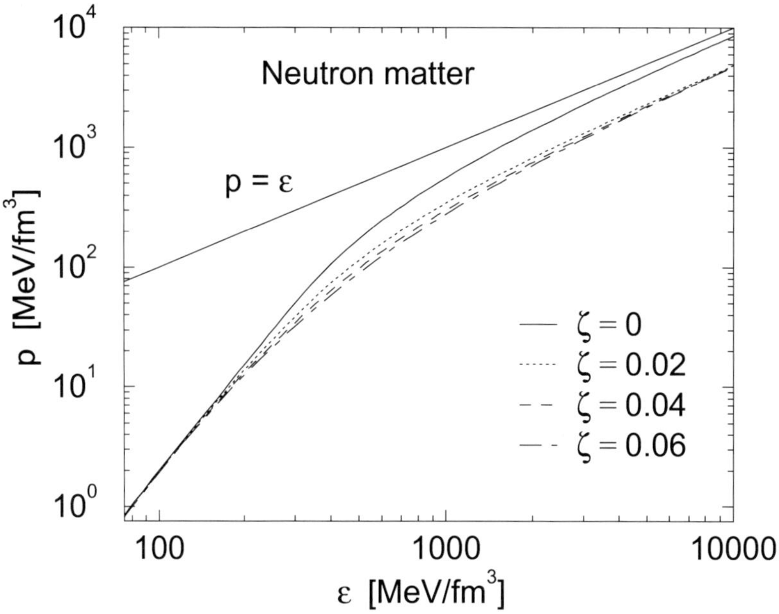

Figure 2. Equation of state of neutron matter.

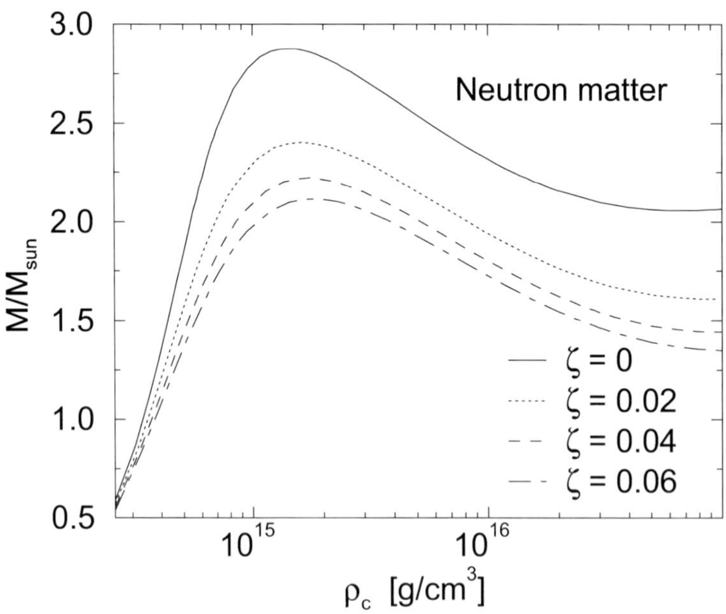

Figure 3. Neutron star masses for different nonlinear couplings ζ. Pure neutron matter is assumed.

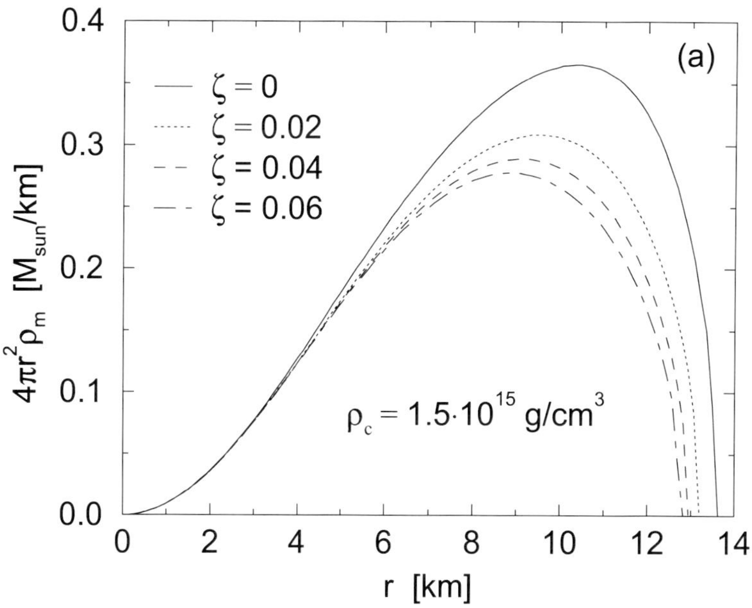

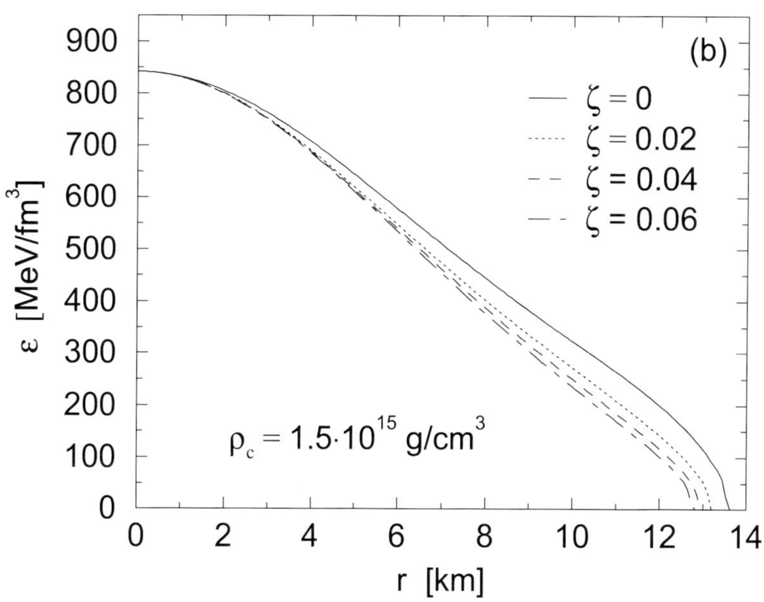

Figure 4. Density distributions of neutron stars as a function of the radius, calculated with different parameters ζ. The central mass density $\rho_c = 1.5 \times 10^{15}$ g/cm^3 is the same for all curves. Part (a) shows the radial mass densities and part (b) the corresponding energy densities. The stars acquire most of their mass from the region $100 \lesssim \mathcal{E} \lesssim 500$ MeV/fm^3.

We now return to the general form in Eq. (38) and allow nonlinear self-interactions of the ρ mesons. In contrast to the isoscalar coupling ζ, the new quantity ξ does not enter in the calculation of our five "standard" equilibrium properties of nuclear matter, and the other parameters are determined independently. This is true even for the symmetry energy, because the new coupling ξ first appears at order $(N-Z)^4$ in an expansion around symmetric nuclear matter. In principle, ξ could be constrained by fits to liquid-drop expansions of the energy, but in most such fits this parameter is set to zero. (We found only one nonzero value in the literature [92].) Thus, at present, contributions to the symmetry energy beyond terms of order $(N-Z)^2$ are practically unconstrained.

Fig. 5 shows the energy density of nuclear matter as a function of the proton fraction y for two different densities. The curves are calculated for various values of ξ with ζ held fixed. For clarity, we plot the fractional shift in $\mathcal{E}$ relative to its value with $\xi = 0$. At normal nuclear density, results for different ξ are virtually indistinguishable, but at high density, the curves differ by a few percent when the proton fraction becomes very small. This demonstrates that it is possible to generate families of models that reproduce identical properties of nuclear matter at low and normal densities, but which generate different predictions at high densities.

According to the analysis in the preceding section, the EOS asymptotically approaches the massless Fermi gas limit for both symmetric and neutron matter. The high-density expansion for neutron matter corresponding to Eq. (45) is

$$\lim_{\mathcal{E}\to\infty} p = \frac{1}{3}\mathcal{E} + \frac{2\pi}{3}\frac{\left[\frac{1}{c_{\rm v}^2}\left(\frac{2}{\pi^2\zeta}\right)^{2/3} + \frac{1}{c_\rho^2}\left(\frac{1}{\pi^2\xi}\right)^{2/3}\right]}{\left[1+\left(\frac{2}{\pi^2\zeta}\right)^{1/3} + \left(\frac{1}{\pi^2\xi}\right)^{1/3}\right]^{1/2}}\,\mathcal{E}^{1/2} + O(\mathcal{E}^0)\,. \qquad (46)$$

In contrast to the previous case, the asymptotic limit is now approached from above. Moreover, the analytical form has changed, since Eq. (46) indicates a series in powers of $\mathcal{E}^{1/2}$.

The dependence of the neutron star masses on ξ follows the same trend as found for ζ. This can be gleaned from Fig. 6, which shows results for pure neutron matter. For fixed ζ, the maximum mass decreases with increasing ξ, although the effect of this new parameter is smaller (less than a 10% change in the maximum mass), given the limitation on parameter values imposed by naturalness.

A more complete picture of $M_{\rm max}$ is given in Fig. 7, where the variations with both ζ and ξ are shown. It is apparent that parameter sets that yield identical properties near nuclear equilibrium can still generate values of the maximum mass that differ by as much as one solar mass. We find variations between $M_{\rm max} = 2.9\,M_\odot$ and $M_{\rm max} = 1.9\,M_\odot$ for stars composed of pure neutron matter. The masses are most sensitive to changes at small values of the couplings, particularly for $\zeta \approx 0$, which can be related to the nonanalytic form of the EOS in terms of ζ and ξ. [See Eqs. (44)–(46).] As seen earlier, the effect of ξ is smaller than that of ζ.

So far, we have considered pure neutron matter, which gives only a qualitative picture of neutron star properties. For a more realistic description, it is necessary to consider beta-stable matter, *i.e.*, matter composed of neutrons, protons, and electrons in beta-decay equilibrium. This situation was realized by adding the contribution of free, relativistic elec-

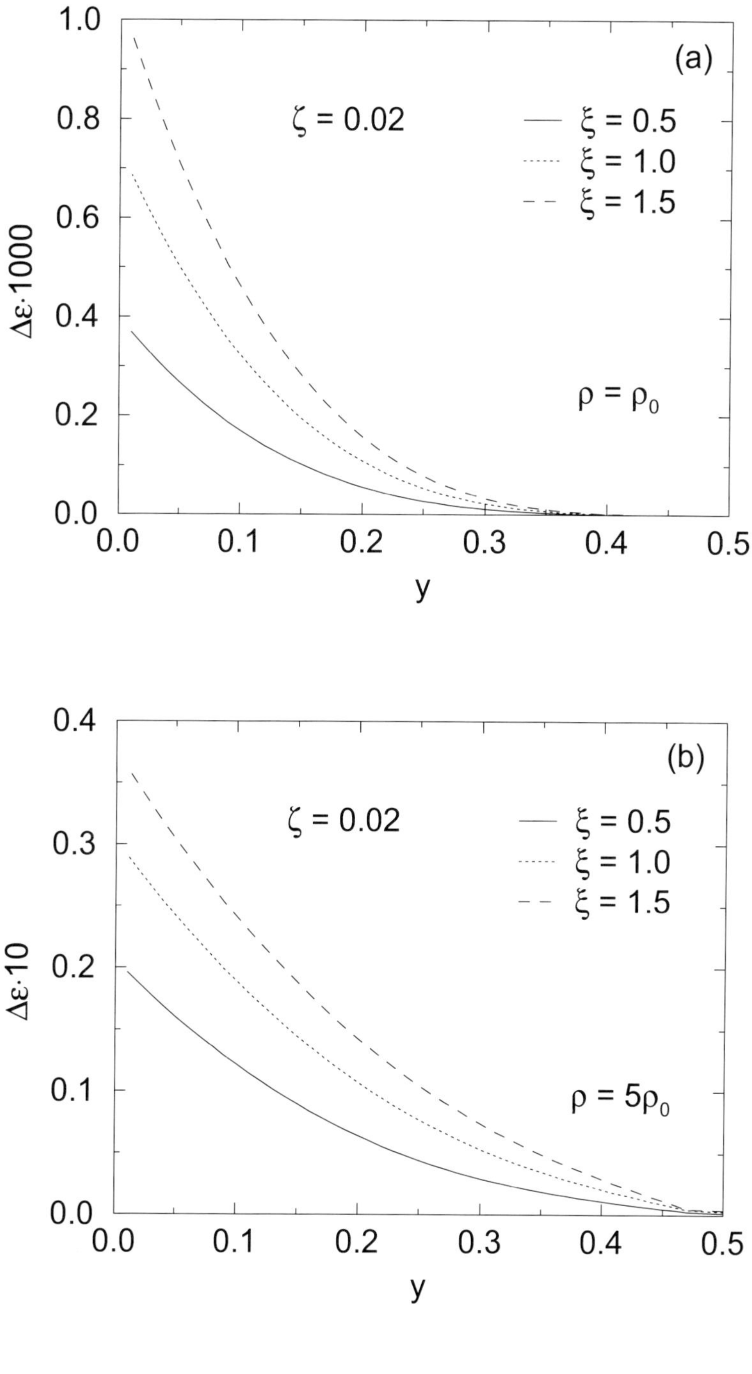

Figure 5. Energy difference $\Delta\mathcal{E} = [\mathcal{E}(\xi = 0) - \mathcal{E}(\xi)]/\mathcal{E}(\xi = 0)$ of nuclear matter as a function of the proton fraction y calculated at constant baryon density. In part (a) the density is fixed at its equilibrium value ρ_0 and part (b) shows the results at $5\rho_0$. Note the different vertical scales in (a) and (b).

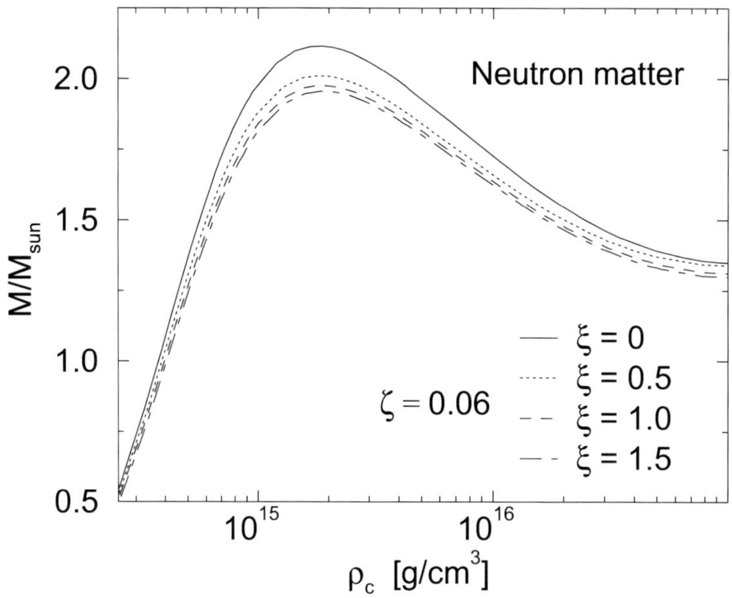

Figure 6. Neutron star mass for different values of ξ at fixed ζ. Pure neutron matter is assumed.

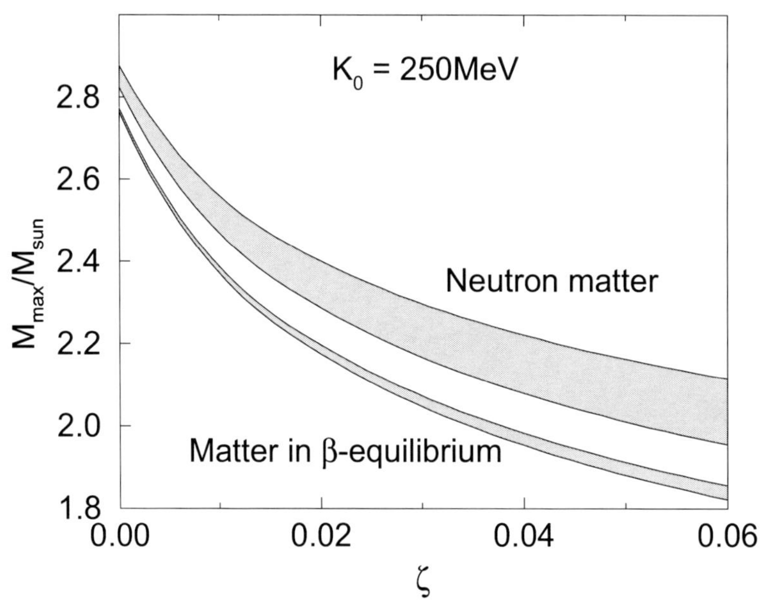

Figure 7. Maximum neutron star mass as function of ζ and ξ. Results for pure neutron matter and for matter in β-equilibrium are displayed. The shaded areas show the mass range obtained when ξ is varied; the upper boundaries correspond to $\xi = 0$ and the lower boundaries to $\xi = 1.5$.

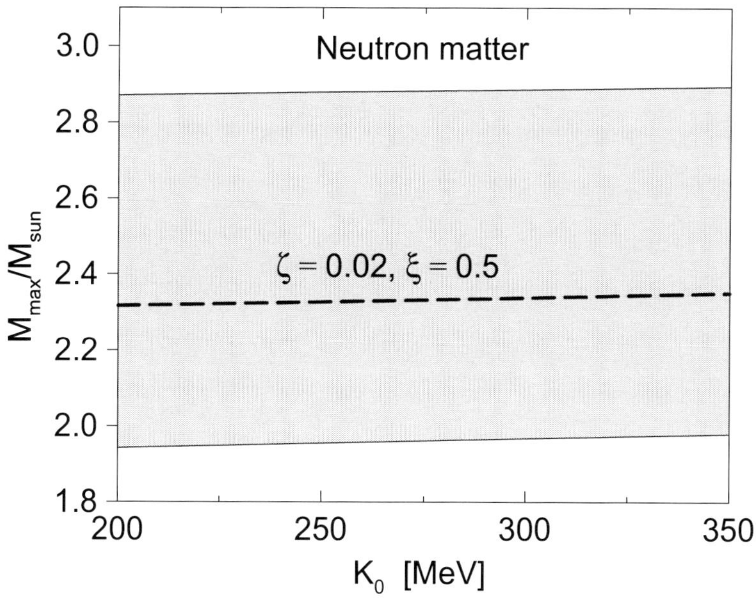

Figure 8. Maximum neutron star mass as a function of the compression modulus K_0. (All other nuclear matter inputs are held fixed.) The shaded area marks the covered range of masses. The upper boundary corresponds to $\zeta = 0$, $\xi = 0$ and the lower boundary to $\zeta = 0.06$, $\xi = 1.5$. For fixed values of the nonlinear couplings, the mass changes marginally with the compression modulus. This can be seen from the dashed line inside the shaded area, which corresponds to $\zeta = 0.02$, $\xi = 0.5$

trons to the nuclear EOS and by imposing the equilibrium condition

$$\mu_n = \mu_p + \mu_e \, .$$

The TOV equation is then integrated under the constraint of total charge neutrality . Maximum neutron star masses for beta-stable matter are also shown in Fig. 7. The dependence on the isoscalar coupling ζ is similar to that obtained earlier (the maximum mass varies between $M_{\max} = 2.8\, M_{\odot}$ and $M_{\max} = 1.8\, M_{\odot}$), but the influence of the isovector coupling ξ is less drastic, since the matter becomes significantly more symmetric in the region that gives the largest contribution to the mass.

In Fig. 8, we examine the dependence of the maximum mass on the compression modulus at equilibrium, which has traditionally been a parameter that has been difficult to deduce from empirical nuclear properties. (Relativistic mean-field models with $200 \lesssim K_0 \lesssim 350\,\mathrm{MeV}$ can produce accurate nuclear binding-energy systematics and surface energetics [13].) The shaded band shows the total predicted variation in maximum mass when both ζ and ξ are varied within the bounds imposed by naturalness. The dashed curve shows the predicted variation when K_0 is varied at fixed ζ and ξ. Evidently, the variations in the maximum mass arising from the possibility of nonlinear vector meson interactions is much greater than that arising from the uncertainty in the nuclear compressibility.

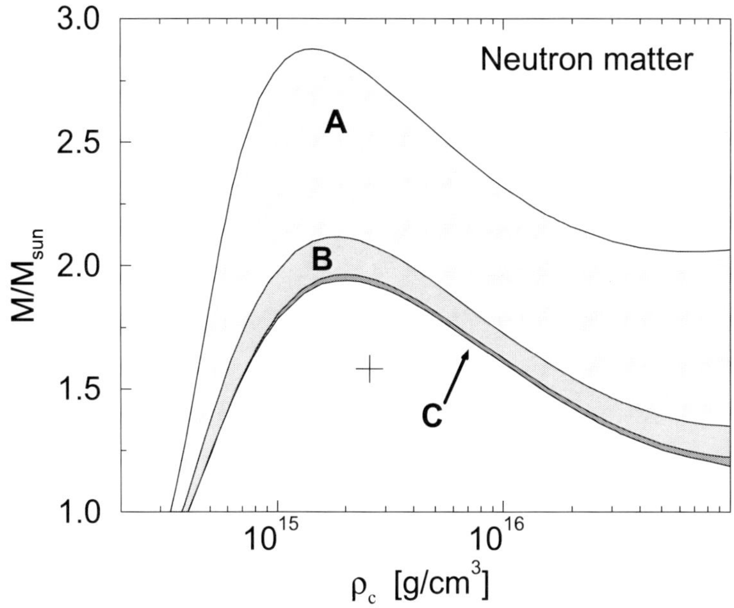

Figure 9. Neutron star masses for models with different nonlinear couplings for the neutral vector meson. The uppermost curve corresponds to the Walecka model, including cubic and quartic couplings for the scalar meson. Region A shows the range of masses obtained when the quartic vector coupling ζ is turned on. Regions A and B correspond to models with up to sixth-order terms, and regions A, B, and C include an eighth-order term. The values of the nonlinear couplings are chosen within a natural range, as described in the text. The cross indicates the maximum mass ($1.58\ M_\odot$) obtained for beta-stable matter in a calculation that includes the ζ, ζ', ζ'', and ξ couplings (see the text).

These results raise the interesting question of whether the only significant nonlinearity is the quartic, isoscalar vector interaction. In other words, once one takes $\zeta \neq 0$ to soften the EOS at high densities, does the addition of further nonlinearities produce only small effects? To examine this question, we extend the model of Eq. (38) to include sixth-order and eighth-order terms involving the vector-isoscalar meson:

$$\Delta\mathcal{V}' = -\frac{\zeta'}{6!}\, g_{\rm v}^6 (V_\mu V^\mu)^3 - \frac{\zeta''}{8!}\, g_{\rm v}^8 (V_\mu V^\mu)^4\ . \tag{47}$$

Thus $\zeta'/6!$ and $\zeta''/8!$ are the relevant ratios to be included in the parameter list given earlier, and we will initially set ξ to zero and examine the consequences of varying ζ, ζ', and ζ'' within the bounds imposed by naturalness.

Figure 9 shows neutron star masses as higher-order vector nonlinearities are included sequentially. Evidently, the quartic interactions are the most important, producing a roughly 30% variation in the maximum. The effects of the sixth-order term are quite modest (roughly 10%), while the eighth-order contributions are essentially negligible (roughly 2%). Here the parameters are varied within the natural ranges $0 \leq \zeta \leq 0.06$, $0 \leq \zeta' \leq 1.2$, and $0 \leq \zeta'' \leq 60$. Thus we have the encouraging result that once the W^4 interaction has been accurately calibrated, contributions from higher-order nonlinearities are relatively unimportant. To indicate the most extreme reduction in maximum mass possible in the present model, the cross in Fig. 9 shows $M_{\max} = 1.58\, M_\odot$, which is obtained for *beta-stable matter* when the couplings $\zeta = 0.06$, $\zeta' = 1.2$, $\zeta'' = 60$, and $\xi = 1.5$ are included. Note that this value of $M_{\max}$ is only slightly ($\approx 10\%$) larger than that of the most massive observed neutron stars at the turn of the millennium [57].

Our analysis to this point has revealed significant model and parameter dependence in the high-density EOS. It is therefore of interest to see if these variations influence predictions arising from other relevant dynamics in systems with high densities. As an example, we study the effect of the high-density hadronic EOS on the existence of quark-matter cores in neutron stars. We adopt a simple two-phase model [1, 66] based on a first-order (van der Waals) phase transition between the hadronic and quark phase. Modern calculations show that the phase transitions are actually "crossovers" for physical quark masses at zero density [94, 95] and that they remain crossovers at finite densities [96]. The zero density crossover temperature appears to be roughly 160 MeV. Nevertheless, the true behavior of the transition at the finite density relevant for neutron stars (and indeed whether one actually exists) is unknown at present. Until more reliable information is available, one must resort to a separate description of the quark phase and the hadronic phase. Moreover, whereas a detailed description involving beta-stable matter requires a careful treatment of the phase transition in systems with two conserved charges (baryon number and isospin) [49, 93], here we will be satisfied with a qualitative discussion based on pure neutron matter. This is certainly reasonable, given the large uncertainties we have already found in the hadronic EOS at high density.

We adopt the simple EOS involving massless u and d quarks given by

$$p = \frac{1}{3}\mathcal{E} - \frac{4}{3}b\,, \tag{48}$$

where the confinement property of QCD (or alternatively, the anomaly in the trace of the energy-momentum tensor) is modeled by a positive constant b, which represents the energy per unit volume in the vacuum.

We return to the hadronic model of Eq. (38), which leads to the high-density expansion in Eq. (46). Our discussion is based on the simple observation that independent of the actual nature of the transition, it is driven purely by the energetics in the two phases.

By comparing Eq. (48) with the expansion in Eq. (46), one observes that in the quark phase, the limit $p = \mathcal{E}/3$ is approached from below, whereas in neutron matter, for the general case $\zeta > 0$ and $\xi > 0$, the limit is approached from above. To decide whether a transition takes place, it is necessary to compare the energy/baryon in both phases. For the quark phase it is expressed as [66]

$$\mathcal{E}/\rho = \frac{3}{4}\pi^{2/3} f(\alpha_s)\rho^{1/3} + b/\rho\,, \tag{49}$$

with

$$f(\alpha_s) = (1 + 2^{4/3})(1 + \frac{2\alpha_s}{3\pi})\,,$$

which includes the lowest-order contribution in α_s (the exchange energy). As discussed in the preceding section, in hadronic models that are characterized by Eqs. (44) and (45), the quadratic terms dominate at high densities, so that

$$\lim_{\mathcal{E}\to\infty} \mathcal{E}/\rho \propto \rho \,, \tag{50}$$

for the special case $\zeta \geq 0$ and $\xi = 0$. This is also true if $\zeta = 0$ and $\xi \geq 0$. Thus, at sufficiently high densities, neutron matter always has higher energy compared to the quark phase [66], and the two phases can be connected by a Maxwell construction, which signals the transition from hadron to quark matter.

The situation is different in the general case $\zeta > 0$, $\xi > 0$. For the asymptotic form of the energy corresponding to Eq. (46), one obtains

$$\lim_{\mathcal{E}\to\infty} \mathcal{E}/\rho = \frac{3}{4}(3\pi^2)^{1/3}\left[1 + \left(\frac{2}{\pi^2\zeta}\right)^{1/3} + \frac{1}{2}\left(\frac{1}{\pi^2\xi}\right)^{1/3}\right]\rho^{1/3} + O(\rho^{-1/3}) \,, \tag{51}$$

which is, up to the prefactor, the same leading behavior as in Eq. (49) for the quark phase. Therefore a phase transition is possible, *i.e.*, neutron matter has higher energy, only if

$$3^{1/3}\left[1 + \left(\frac{2}{\pi^2\zeta}\right)^{1/3} + \frac{1}{2}\left(\frac{1}{\pi^2\xi}\right)^{1/3}\right] > f(\alpha_s) \,. \tag{52}$$

This remarkable observation implies that for sufficiently large values of the nonlinear couplings, the matter remains in the hadron (neutron) phase, at least in the simple model discussed here. More generally, one observes that independent of the asymptotic form, increasing the couplings ζ and ξ increases the density at which the phase transition occurs (if it does), since the hadronic EOS becomes softer as the nonlinear couplings increase.

The different possibilities are illustrated in Fig. 10. The curve labeled (a) corresponds to Eq. (50); the asymptotic behavior is clearly different from the quark EOS. The phase transition occurs at roughly $3\rho_0$, and the Maxwell construction is indicated by the dotted line. In the situation described by curve (b), neutron matter and quark matter have a similar asymptotic behavior, but the condition (52) remains true, and the two curves cross, leading to a phase transition at roughly $6\rho_0$. Finally, curve (c) lies completely below the quark EOS and a transition is not possible; neutron matter is stable at all densities.

If one introduces interaction terms of higher than fourth order in the fields, for example, $(V_\mu V^\mu)^3$, the hadronic energy is dominated by the Fermi gas contribution, and Eq. (51) must be replaced by

$$\lim_{\mathcal{E}\to\infty} \mathcal{E}/\rho = \frac{3}{4}(3\pi^2)^{1/3}\rho^{1/3} + O(\rho^{-1/3}) \,. \tag{53}$$

Since $3^{1/3} < f(\alpha_s)$, no transition is possible in the asymptotic regime for any choice of hadronic parameters in this case. On the other hand, the transition regime also depends on the model and parameters used for the description of the quark phase. In our model, the vacuum constant b and strong coupling α_s can be used to shift the transition point substantially [97, 98], so that a transition may occur outside the asymptotic regime. It is clear, however, that the strong model dependence in the hadronic EOS introduces large uncertainties in any attempted prediction of these values.

The consequences for neutron stars in this model follow straightforwardly. Nonlinear vector meson interactions soften the hadronic EOS, which lowers the maximum neutron

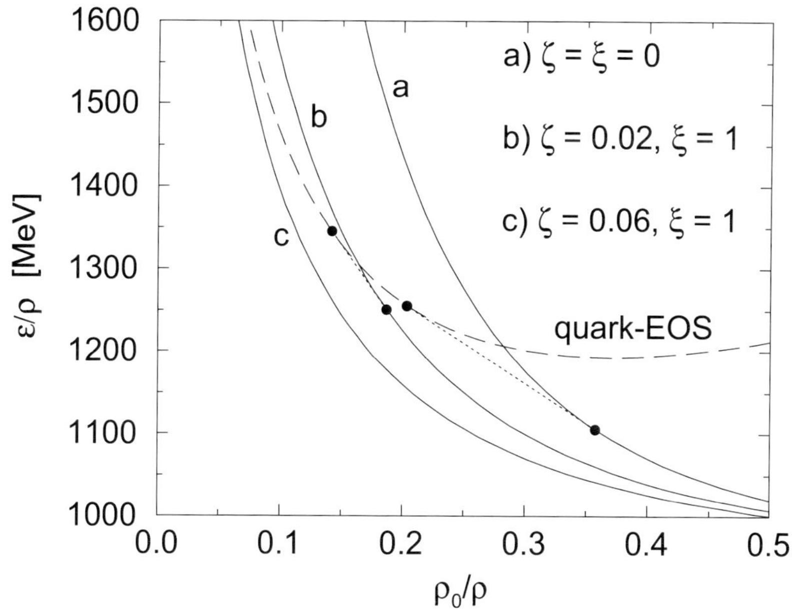

Figure 10. Energy per baryon for neutron matter (solid) and quark matter (dashed). The quark matter results are calculated using $b = 120$ MeV/fm^3 and $\alpha_s = 0.4$ [97].

star mass and increases the density of the transition to quark matter. In contrast, a stiff hadronic EOS lowers the density of the phase transition, and since the quark matter EOS is soft, also tends to decrease the maximum star mass. Thus it may be impossible to decide, from neutron star masses alone, whether quark matter cores exist in neutron stars, and similar conclusions may be drawn about other exotic phenomena that soften the EOS. (The situation is complicated further by the continuous nature of the transition when two conserved charges are involved, which is the more physical case [49, 93].)

5. Recent Developments in Observation and Theory

There has been significant progress in the study of neutron stars since the original version of this paper was written [36]. Here we will describe some advances in the last fifteen years within the framework of relativistic mean-field descriptions of the high-density EOS.

The first theoretical advance appeared in early 2001 [85], which studied the relationship between the neutron-rich skin of a heavy nucleus and the onset of a neutron star crust . These authors added to the lagrangian nonlinear couplings between the isovector and isoscalar mesons of both the Lorentz scalar and vector type. The explicit potential chosen was

$$\sum_{i,j,k} a_{ijk}\,\phi^i\,(V_\mu V^\mu)^j\,(\mathbf{b}_\mu\cdot\mathbf{b}^\mu)^k = -\tfrac{\kappa}{3!}\,(g_s\phi)^3 - \tfrac{\lambda}{4!}\,(g_s\phi)^4 + \tfrac{\zeta}{4!}\,g_v^4(V_\mu V^\mu)^2$$
$$+\tfrac{\xi}{4!}\,g_\rho^4(\mathbf{b}_\mu\cdot\mathbf{b}^\mu)^2 + g_\rho^2\,(\mathbf{b}_\mu\cdot\mathbf{b}^\mu)\,[\Lambda_4 g_s^2\phi^2 + \Lambda_v g_v^2\,V_\mu V^\mu]\,. \tag{54}$$

The purpose of these new couplings was to allow for a tuning of the density dependence of the nuclear symmetry energy, without upsetting the accurate fits to other nuclear observables. (Note that the "other" observables were the ones used in the previous sections of this chaper.) The conclusion of this work was that it was possible to produce an EOS with a "softer" symmetry energy (one that rises more slowly with the nuclear density) than in previous studies. This implied that the thicker the neutron skin in ^{208}Pb, the faster the electron fraction Y_e in a neutron star rises with density. Moreover, an inverse correlation was found between the neutron-skin thickness and the density of a phase transition from uniform to nonuniform neutron-rich matter. This suggests that an accurate measurement of the neutron-skin thickness in ^{208}Pb could provide important information on the thickness of the neutron-star crust.

Following these correlations further, Horowitz and Piekarewicz [86] derived a new relation between the neutron skin of a heavy nucleus and the radius of a neutron star: the larger the neutron skin, the larger the stellar radius. By varying the density dependence of the symmetry energy (as before), one can accurately reproduce a variety of nuclear ground-state observables with various neutron skins in heavy nuclei; moreover, scalar self-interactions reduce the compression modulus K_0 to a value consistent with the isoscalar giant monopole resonance (GMR) in ^{208}Pb [99], while omega-meson self-interactions are also essential for softening the EOS at high density. Apparently, a softer symmetry energy leads to larger central densities in the interiors of neutron stars and hence to smaller stellar radii [86]. Although the correlation between the neutron skin and the neutron star radius can put constraints on the high-density EOS, the radius of a neutron star is not uniquely constrained by the neutron skin. This is because the stellar radius depends on the EOS at much higher densities than the neutron skin.

In a deeper look at the role of nuclear giant resonances and the calibration of the nuclear EOS, Piekarewicz [100] showed that by looking at the isoscalar-monopole strength in ^{90}Zr and ^{208}Pb, and the isovector-dipole strength in ^{208}Pb, one finds that the isoscalar-monopole strength in ^{208}Pb is sensitive to the density dependence of the symmetry energy, while this is not true in ^{90}Zr. Thus it is more appropriate to use ^{90}Zr (with a small neutron-proton asymmetry) to constrain the compression modulus of symmetric nuclear matter (K_0). In contrast, the sensitivity of ^{208}Pb to the density dependence of the symmetry energy is used to constrain the size of the neutron skin.

By starting with an accurately calibrated relativistic mean-field parametrization, Todd-Rutel and Piekarewicz [88] computed ground-states of nuclei, their linear response, and the structure of neutron stars. They used an effective potential of the type in Eq. (54), except with $\xi = 0$ and $\Lambda_4 = 0$. This new calibration produces an EOS that is considerably softer than previous equations of state *both for symmetric nuclear matter and for the symmetry energy*. This softness is apparently required to accurately describe several different giant resonances with different neutron–proton ratios. Notice that this model introduces two new parameters, ζ and $\Lambda_{\rm v}$, relative to old-fashioned mean-field parameterizations, such as those called NLi in Ref. [2]. The first parameter is used to soften the EOS of symmetric matter, and the second softens the symmetry energy. The necessity of these new parameters to describe new calibration data is consistent with the counting conclusions of Ref. [5].

As further validation of the newly calibrated model, Piekarewicz [101] applied it to EOS observables that were not used in the calibration procedure. These new comparisons involved dilute neutron gases, heavy-ion experiments that probe both the low- and high-density domain of the EOS, and astronomical observations that limit the masses and radii of neutron stars. The new model completed all of the tests successfully. Piekarewicz draws the strong conclusion that to adequately test the validity and applicability of relativistic mean-field models, one must *simultaneously* reproduce many diverse observables that probe the EOS over a wide dynamic range. Only then can one believe in the emergence of exotic phenomena.

In response to recent astrophysical measurements of the pressure of cold matter above nuclear-matter equilibrium density, the newly calibrated model was used to compute the EOS of neutron-star matter [102]. The effective potential is of the type in Eq. (54), except with $\xi = 0$ and $\Lambda_4 = 0$. The uniform stellar core was assumed to consist of nucleons and leptons in beta equilibrium, and no exotic degrees of freedom were considered. The conclusion was that the predicted radius for a given neutron-star mass was consistently larger than observations.

In response to this result, the group has recently introduced a new relativistic effective interaction that is simultaneously constrained by the properties of finite nuclei, giant resonances, and *neutron star properties* [103]. The form of the effective interaction is the same as before. These authors could efficiently tune the neutron-skin thickness of ^{208}Pb and the maximum star mass. They compared their results to a recent X-ray measurement [104] and found reasonable agreement for both maximum stellar mass and implied stellar radii.

Finally, a most exciting measurement of a two-solar-mass neutron star has been made using Shapiro delay , an effect of general relativity in the curved spacetime near a massive body [105]. Radio-timing observations of the binary millisecond pulsar J1614−2230 imply a pulsar mass of $(1.97 \pm 0.04)M_{\odot}$. This large mass rules out almost all currently proposed hyperon or boson condensate equations of state [55, 58, 61]. Moreover, quark matter can support a star this massive only if the quarks are strongly interacting [60] and are therefore not "asymptotically free."

This brings us to a fundamental question involving the role of quarks in the high-density neutron-star EOS. It is *obvious* that there are quarks in the cores of compact stars; but are they confined into hadrons? What one needs are three things:

- A systematic and controlled way to describe QCD interactions at high density;
- The quarks interact in some way that does not confine the color degrees of freedom into singlets (hadrons);
- There must be some *observational* way to tell the difference between confined and unconfined quarks.

6. Summary

In this paper we study the equation of state of nuclear and neutron-star matter based on relativistic mean-field theory. Our starting point is an effective action (or energy functional) containing Dirac nucleons and local scalar and vector fields. These fields are interpreted as relativistic Kohn–Sham potentials, and nonlinear interactions between the fields are introduced to parameterize the density dependence of the energy functional. We calibrate the energy functional by observing that at normal nuclear densities, the ratios of the mean fields to the nucleon mass are small, and thus the nonlinear interactions can be truncated at some low order in the fields. The unknown parameters can then be fit to properties of nuclear matter near equilibrium that are known to be characteristic of the observed bulk and single-particle properties of nuclei.

We then extrapolate the resulting equation of state into the neutron-rich, high-density regime to calculate the properties of neutron stars. Two problems arise in the extrapolation: First, even with a significant truncation of the energy functional, the unknown parameters are underdetermined. Thus there exist families of parameters that reproduce exactly the same nuclear matter properties near equilibrium, but which produce potentially different high-density equations of state. Second, terms omitted from the functional because they are negligible at normal density may become important at densities relevant for neutron stars. This is true even if we assume that the coupling parameters are "natural", which means that they are all of roughly the same size when expressed in appropriate dimensionless ratios.

Our basic goal is to determine, in light of these two problems, whether the calibration at equilibrium nuclear matter density is sufficient to predict a maximum neutron star mass within a reasonably small range. This is relevant in view of recent calculations that hope to see evidence for "new" physics in neutron stars (such as quark cores, strange matter, or kaon condensates) based on the need for a softer high-density equation of state than that provided by neutrons, protons, and electrons alone. These calculations assume that the high-density behavior of these more mundane components is well known, and in particular, that the contributions of many-nucleon forces are negligible [49–54]. These many-body, density-dependent forces are precisely the ones that are difficult to calibrate using observed nuclear properties; the question is whether one can build nuclear equations of state with different types of many-body forces that all reproduce the observed properties near equilibrium, but which yield significantly different results at high density.

By beginning with a meson self-interaction potential containing arbitrary powers of scalar-isoscalar (σ), vector-isoscalar (ω), and vector-isovector (ρ) fields, we show that the meson nonlinearities can have a profound effect on the high-density equation of state. In models where the vector mesons enter the potential at most quadratically, the equation of state is stiff and asymptotically approaches $p = \mathcal{E}$. (The Walecka model is a special case.) In models where the vector fields enter with high powers, these fields become negligible at high density, and the asymptotic equation of state resembles that of a free, relativistic gas: $p = \mathcal{E}/3$. The intermediate case occurs when the vector fields enter quartically; the asymptotic equation of state is still soft ($p = \mathcal{E}/3$), but the approach to the asymptotic limit is determined by the coupling parameters.

We illustrate these results using specific models containing quartic ω and ρ meson couplings and also sixth- and eighth-order ω couplings. All models are calibrated to exactly the same nuclear properties at equilibrium, for all choices of parameters. We find that by far the most important coupling constant is that of the quartic ω term; even when restricted by the requirements of naturalness, variations in this parameter can produce variations of nearly one solar mass in the predicted maximum neutron star mass. (This is true for both pure neutron matter and for beta-stable matter.) This uncertainty is clearly relevant on the scale on which one hopes to identify new, exotic effects. Moreover, increasing the strength of the quartic coupling softens the equation of state, which is precisely the effect sought from the exotica. The impact of a quartic ρ meson coupling is smaller, and its effects are only appreciable in stars made of pure neutron matter; the maximum masses of stars computed with beta-stable matter show little change when this parameter is varied within the bounds imposed by naturalness. Similarly, sixth-order ω interactions have only a modest effect on the predicted maximum mass, and by the time the eighth-order terms are included, the high-density equation of state is already so soft that these terms are negligible.

We emphasize that the importance of these many-body effects is not limited to the domain of relativistic mean-field theories; equations of state based on nonrelativistic potentials [48] use interactions that are calibrated to few-body systems and that are insensitive to possible six- or eight-body forces that may be relevant at high density. This is especially important because a mean-field calculation with four-component spinors and just *two-body* Lorentz scalar forces already implicitly contains an infinite string of many-body forces if it is recast in terms of two-component spinors.

To illustrate these difficulties more concretely, we also study the role of the hadron/quark phase transition in a simple two-phase model with a first-order transition. We find that with couplings well within the bounds of naturalness, it is possible to push the phase transition to arbitrarily high density, and even to make it disappear altogether. Although the absence of a phase transition is probably unrealistic and could serve to exclude some values of the hadronic parameters, there are still too many uncertainties on both sides of the transition (as well as in the nature of the transition itself!) to make any definitive statements.

We therefore conclude from these initial investigations that methods for calibrating the nuclear equation of state that rely solely on static nuclear properties cannot constrain the predicted maximum star mass well enough to make reliable statements about the existence of "new" physics beyond the dynamics of neutrons, protons, and electrons. We show that the uncertainties arising from an incomplete knowledge of the hadronic many-body forces are much larger than those arising from an imperfect knowledge of the properties of nuclear matter near equilibrium, such as the nuclear matter compression modulus. Even the old question of the role of the hadron/quark phase transition is problematic, since the high-density hadronic equation of state can be made essentially as soft as desired by the addition of nonlinear interactions that are still consistent with equilibrium nuclear matter properties.

Subsequent calculations have come to a partial resolution of these problems. One retains the quartic isoscalar interaction and includes an additional quartic isoscalar-isovector term. The key is to use nuclear dynamics, for example, observed properties of giant resonances, to constrain these two parameters, and the well-measured giant resonances in ^{90}Zr and ^{208}Pb are sufficient. Compared to earlier calibrations, one finds a softer EOS for symmetric nuclear matter, which is determined primarily by the quartic isoscalar term, and also a softer symmetry energy, which follows from a nonzero quartic isoscalar-isovector term. With this new calibration procedure, the neutron-star EOS is consistent with both the maximum stellar mass and the implied stellar radii, as deduced from recent X-ray measurements [103, 104].

To summarize, precise predictions of the properties of neutron stars require accurate calibrations of the nuclear equation of state that rely on both static and dynamical observables. It is especially important to have the high-density behavior of the "standard" components (neutrons, protons, and electrons) under control before one can make reliable statements about the existence of "new" physics, and significant progress has been made recently toward that goal. In particular, with the observation of a two-solar-mass neutron star, one can now draw the firm conclusion that soft neutron-star equations of state are ruled out [105, 106].

Acknowledgments

The bulk of the material in this chapter was originally written in collaboration with H. Müller [36]. We thank R. J. Furnstahl, J. Piekarewicz, H.-B. Tang, and J. D. Walecka for useful comments. This work was supported in part by the Department of Energy under Contract No. DE–FG02–87ER40365.

References

[1] B. D. Serot and J. D. Walecka, *Adv. Nucl. Phys.* 16 (1986) 1 and references therein.

[2] B. D. Serot, *Rep. Prog. Phys.* 55 (1992) 1855.

[3] R. J. Furnstahl, B. D. Serot and H.-B. Tang, *Nucl. Phys.* A 615 (1997) 441; A 640 (1998) 505 (E).

[4] B. D. Serot and J. D. Walecka, *Int. J. Mod. Phys.* E 6 (1997) 515 and references therein.

[5] R. J. Furnstahl and B. D. Serot, *Nucl. Phys.* A 671 (2000) 447.

[6] R. J. Furnstahl, *Int. J. Mod. Phys.* B 17 (2003) 5111.

[7] R. J. Furnstahl, in: Lecture Notes in Physics, vol. 641, G. A. Lalazissis, P. Ring and D. Vretenar, eds. (Springer, Berlin, Heidelberg, 2004), p. 1.

[8] B. D. Serot, in: Lecture Notes in Physics, vol. 641, G. A. Lalazissis, P. Ring and D. Vretenar, eds. (Springer, Berlin, Heidelberg, 2004), p. 31.

[9] J. D. Walecka, *Ann. Phys.* (N.Y.) 83 (1974) 491.

[10] J. Boguta and A. R. Bodmer, *Nucl. Phys.* A 292 (1977) 413.

[11] B. D. Serot, *Phys. Lett.* 86 B (1979) 146; 87 B (1979) 403 (E).

[12] R. J. Furnstahl, H.-B. Tang and B. D. Serot, *Phys. Rev.* C 52 (1995) 1368.

[13] R. J. Furnstahl, B. D. Serot and H.-B. Tang, *Nucl. Phys.* A 598 (1996) 539.

[14] B. D. Serot, *Ann. Phys. (N.Y.)* 322 (2007) 2811.

[15] P. G. Reinhard, M. Rufa, J. Maruhn, W. Greiner and J. Friedrich, *Z. Phys.* A 323 (1986) 13.

[16] R. J. Furnstahl, C. E. Price and G. E. Walker, *Phys. Rev.* C 36 (1987) 2590.

[17] R. J. Furnstahl and C. E. Price, *Phys. Rev.* C 40 (1989) 1398.

[18] A. R. Bodmer and C. E. Price, *Nucl. Phys.* A 505 (1989) 123.

[19] Y. K. Gambhir, P. Ring and A. Thimet, *Ann. Phys.* (N.Y.) 198 (1990) 132.

[20] R. J. Furnstahl and B. D. Serot, *Phys. Rev.* C 47 (1993) 2338.

[21] D. G. Boulware, *Ann. Phys.* (N.Y.) 56 (1970) 140.

[22] S. Weinberg, *Phys. Rev.* 166 (1968) 1568.

[23] R. Dashen and M. Weinstein, *Phys. Rev.* 183 (1969) 1261.

[24] S. Weinberg, Physica A 96 (1979) 327.

[25] S. Weinberg, *Phys. Lett.* B 251 (1990) 288.

[26] S. Weinberg, *Nucl. Phys.* B 363 (1991) 3.

[27] S. Weinberg, *Phys. Lett.* B 295 (1992) 114.

[28] J. Gasser and H. Leutwyler, *Ann. Phys.* (N.Y.) 158 (1984) 142.

[29] J. Gasser and H. Leutwyler, Nucl. Phys. B 250 (1985) 465, 517, 539.

[30] U.-G. Meißner, *Phys. Rep.* 161 (1988) 213.

[31] H. Leutwyler, *Ann. Phys.* (N.Y.) 235 (1994) 165.

[32] J. L. Friar, D. G. Madland and B. W. Lynn, *Phys. Rev.* C 53 (1996) 3085.

[33] A. R. Bodmer, *Nucl. Phys.* A 526 (1991) 703.

[34] S. Gmuca, *Z. Phys.* A 342 (1992) 387.

[35] S. Gmuca, *Nucl. Phys.* A 547 (1992) 447.

[36] H. Müller and B. D. Serot, *Nucl. Phys.* A 606 (1996) 508.

[37] R. M. Dreizler and E. K. U. Gross, Density functional theory (Springer, Berlin, 1990).

[38] C. Speicher, R. M. Dreizler and E. Engel, *Ann. Phys.* (N.Y.) 213 (1992) 312.

[39] R. N. Schmid, E. Engel and R. M. Dreizler, *Phys. Rev.* C 52 (1995) 164.

[40] A. H. MacDonald and S. H. Vosko, *J. Phys.* C 12 (1979) 2977.

[41] W. Kohn and L. J. Sham, *Phys. Rev.* A 140 (1965) 1133.

[42] W. Kohn, *Rev. Mod. Phys.* 71 (1999) 1253.

[43] B. D. Serot and J. D. Walecka, in: 150 Years of quantum many-body theory, R. F. Bishop, K. A. Gernoth, N. R. Walet, eds. (World Scientific, Singapore, 2001), p. 203.

[44] G. A. Lalazissis, P. Ring and D. Vretenar, eds., Extended density functionals in nuclear structure physics (Springer, Berlin, Heidelberg, 2004).

[45] M. A. Huertas, *Phys. Rev.* C 66 (2002) 024318; 67 (2003) 019901 (E).

[46] M. A. Huertas, *Acta Phys. Polon.* B 34 (2003) 4269.

[47] M. A. Huertas, *Acta Phys. Polon.* B 35 (2004) 837.

[48] B. Friedman and V. R. Pandharipande, *Nucl. Phys.* A 361 (1981) 502.

[49] N. K. Glendenning, *Phys. Rev.* D 46 (1992) 1274.

[50] T. Maruyama, H. Fujii, T. Muto, T. Tatsumi, *Phys. Lett.* B 337 (1994) 19.

[51] R. Knorren, M. Prakash and P. J. Ellis, *Phys. Rev.* C 52 (1995) 3470.

[52] M. Prakash, J. Cooke and J. M. Lattimer, *Phys. Rev.* D 52 (1995) 661.

[53] M. Prakash, I. Bombaci, M. Prakash, P. J. Ellis, J. M. Lattimer and R. Knorren, *Phys. Rep.* 280 (1997) 1.

[54] T. Maruyama, H. Shin, H. Fujii and T. Tatsumi, *Prog. Theor. Phys.* 102 (1999) 809.

[55] N. K. Glendenning and J. Schaffner-Bielich, *Phys. Rev. Lett.* 81 (1998) 4564.

[56] J. M. Lattimer and M. Prakash, *Astrophys. J.* 550 (2001) 426.

[57] J. M. Lattimer and M. Prakash, *Science* 304 (2004) 536.

[58] B. D. Lackey, M. Nayyar and B. J. Owen, *Phys. Rev.* D 73 (2006) 024021.

[59] F. Özel, *Nature* 441 (2006) 1115.

[60] M. Alford, D. Blaschke, A. Drago, T. Klähn, G. Pagliara and J. Schaffner-Bielich, *Nature* 445 (2007) E7.

[61] J. M. Lattimer and M. Prakash, *Phys. Rep.* 442 (2007) 109.

[62] G. Velo and D. Zwanzinger, *Phys. Rev.* 188 (1969) 2218.

[63] G. Velo, Nucl. Phys. B 65 (1973) 427.

[64] V. I. Ogievetskij and I. V. Polubarinov, *Ann. Phys.* (N.Y.) 25 (1963) 358.

[65] J. C. Collins and M. J. Perry, *Phys. Rev. Lett.* 34 (1975) 1353.

[66] G. Baym and S. A. Chin, *Phys. Lett.* 62 B (1976) 241.

[67] N. K. Glendenning, *Ap. J.* 293 (1985) 470.

[68] N. K. Glendenning, *Nucl. Phys.* A 493 (1989) 521.

[69] J. Ellis, J. I. Kapusta and K. A. Olive, *Nucl. Phys.* B 348 (1991) 345.

[70] N. K. Glendenning, Compact stars, nuclear physics, particle physics and cosmology (Springer, New York, 1997).

[71] D. B. Kaplan and A. E. Nelson, *Phys. Lett.* B 175 (1986) 57.

[72] D. B. Kaplan and A. E. Nelson, *Phys. Lett.* B 179 (1986) 409.

[73] H. D. Politzer and M. B. Wise, *Phys. Lett.* B 273 (1991) 156.

[74] V. Thorsson, M. Prakash and J. M. Lattimer, *Nucl. Phys.* A 572 (1994) 693.

[75] A. L. Fetter and J. D. Walecka, Quantum theory of many-particle systems (McGraw–Hill, New York, 1971). Reprinted by Dover Publications, Mineola, New York (2003).

[76] C. Itzykson and J.-B. Zuber, Quantum field theory (McGraw–Hill, New York, 1980), sec. 11–4–2.

[77] J. Iliopulos, C. Itzykson and A. Martin, *Rev. Mod. Phys.* 47 (1975) 165.

[78] R. J. Furnstahl, H.-B. Tang and B. D. Serot, *Phys. Rev.* C 52 (1995) 1368.

[79] C. J. Horowitz and B. D. Serot, *Nucl. Phys.* A 399 (1983) 529.

[80] C. J. Horowitz and B. D. Serot, *Nucl. Phys.* A 464 (1987) 613; A 473 (1987) 760 (E).

[81] H. Georgi, Weak interactions and modern particle theory (Benjamin/Cummings, Menlo Park, CA, 1984).

[82] E. K. Heide, S. Rudaz and P. J. Ellis, *Nucl. Phys.* A 571 (1994) 713.

[83] J. McIntire, Y. Hu and B. D. Serot, *Nucl. Phys.* A794 (2007) 166.

[84] Y. Hu, J. McIntire and B. D. Serot, *Nucl. Phys.* A794 (2007) 187.

[85] C. J. Horowitz and J. Piekarewicz, *Phys. Rev. Lett.* 86 (2001) 5647.

[86] C. J. Horowitz and J. Piekarewicz, *Phys. Rev.* C 64 (2001) 062802(R).

[87] J. Piekarewicz, *Phys. Rev.* C 69 (2004) 041301(R).

[88] B. G. Todd-Rutel and J. Piekarewicz, *Phys. Rev. Lett.* 95 (2005) 122501.

[89] R. J. Furnstahl, B. D. Serot and H.-B. Tang, *Nucl. Phys.* A 618 (1997) 446.

[90] B. D. Serot, in: Frontiers in field theory, O. Kovras, ed. (Nova Science, New York, 2005), p. 89.

[91] S. Weinberg, Gravitation and cosmology (Wiley, New York, 1972).

[92] W. D. Myers and W. J. Swiatecki, *Nucl. Phys.* A 601 (1996) 141.

[93] H. Müller and B. D. Serot, *Phys. Rev.* C 52 (1995) 2072.

[94] Y. Aoki, G. Endrõdi, Z. Fodor, S. D. Katz, and K. K. Szabó, *Nature* 443 (2006) 675.

[95] S. Ejiri, F. Karsch, E. Laermann, C. Miao, S. Mukherjee, P. Petreczky, C. Schmidt, W. Soeldner and W. Unger, *Phys. Rev.* D 80 (2009) 094505.

[96] C. DeTar, L. Levkova, S. Gottlieb, U. M. Heller, J. E. Hetrick, R. Sugar and D. Toussaint, *Phys. Rev.* D 81 (2010) 114504.

[97] B. D. Serot and H. Uechi, *Ann. Phys.* (N.Y.) 179 (1987) 272.

[98] H. A. Bethe, G. E. Brown and J. Cooperstein, *Nucl. Phys.* A 462 (1987) 791.

[99] D. H. Youngblood, H. L. Clark and Y.-W. Liu, *Phys. Rev. Lett.* 82 (1999) 691.

[100] J. Piekarewicz, *Phys. Rev.* C 69 (2004) 041301(R).

[101] J. Piekarewicz, *Phys. Rev.* C 76 (2007) 064310.

[102] F. J. Fattoyev and J. Piekarewicz, *Phys. Rev.* C 82 (2010) 025805.

[103] F. J. Fattoyev, C. J. Horowitz, J. Piekarewicz, and G. Shen, *Phys. Rev.* C 82 (2010) 055803.

[104] A. W. Steiner, J. M. Lattimer and E. F. Brown, *Ap. J.* 722 (2010) 33.

[105] P. B. Demorest, T. Pennucci, S. M. Ransom, M. S. E. Roberts and J. W. T. Hessels, *Nature* 467 (2010) 1081.

[106] F. Özel, *Nature* 445 (2007) E8.

In: Neutron Stars
Editors: Hiroshi Uechi et al, pp. 73-92
ISBN: 978-1-62081-123-8

Chapter 3

QUANTUM HADRODYNAMICS: EVOLUTION AND REVOLUTION*

R.J. Furnstahl[1,†] and Brian D. Serot[2,‡]
[1]Department of Physics, The Ohio State University,
Columbus, OH 43210
[2]Department of Physics and Nuclear Theory Center
Indiana University, Bloomington, IN 47405

Abstract

The underlying philosophy and motivation for quantum hadrodynamics (QHD), namely, relativistic field theories of nuclear phenomena featuring manifest covariance, have evolved over the last quarter century in response to successes, failures, and sharp criticisms. A recent revolution in QHD, based on modern effective field theory and density functional theory perspectives, explains the successes, provides antidotes to the failures, rebuts the criticisms, and focuses the arguments in favor of a covariant representation.

1. Introduction

Quantum hadrodynamics (QHD) refers to relativistic field theories for nuclei based on hadrons, in which the representation is manifestly covariant. The distinguishing empirical feature of covariant QHD is the presence of large (several hundred MeV), isoscalar, Lorentz scalar and vector mean fields (optical potentials) in nuclear matter at normal nuclear densities. During the last quarter century, QHD calculations have had numerous successes, but also apparent failures that arose when the dynamics of the quantum vacuum were computed with the same degrees of freedom used to describe the valence-nucleon physics. Moreover, there have been sharp criticisms over the years (some of which appeared in earlier issues of *Comments*) based on the relevance of a covariant approach to the nuclear many-body

*First published as Comments Mod. Phys. 2 (2000) A23.
†E-mail address: furnstahl.1@osu.edu
‡E-mail address: serot@indiana.edu

problem at observable densities, the use of relativistic, renormalizable quantum field theories with nucleon fields to describe composite hadrons and the quantum chromodynamics (QCD) vacuum, and the apparent lack of pion dynamics and chiral symmetry in the empirically successful calculations.

The underlying philosophy of QHD has evolved over the past twenty-five years in response to these successes, failures, and criticisms. Ultimately, the original motivation for renormalizable QHD lagrangians yielded to a more general approach based on the modern ideas of nonrenormalizable, effective relativistic field theories. This change in QHD philosophy is nothing short of a "revolution" that occurred during a two- or three-year span in the mid 1990s. The new approach, based on effective field theory (EFT) and density functional theory (DFT), allows us to understand the successful mean-field calculations of nuclear properties and how chiral symmetry works in QHD. It provides antidotes to earlier failures through a consistent, systematic, covariant treatment of the nuclear many-body problem, and it provides rebuttals to the previous criticisms.

In this *Comment*, we trace the QHD evolution and revolution, with special attention to how past criticisms and deficiencies have been nullified.

2. Evolution

2.1. Original Motivation

Calculations of nuclear many-body systems based on relativistic hadronic field theories have existed for many years; one can trace their history back at least as far as the seminal work of Schiff in 1951 [1]. An important advance was made roughly 25 years ago, when Walecka enumerated the philosophy underlying a covariant description of nuclear matter, together with the rules for calculating within such a framework and a systematic program for investigating the observable consequences of the approach [2].[1]

To paraphrase the motivation presented so clearly by Walecka: to discuss neutron stars, it is necessary to have an equation of state that describes matter from observed terrestrial densities upward; a consistent theory should include mesonic degrees of freedom explicitly to allow for extrapolation to high densities; as the density of the matter is increased, relativistic propagation of the nucleons (and the retarded propagation of the virtual mesons) must be included; and causal restrictions on the propagation of excitation modes of the interacting system must be automatically contained in the theory. The basic conclusion following from these ideas is that, "The only consistent approach ... which meets these [objectives] is ... a local, relativistic, many-body quantum field theory."

The systematic program to develop QHD as defined by Walecka focused on the nuclear many-body problem and relied on *renormalizable* lagrangian densities to define the models. This restriction was motivated by the desire to extrapolate away from the empirical calibration data in a manner that did not introduce any new, unknown parameters. Perhaps more importantly, the faithful pursuit of this framework would reveal whether renormalizable QHD was feasible and practical or not [3, 4, 5].

[1] At the time of Walecka's original paper, several other groups were studying relativistic, hadronic, field-theoretic approaches to the nuclear many-body problem, using similar but slightly different philosophies. See Ref. [3] for discussions of other approaches and results.

The original model contained neutrons, protons, and isoscalar, Lorentz scalar and vector mesons. In the nonrelativistic (Yukawa potential) limit (which was never actually used in the calculations), single-meson exchange generates basic observed features of the NN interaction: a strong, short-range repulsion and a medium-range attraction. It was assumed by fiat that the possible (renormalized) nonlinear couplings between the scalar fields were zero, for simplicity, although this restriction was relaxed quite early in the development by other practitioners [6]. Moreover, no attempt was made to reconcile the model with the spontaneously broken, approximate $SU(2)_L \times SU(2)_R$ chiral symmetry of hadronic interactions.

The model contains large couplings, so a practical, nonperturbative approximation is needed as a starting point to describe the nuclear equation of state (EOS). Walecka argued that at high enough density, fluctuations in the meson fields could be ignored, and they could be replaced by their classical expectation values or *mean fields*. He also assumed that these conditions were sufficiently valid at ordinary nuclear density, so that the model could be calibrated in the mean-field approximation, and that the mean-field contributions would dominate the high density (e.g., neutron-star) EOS. Corrections to the mean-field approximation were calculated, and it was indeed found that in the context of renormalizable, Walecka-type models, the "stiff" EOS predicted by the mean-field theory (MFT), in which the pressure approaches the total energy density from below, becomes increasingly accurate as the nuclear density increases.

The hope was that nonrenormalizable and vacuum (short-range) effects, which had not been included in the corrections noted above, would be small enough to be described adequately by the long-range degrees of freedom of a renormalizable field theory, through the systematic evaluation of quantum loops. This hope was not fulfilled by explicit calculation of short-range loop effects. Furthermore, enlarging the nonlinear meson self-couplings to allow for effective, nonrenormalizable terms alters the high-density nuclear EOS qualitatively [7].

2.2. Successes

At the same time as the neutron-matter EOS was being studied, the MFT model was applied to the bulk and single-particle properties of doubly magic nuclei. The most important conclusion of this early work was that calibration to the empirical equilibrium point of ordinary nuclear matter produces scalar and vector mean fields of roughly several hundred MeV at equilibrium density. When extended to finite nuclear systems, the resulting single-particle spin-orbit potential is roughly the same size as the observed spin-orbit potential, and thus the relativistic MFT *predicts the existence of the nuclear shell model*, without any *ad hoc* adjustments to the spin-orbit force. Moreover, MFT models that incorporated nonlinear (i.e., cubic and quartic) scalar field couplings reproduced bulk and single-particle nuclear observables as well as or better than any other concurrent models.

Relativistic MFT calculations have been performed for nuclei throughout the Periodic Table, with similarly realistic results and predictions; the reader is directed to the cited review articles for discussions of these numerous calculations [8, 9, 10, 11, 12, 13, 14]. When the MFT nuclear densities were folded with the free NN scattering matrix to compute proton–nucleus scattering observables (this is called the relativistic impulse approximation or RIA [15, 16, 17, 18, 19]), excellent descriptions of existing data and predictions for upcoming data were found—far superior to nonrelativistic calculations at the same level of approximation.

The vast majority of successful QHD predictions rely on the important observation that there are large, isoscalar, Lorentz scalar and vector mean fields in nuclei. Moreover, the successful calculations explicitly include only valence nucleons and long-range, many-body dynamics; the QHD degrees of freedom are designed precisely to describe this type of dynamics. It is also possible to study nuclear excited states in a random-phase approximation (RPA) that involves only long-range dynamics, but that still maintains the underlying symmetries of the lagrangian [20].

To explicitly include two-nucleon correlations, one uses the so-called Dirac–Brueckner–Hartree–Fock (DBHF) theory . With a covariant NN kernel that is fitted to two-body data and that contains large Lorentz scalar and vector (and pionic) components, one can simultaneously reproduce the nuclear matter equilibrium point at the two-hole-line level [21, 22, 23, 24]. Moreover, although the correlation corrections produce changes in the MFT binding energy that are of the same order as the binding energy itself, *the corrections to the large MFT scalar and vector self-energies (optical potentials) are small* [25, 26]. Nevertheless, to our knowledge, numerous approximations in the DBHF approach have never been quantitatively tested, and the systematic inclusion of contributions from the quantum vacuum was an unsolved problem [13].

2.3. Difficulties

Along with the numerous successes, there were also various difficulties that can be traced to a common source: the requirement that the QHD models be *renormalizable* . The difficulties fall into two basic classes: those arising from the computation of quantum loops (short-range physics) using long-range degrees of freedom, and those arising from attempting to maintain both renormalizability and chiral symmetry simultaneously.

Although the relativistic MFT results are encouraging, the QHD program sought a more complete description of the nuclear many-body system, which *requires* the development of reliable techniques to extend these calculations. Quantum loops are important for several reasons: loops ensure the unitarity of scattering amplitudes, baryon loops containing valence nucleons incorporate familiar many-body effects, loops introduce effects arising from the modification of the quantum vacuum in the presence of valence nucleons, and meson loops in particular generate contributions to the extended structure of the nucleon.

Not all loops in QHD are problematic. For example, loops involving fluctuations of the pion field generate long-range effects (since the pion is light) and should be accurately described by explicit calculation within the QHD model (provided that approximate chiral symmetry is maintained). The strong, mid-range NN attraction arises from pion rescattering loops when two pions in the scalar, isoscalar channel are exchanged between nucleons. These long-range dynamical effects are much more efficiently described with hadrons than with QCD quarks and gluons; in fact, most QHD models go one step further and simulate the mid-range NN attraction using a Yukawa coupling to an explicit scalar, isoscalar field.

Problems arise with loops when one attempts to describe short-range dynamics using the heavier QHD degrees of freedom (nucleons and non-Goldstone bosons). In a renormalizable theory, a finite result can be obtained for the Casimir effect, and its addition to the MFT produces what is usually called the relativistic Hartree approximation (RHA). Although the new contributions are finite, they degrade the agreement of the nuclear predictions with experiment, particularly when one examines spin-orbit splittings for single-particle levels near the Fermi surface. These results imply that the QHD treatment of the quantum vacuum at the one-baryon-loop level is, at best, inadequate; although higher-order corrections might reduce the size of the one-loop terms, this can occur only through sensitive cancelations between relatively large contributions.

In fact, contributions from higher-loop terms within the renormalizable QHD framework do not improve the situation. Explicit calculations of the nuclear matter energy density at the two-loop level found enormous contributions that altered the description of the nuclear ground state qualitatively [27]. The conclusion was that the loop expansion does not provide a reliable approximation scheme in renormalizable QHD.

Similarly, vacuum contributions in the summation of ring diagrams produced unphysical poles at spacelike momenta in the meson propagators (sometimes called " ghosts"), which signal either an inconsistency in the QHD framework or an inadequate level of approximation. It was proposed that vertex corrections within the theory could solve these problems [28, 29], but complete calculations involving vertex insertions proved to be impractical, and to our knowledge, no systematic, reliable approximation scheme for incorporating both long-range and short-range loop effects in renormalizable QHD theories has ever been found.

The second class of difficulties arises when one tries to embed the approximate, spontaneously broken, $SU(2)_L \times SU(2)_R$ chiral symmetry of QCD in a renormalizable QHD theory. The original Walecka model made no mention of chiral symmetry, but there were chiral models in use at that time based on the well-known Sigma model of Schwinger [30] and of Gell-Mann and Lévy [31]. In fact, if one takes the original Walecka model, adds massless pions that couple to nucleons with a pseudoscalar (γ_5) Yukawa coupling, and demands that the theory be Lorentz covariant, parity invariant, isospin and chiral invariant, and renormalizable, one is led to a lagrangian that is simply the Sigma model with an additional isoscalar vector meson. The nonzero nucleon and scalar masses are generated through the familiar spontaneous symmetry breaking, and given the similarity to the Walecka-model lagrangian, it is natural to identify the scalar field (which is the chiral partner of the pion) with the scalar field in the Walecka model.

The MFT for the chiral model can be motivated precisely as before, except that there are now cubic and quartic scalar self-couplings that are not free parameters, but that are specified by the chiral symmetry. Unfortunately, the assumption that the chiral scalar field is the same as the Walecka scalar leads to dire consequences. First, it is impossible to reproduce the empirical nuclear matter equilibrium point in the MFT. Including quantum loops does not help, since one either generates extremely large contributions or arrives at uncertain results due to the appearance of unphysical poles in the meson propagators. An extensive mean-field analysis shows that it is impossible to generate realistic results for finite nuclei within this framework (see Refs. [32, 33, 34]). The conclusion is that the standard form of spontaneous chiral symmetry breaking, implemented in a model with a linear realization of the symmetry, *cannot* produce successful nuclear phenomenology at the mean-field level, if the chiral scalar field is identified with the scalar field in the Walecka model. This failure of the Sigma model is evidence that the simultaneous constraints of renormalizability and linear chiral symmetry are too restrictive.

Some progress was made on this problem by following the work of Weinberg and by making field transformations of the nucleon, pion, and scalar fields [35, 36]. In addition to changing the form of the pion–nucleon interaction to include a pseudovector ($\gamma_\mu\gamma_5$) coupling, the transformation allows the introduction of a *new* scalar, isoscalar field that is not the chiral partner of the pion. This field plays the same role as in the Walecka model: it simulates important $\pi\pi$ and NN interactions that must be included from the outset to generate a realistic description of nuclear matter and nuclei.

A profound change has occurred, however, because in contrast to the original proposal of QHD as a renormalizable field theory, we are now forced to consider the new scalar field as an effective degree of freedom and the new chiral lagrangian as a nonrenormalizable, effective lagrangian. Thus the earlier philosophy of QHD must be generalized to include nonrenormalizable, effective field theories.

3. Revolution

3.1. EFT/DFT Perspective

The revolution in QHD started with the reinterpretation of QHD lagrangians as nonrenormalizable EFT lagrangians [37]. An effective lagrangian consists of known long-range interactions constrained by symmetries and a complete set of generic short-range interactions. The division between long and short is characterized by the breakdown scale Λ of the EFT. While it is not possible at present to derive an effective hadronic theory directly from the underlying QCD, the EFT perspective implies that this is not necessary. If one constructs a general lagrangian that respects the symmetries of QCD: Lorentz covariance, parity conservation, time-reversal and charge-conjugation invariance, (approximate) isospin symmetry, and spontaneously broken chiral symmetry, then the EFT is a general parametrization of observables below the breakdown scale.

The EFT perspective, with the freedom to redefine and transform fields, implies that *there are infinitely many representations of low-energy QCD physics*. But they are not all equally efficient or physically transparent. One of the possible choices is between Lorentz covariant and nonrelativistic formulations. (In the context of EFT, these can be related by the heavy-baryon expansion [38].) Recent developments in baryon chiral perturbation theory support the consistency (and utility) of a covariant EFT, with Dirac nucleon fields in a Lorentz invariant effective lagrangian density [39, 40]. A similar framework underlies QHD approaches to nuclei.

For QHD, we identify Λ with the scale of non-Goldstone-boson physics (roughly 600 MeV). At momenta small compared to Λ, short-distance physics (such as the substructure of nucleons) is only partially resolved and so may be incorporated into the coefficients of operators organized as a derivative expansion. The coefficients of these short-range terms may eventually be derived from QCD, but at present, they must be fitted by matching calculated and experimental observables. In principle, there are an infinite number of (nonrenormalizable) terms, but in practice, the lagrangian or energy functional can be truncated to work to a given precision [41]. The EFT is useful if this truncation can be made at low enough order that the number of free parameters is not prohibitive.

In QHD, the only *essential* hadronic degrees of freedom are the nucleons and pions. The long-range pion–pion and pion–nucleon interactions are included in a nonlinear realization of chiral symmetry, which avoids dynamical assumptions inherent in linear representations. These interactions can be written down systematically (given a power-counting scheme) [41]. Low-mass vector mesons are typically included for phenomenological reasons, but are not required since their masses are of the order of Λ; they are absent from point-coupling models, for example. In descriptions of NN scattering and of nuclear structure and reactions, the heavy bosons carry *spacelike* four-momenta and are "off the mass shell"; they therefore serve simply as a convenient way to parameterize the NN interaction in exchange channels with vector quantum numbers. This explains why it is useful to introduce collective degrees of freedom with other quantum numbers, such as a Delta baryon (with spin and isospin of 3/2) to incorporate important pion–nucleon interactions. Because one must always truncate the lagrangian, these degrees of freedom can be efficient in the many-body problem *whether or not they are actually observed as hadronic resonances*.

A scalar, isoscalar mean-field in nuclei is an efficient way to include implicitly the effects of pion exchange that are the most important for describing bulk nuclear properties. Because chiral symmetry is realized nonlinearly, one can *add* a light scalar, isoscalar, chiral-singlet field to the theory and give it a Yukawa coupling to the nucleon, just as in the Walecka model. Nonlinear self-interactions of this new scalar must be included, with adjustable couplings that arise in part from the nucleon substructure. Since the expectation value of the pion field in nuclear matter vanishes at the mean-field level, one makes the remarkable observation that the MFT of Walecka-type QHD models is the same as the MFT of the chiral EFT model! Thus the MFT of Walecka-type models *is consistent with chiral symmetry* , provided we think in terms of a nonlinear realization of the symmetry. The light scalar, isoscalar field, which is *not* the chiral partner of the pion, plays the same role as in the Walecka model: it simulates important $\pi\pi$ and NN interactions that must be included from the outset to generate a realistic description of nuclear matter and nuclei.

To make systematic calculations, the EFT approach exploits the separation of scales in physical systems, with the ratios of scales providing expansion parameters. A connection between appropriate QCD scales and nuclear phenomenology is made by applying Georgi and Manohar's Naive Dimensional Analysis (NDA) and naturalness [42, 43]. These principles prescribe how to count powers of the pion decay constant $f_\pi \approx 94\,\text{MeV}$ and a larger mass scale Λ in effective lagrangians or energy functionals. The mass scale Λ is associated with the new physics beyond the pions: the non-Goldstone boson masses or the nucleon mass. The signature of these low-energy QCD scales in the coefficients of a relativistic point-coupling model was first pointed out by Friar, Lynn, and Madland [44]. Subsequent analyses have extended and supplemented this idea, testing it in nonrelativistic mean-field models as well as in different types of relativistic models. Estimates of contributions to the energy functional from individual terms, based on NDA power counting, are *quantitatively* consistent with direct, high-quality fits to bulk nuclear observables [45, 46]. Naturalness based on NDA scales has proved to be a very robust concept: nuclei know about these scales!

The successes of QHD mean-field phenomenology are, at first, rather mysterious from the EFT perspective alone, since the Hartree approximation is only the finite-density counterpart of the Born approximation at zero density. The density functional theory (DFT) perspective explains the successes of mean-field models and provides a new context for EFT power counting.

Conventional density functional theory is based on energy functionals of the ground-state density of a many-body system, whose extremization yields a variety of ground-state properties. In a covariant generalization of DFT applied to nuclei, these become functionals of the ground-state scalar density ρ_{s} as well as the baryon current B_μ. Relativistic mean-field models are analogs of the Kohn–Sham formalism of DFT [47, 48, 49], with local scalar and vector fields $\Phi(\mathbf{x})$ and $W(\mathbf{x})$ appearing in the role of relativistic Kohn–Sham potentials [14]. The mean-field models approximate the exact functional, which includes all higher-order correlations, using powers and gradients of auxiliary meson fields or nucleon densities.

The scalar and vector potentials are determined by extremizing the energy functional, which gives rise to a Dirac single-particle hamiltonian. The isoscalar part (for spherical nuclei) is

$$h_0 = -i\,\boldsymbol{\nabla}\cdot\boldsymbol{\alpha} + \beta\Big(M - \Phi(r)\Big) + W(r)\,, \tag{1}$$

where M is the nucleon mass and we define $M^* \equiv M - \Phi$. It is not necessary that Φ is simply proportional to a scalar meson field ϕ. In fact, Φ could be proportional to ϕ (as in the original QHD models) or could be expressed as a sum of scalar and vector densities (as in relativistic point-couplings models) or could be a nonlinear function of ϕ.

Density functional theory can provide a framework for the systematic incorporation of correlation effects, *which are included exactly if the correct functional is identified*. Mean-field models approximate this functional with powers and gradients of fields or densities, with the truncation determined by power counting. The inclusion of vacuum contributions, which was such a difficulty before, now becomes simple with the realization that all necessary counterterms are already present. The convergence of the EFT/DFT expansion is reasonable, but is slow enough that there are still too many terms to calibrate accurately by fitting to nuclear data. We rely on existing phenomenology as a guide to truncating the lagrangian most efficiently in light of ill-determined coefficients. In addition, while it is known that correlation corrections modify the scalar and vector self-energies by only a small amount ("Hartree dominance "), the mean-field energy functional omits possible non-analytic terms; a combination of EFT and DFT may show us how to systematically include them [50, 51].

3.2. Antidotes, Rebuttals, and Reinterpretations

In this section, we revisit past criticisms of QHD, many taken from earlier *Comments* [52, 53, 54] and others commonly expressed at conferences or elsewhere in the literature. We present without attribution a series of paraphrased statements in **boldface** that criticize different aspects of QHD and its predictions. Each statement is followed by a resolution based on the modern EFT/DFT perspective of QHD. We find that each criticism is either addressed and answered, revealed to be incorrect, or rendered moot.

Nuclei are nonrelativistic systems because corrections to the kinetic energy are small. Relativistic phenomenology for nuclei has often been motivated by the need for relativistic kinematics when extrapolating to extreme conditions of density, temperature, or momentum transfer. Unfortunately, this motivation obscures the issue of Lorentz covariant vs. nonrelativistic approaches for nuclei under ordinary conditions. Relativistic kinematic corrections are indeed small for ordinary nuclear systems. The important aspect of relativity in these systems is *not* that a nucleon's momentum is comparable to its rest mass, but that maintaining covariance allows scalars to be distinguished from the time components of four-vectors. This distinction is easy to see by expanding the self-consistent nuclear matter energy density $\mathcal{E}$ in powers of the Fermi momentum $k_{\rm F}$ (see Ref. [14], p. 554):[2]

$$\begin{aligned}\mathcal{E}/\rho_{\rm B} &= M + \left[\frac{3k_{\rm F}^2}{10M} - \frac{3k_{\rm F}^4}{56M^3} + \frac{k_{\rm F}^6}{48M^5} - \frac{15k_{\rm F}^8}{1408M^7} + \frac{21k_{\rm F}^{10}}{3328M^9} + \cdots\right] \\ &+ \frac{g_{\rm V}^2}{2m_{\rm V}^2}\rho_{\rm B} - \frac{g_{\rm S}^2}{2m_{\rm S}^2}\rho_{\rm B} + \frac{g_{\rm S}^2}{m_{\rm S}^2}\frac{\rho_{\rm B}}{M}\left[\frac{3k_{\rm F}^2}{10M} - \frac{36k_{\rm F}^4}{175M^3} + \frac{16k_{\rm F}^6}{105M^5} - \frac{64k_{\rm F}^8}{539M^7} + \cdots\right] \\ &+ \left(\frac{g_{\rm S}^2\rho_{\rm B}}{m_{\rm S}^2 M}\right)^2\left[\frac{3k_{\rm F}^2}{10M} - \frac{351k_{\rm F}^4}{700M^3} + \cdots\right] + \left(\frac{g_{\rm S}^2\rho_{\rm B}}{m_{\rm S}^2 M}\right)^3\left[\frac{3k_{\rm F}^2}{10M} - \cdots\right] + \cdots . \qquad (2)\end{aligned}$$

The corrections to the nonrelativistic kinetic energy (contained in the first term in brackets) are indeed small at equilibrium density, but the velocity dependence inherent in the Lorentz scalar interaction introduces significant corrections of higher than linear order in $\rho_{\rm B}$. The leading correction in each order is *repulsive*, so these corrections are important in establishing the equilibrium point.

In the nuclear medium, a covariant treatment implies distinct scalar and four-vector nucleon self-energies or optical potentials. The relevant question is: What are their natural mean values? QHD phenomenology implies several hundred MeV in the center of a heavy nucleus.

The success of nonrelativistic approaches shows that covariant approaches are wrong or unnecessary. Historically, the successes of nonrelativistic nuclear phenomenology have been cited to cast doubt on the relevance of large scalar and vector potentials. But in a nonrelativistic treatment of nuclei, the distinction between a potential that transforms like a scalar and one that transforms like the time component of a four-vector is lost. Because the leading-order contributions of these two types are opposite in sign, an underlying large scale characterizing individual covariant potentials would be *hidden* in the nonrelativistic central potential. (The scalar and vector terms add constructively in the nonrelativistic spin-orbit potential, producing an uncharacteristically large result.) Furthermore, the EFT expansion implies that even potentials as large as 300 to 400 MeV are sufficiently smaller than the nucleon mass that a nonrelativistic expansion should converge, if not necessarily optimally. *Thus the success of nonrelativistic nuclear phenomenology provides little direct evidence about covariant potentials.*

The primary focus of nuclear theory is to fit the two-nucleon data and then to solve the many-body problem. The EFT coefficients are fixed by *any* sufficient data set; NN data has no special significance. Indeed, a density functional is best determined by finite-density data. Furthermore, NN data by itself *cannot* be sufficient. Many-body forces are inevitable [55, 56, 57, 51], and their size can be estimated and shown to be non-negligible at ordinary

[2]Here $\rho_{\rm B} = 2k_{\rm F}^3/3\pi^2$ is the baryon density, and g_i and m_i denote the scalar and vector couplings and masses, respectively. Nonlinear meson interactions are omitted for brevity.

densities (verified phenomenologically [41, 58]). In a mean-field density functional to be used for medium to heavy nuclei, at least one three-body and two four-body parameters are necessary [46].

Why use a field theory? Potential models are easier. Relativistic quantum field theory based on a local lagrangian density provides a general parametrization of experimental observables consistent with the essential physics of the strong interaction: quantum mechanics, special relativity, unitarity, causality, cluster decomposition, and the intrinsic symmetries of QCD [59]. Consequently, we can describe the physics of low-energy QCD (e.g., ordinary nuclei) using an effective field theory with hadronic degrees of freedom. *Field theory* offers advantages over conventional *potential models* by systematically accommodating relativistic corrections and by allowing the construction of complete, consistent operators to describe interactions with external probes.

It is important to find specific observables that distinguish between relativistic and nonrelativistic theory. This pursuit will not be fruitful. There are field transformations that connect relativistic (covariant) and nonrelativistic theories, with the ratios of the fields to the nucleon mass acting as the parameters controlling truncation (see Refs. [60] and [11]). These parameters are small enough that a nonrelativistic approach should reproduce relativistic results, although not necessarily at the same level of approximation. The more appropriate question is: What is the most efficient representation?

Large potentials are an artifact of a relativistic formulation. We argue that the large potentials used in a covariant description of nuclear phenomenology are manifestations of the underlying mass scales of low-energy QCD, which are hidden in nonrelativistic treatments [61]. These QCD mass scales are inescapable if one considers the ${}^1S_0$ NN phase shift (which becomes repulsive at about 250 MeV laboratory kinetic energy) together with the singlet scattering length of roughly $(8\,\mathrm{MeV})^{-1}$ that signals an almost-bound state near zero energy [62].

It is more efficient to work with a nonrelativistic theory because large cancelations are built in. If there were an approximate *symmetry* that enforced the cancelation between scalar and vector contributions, then it would be desirable to build the cancelation into any EFT lagrangian or energy functional. (Chiral symmetry alone does *not* lead to scalar-vector fine tuning.) However, if the cancelation is accidental or of unknown origin, hiding the underlying scales may be counterproductive. We argue that nuclei naturally fall into the second category, with the relevant scales set not by the nonrelativistic binding energy and central potential (tens of MeV), but by the large covariant potentials (hundreds of MeV). The signals of large underlying scales are patterns in the data that are simply and efficiently explained by large covariant potentials, but which require more complicated explanations in a nonrelativistic treatment [61]. Examples of these are:

- The spin-orbit force , which appears automatically with the observed strength in a covariant formulation, but is not fully reproduced in even the most sophisticated nonrelativistic calculations [63].
- Medium-energy proton–nucleus spin observables, which are reproduced by the relativistic impulse approximation with intuitive real optical potentials [15, 16, 17, 18], while nonrelativistic treatments require full-folding and medium effects [64, 65, 66, 67], and have nonintuitive potentials that change *qualitatively* with projectile energy.

- The energy dependence of the optical potential for nucleon–nucleus scattering up to 100 MeV, which is predicted at the relativistic mean-field level from the Lorentz structure of the interaction [3] (and higher-order corrections are small). In conventional nonrelativistic treatments, the energy dependence comes from the nonlocality of exchange corrections in a Hartree–Fock or Brueckner–Hartree–Fock approximation.

- The scalar, isoscalar part of the NN kernel below 1 GeV, which can be studied in an essentially model-independent way. Chiral symmetry, unitarity, and the natural strength of the $\pi\pi$ interaction imply an integrated strength that results in a large scalar single-particle potential [68, 69]. We are not aware of a loophole here.

- The equilibrium of nuclear matter, which is not an ordinary, nonrelativistic Fermi liquid, since it is too dilute and too weakly bound. These characteristics arise in the mean-field energy/particle from an empirically small coefficient of the k_{F}^3 term in a density expansion [see Eq. (2)]. The cancelation between scalar and vector contributions to the nuclear matter binding energy account for this fine tuning.

There is no experimental evidence of large scalar and vector fields. There can be no *direct* experimental verification (or refutation) of any nuclear potentials. The evidence that a natural representation contains large fields, which is achieved only with a covariant formulation, comes from both empirical and theoretical analyses of NN scattering and nuclear properties [23, 46, 61]. As noted above, this manifestation of QCD scales translates in many instances into simpler, more efficient, more compelling explanations of nuclear phenomena than in nonrelativistic formulations.

Empirical support from nuclear properties comes from the study of covariant density functionals fit to nuclei [41]. A good fit to nuclear properties requires the local scalar and vector potentials to be roughly 300 MeV, and the hierarchy of energy contributions follow NDA predictions [46] (see Fig. 3.1). A more subtle argument is that the spread of 15 MeV or more among "realistic" nonrelativistic predictions of the nuclear matter equilibrium binding energy (the "Coester line ") [23] would be difficult to understand as calibration errors ("off-shell effects "), if the underlying scale of the two-body interaction were only 50 MeV. In contrast, large covariant two-body potentials in a relativistic formulation imply sizable three-body contributions in the corresponding nonrelativistic calculation that are consistent with this spread. Finally, independent empirical support comes from fits of a covariant kernel for the NN interaction, which is used to calculate the NN scattering matrix. Every accurate fit has led to large, isoscalar, scalar and vector contributions of comparable magnitude, but of opposite sign, which translate in the medium into single-particle potentials of several hundred MeV at equilibrium density [23].

The pieces of evidence supporting a representation with large nucleon scalar and vector potentials, while not definitive when considered individually, collectively comprise a compelling positive argument [61].

We know that mean-field theory cannot be a correct description of nuclei because important long- and short-range correlations are omitted. Mean-field models are approximate implementations of Kohn–Sham density functional theory [47, 48, 49], which means that correlation effects are included in simple Hartree calculations. Moreover, the

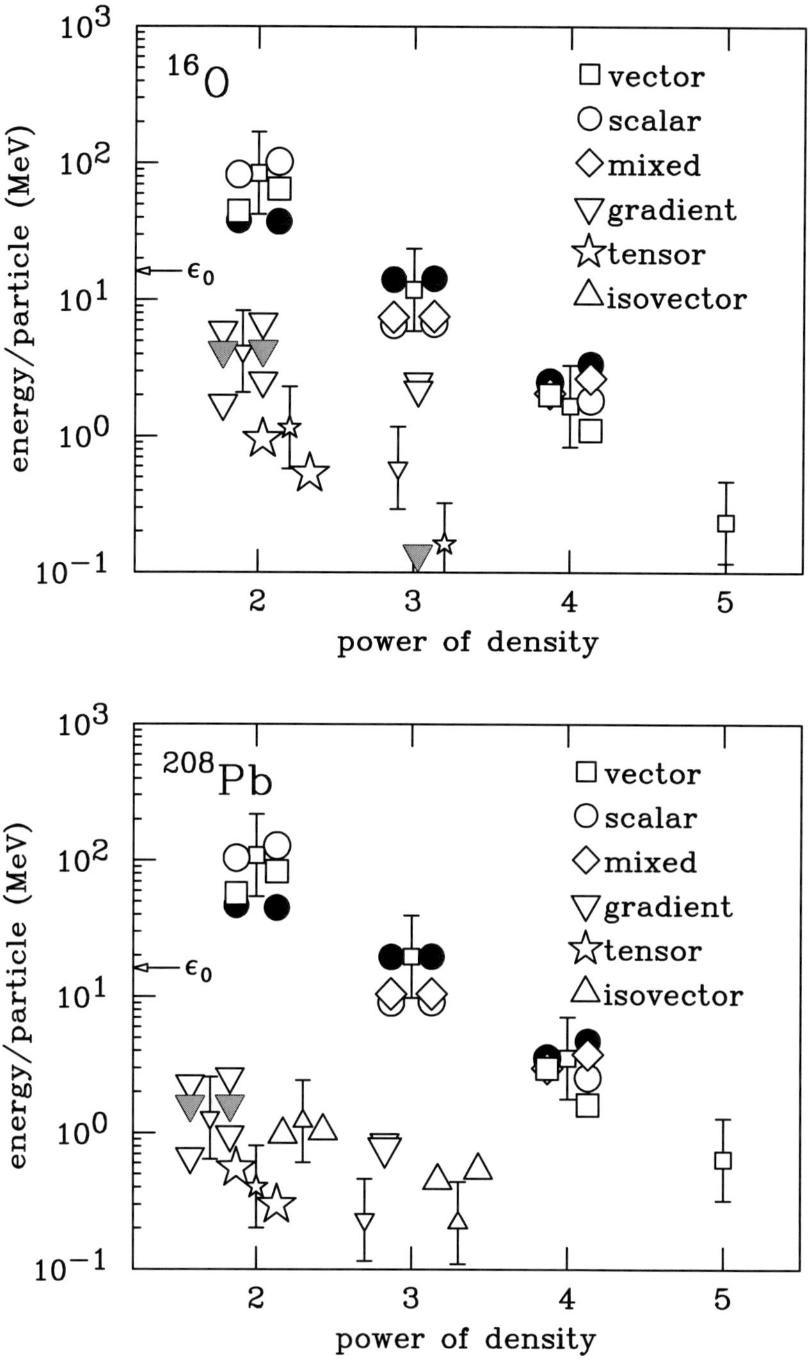

Figure 1. Contributions to the energy/particle in ^{16}O and ^{208}Pb for two covariant, mean-field, point-coupling models [46], one on each side of the error bars. Absolute values are shown. The filled symbols are net values. The small symbols indicate estimates based on NDA, with the error bars corresponding to natural coefficients from 1/2 to 2. The equilibrium binding energy of nuclear matter is ϵ_0.

"Hartree dominance" of the single-particle potentials has been demonstrated, implying that short-range correlation corrections are no more than tens of MeV [25, 26, 3]. The bulk properties of interest in mean-field phenomenology are primarily isoscalar observables that involve low resolution [46], so long-range pionic correlations are of minor importance; for other observables, EFT provides a systematic framework for explicitly including pionic contributions.

The relativistic mean-field approximation may be valid at high densities, but nuclei are low-density systems. The modern view is that the successes of relativistic mean-field theory do not depend on the justification of the mean-field approximation at high density, but on the flexibility of the mean-field density functional near equilibrium density. The combination of EFT and DFT concepts and methods applied to mean-field models of nuclei reveals that:

- NDA provides an organizational principle for the EFT. Power counting and the limited number of bulk nuclear observables explain the success of conventional mean-field models, which contain fewer parameters than the most general EFT models.
- Vacuum effects, chiral symmetry, and nucleon substructure *are all included* in general QHD models.
- Ground-state nuclear properties provide information at low resolution. Models with different degrees of freedom (e.g., four- vs. two-component nucleons or point-coupling vs. meson models) are simply different organizations of the EFT. All are consistent with NDA.

Relativistic many-body calculations have unquantifiable errors. The EFT framework based on NDA and naturalness provides an organizational scheme for truncating a lagrangian or energy functional and for making well-defined error estimates (see Fig. 3.1). Vacuum corrections, which disrupted early attempts at QHD expansion schemes, are innocuous in the EFT approach. (They are automatically absorbed into the coefficients.) Moreover, there are (in principle) no off-shell ambiguities.

QHD calculations apply perturbation theory, which is not sensible with large coupling constants. In fact, QHD does have a sensible expansion, which is *not* in powers of the couplings. We work instead with density functional theory, with NDA power counting identifying reasonable expansion parameters. It is true that the short-distance (ultraviolet) behavior may be incorrect, but the EFT can correct the behavior systematically for low-energy observables using a small number of parameters (verified phenomenologically [46]).

Calculations of magnetic moments in relativistic models are inconsistent with the data due to enhancements from a small effective nucleon mass. Naively, the baryon current of a nucleus with a single valence nucleon with momentum p outside a closed shell is p/M^*, compared to the Schmidt current p/M. However, if the calculation is forced to respect Lorentz covariance and the first law of thermodynamics, the nuclear current is constrained to be p/μ, where $\mu \approx M$ is the chemical potential [70, 71]. Thus there is no enhancement in a consistent relativistic framework.

Relativistic theories have ghosts. No they don't. Ghosts arise from improper treatment of the ultraviolet behavior in renormalizable theories; in EFT, this short-distance behavior is included systematically by fitting a small number of parameters to nuclear observables. Moreover, the long-distance instability known as "Brown–Ravenhall disease" [72] does not arise in EFT, because the EFT framework is not quantum mechanics with a fixed number of particles.

The successes of the relativistic impulse approximation are meaningless because correlation corrections are large. On the contrary, relativistic correlation corrections to the optical potential are small, in contrast to the nonrelativistic framework ("Hartree dominance"). Thus the success of RIA calculations is expected. We emphasize that covariant and nonrelativistic expansions can have very different rates of convergence. The covariant representation appears superior in most instances [64, 67].

G-parity implies that a successful description of NN scattering leads to unphysical predictions in the $\overline{\text{N}}$N sector. Explicit calculations show that absorptive processes dominate the $\overline{\text{N}}$N optical potential [73]. Thus the consequences of *G*-parity are not directly observable, since the transformed NN scattering amplitude is known only at unphysical kinematics for the $\overline{\text{N}}$N system. Moreover, in the EFT framework, there are no on-shell antinucleons; there are only valence nucleons. A simple extension of QHD to the $\overline{\text{N}}$N sector pushes the EFT expansion beyond its breakdown scale Λ, so this extrapolation is suspect.

QHD does not have pions and chiral symmetry. Modern QHD effective lagrangians include pions in a nonlinear realization of chiral symmetry [41]. Confusion about the apparent absence of pions arises because *explicit* pionic contributions do not contribute to mean-field energy functionals, although correlated pionic contributions are *implicitly* contained in the effective scalar field. Long-range pionic contributions can be included systematically, but do not qualitatively change mean-field phenomenology [74, 3, 50].

The factorization of nuclear amplitudes into a product of on-shell, single-nucleon form factors and many-body amplitudes is incorrect. The modern chiral EFT lagrangians of QHD do not require such a factorization. The single-nucleon structure is included explicitly in the lagrangian through a derivative expansion [41]. This produces results that are very similar to the standard "folding" procedure [8].

The anomalous moment of the nucleon is clearly a property of its internal quantum structure; by itself, this precludes the representation of the nucleon as a local field. This is directly refuted by the EFT lagrangian [41], which not only accommodates an anomalous moment, it requires it!

Local meson fields and "point" nucleons provide no possibility for quark substructure. This is simply incorrect. At energy and momentum scales small compared to the underlying QCD scale Λ, details of the quark substructure are not resolved. It follows that the substructure can be incorporated through a systematic expansion of nonlinear and gradient interactions in the effective lagrangian, with the dynamics encoded in the local hadronic couplings. This is the essence of the EFT approach. A clear example of this expansion is the single-nucleon structure included in modern QHD chiral lagrangians [41].

Virtual nucleon–antinucleon Z graphs, which are essential to relativistic phenomenology, should be suppressed because the nucleon has substructure. A local Dirac field for the nucleon does not imply a *physical* point nucleon [3, 41]. Moreover, the virtual $\overline{\text{N}}$N pair is *far off-shell effects off shell*, and the off-shell intermediate states in a particular representation cannot be interpreted in terms of on-shell physics. The EFT framework ensures that any incorrect short-distance dynamics can be corrected systematically with counterterms. Recent formulations of covariant chiral perturbation theory also verify that implicit Z graphs are not a problem, and that consistent power counting is possible [39, 40].

The QHD treatment of the vacuum neglects nucleon substructure, violates N_c counting rules , and relies on unphysical $\overline{\text{N}}$N contributions (Z graphs). The modern QHD treatment of vacuum dynamics has changed this discussion completely. Vacuum contributions in the EFT framework are not calculated explicitly but are implicitly and systematically contained in a small number of fitted parameters. Any physical consequences of hadronic substructure, for example, are automatically included. Furthermore, these implicit contributions are consistent with N_c counting using NDA [75].

Hidden QCD color is relevant for low-energy nuclear physics and is not contained in QHD [52]. The EFT perspective says this objection *must be irrelevant*, because all observable amplitudes are color singlets (since color is confined). Regardless of the underlying QCD picture, the EFT must be valid at sufficiently low energies, without explicitly invoking colored degrees of freedom.

Quantum chromodynamics of quarks and gluons is the fundamental theory of the strong interaction, so we should describe nuclei in terms of quarks. For most ordinary nuclear phenomena, a description based on hadronic degrees of freedom is most appropriate: Hadrons are the particles actually observed in experiments and thus are more efficient. Hadronic calculations can be calibrated using empirical nuclear properties and scattering observables. Hadronic models have historically provided accurate descriptions of NN scattering and the bulk and single-particle properties of nuclei. It is better to *match* the effective hadronic theory to QCD to determine its coefficients and then use the EFT to calculate nuclear structure and reactions.

4. Discussion

4.1. Summary

The hadronic theory of QHD is truly a manifestation of QCD in the strong-coupling regime. It is currently impossible to construct this theory directly from QCD, but the effective field theory perspective shows that we can make progress regardless, without recourse to *ad hoc* models.

The history of QHD from 1974 shows an evolution driven by the successes and difficulties of the original approach. One of the cornerstones of this approach was the necessity for a consistent, microscopic treatment of nuclear systems using hadrons [2]. Although the mean-field theory was phenomenologically successful, the ultimate goal was to improve upon this approximation to incorporate both many-body and short-distance (quantum vacuum) effects. Unfortunately, this goal was sometimes overlooked, because numerous "improvements" degraded the quality of the mean-field-theory results; this often led to the imposition of arbitrary constraints or restrictions on QHD calculations. Fortunately, however, perseverance within the original framework ultimately led to the conclusion that the constraint of renormalizability is too restrictive, and alternatives to this requirement were sought.

The result is the modern viewpoint of QHD based on effective field theory and density functional theory (the "revolution"), which solves the most serious problems while preserving intact the successful predictions for bulk and single-particle nuclear observables. The EFT framework identifies the systematic "organizing principle" behind the successful QHD calculations: energy scales arising from the underlying QCD define the dimensional analysis for terms in the effective lagrangian. Naturalness and the size of the nuclear mean fields allow for a practical expansion and truncation, and also clarify the scope and limitations of QHD. Density functional theory and the Kohn–Sham formalism then explain why the truncated mean-field energy functional can be flexible enough to yield accurate results for certain nuclear observables.

Analyses based on QHD, as defined here, provide a correct description of baryonic systems at sufficiently large distances and low energies. But we argue further that a *covariant* formulation of the dynamics manifests the *true energy scales* of QCD in nuclei and provides an efficient and comprehensive explanation of observed bulk and single-particle systematics.

4.2. Outlook

The development of QHD is far from over, and there are many issues to be addressed. Whereas the majority of existing QHD calculations focus on isoscalar physics, the incorporation of pions using a nonlinear realization of chiral symmetry and the inclusion of the Delta baryon as a collective πN degree of freedom should produce accurate results in the isovector sector; this merits further study. Calculations of excited states using a consistent, conserving approach to the random-phase approximation [20, 76], coupled with new, more accurate data on nuclear breathing modes [77], could lead to a more precise determination of QHD parameters [46]. The situation for nuclear currents and magnetic nuclear form factors at low momentum transfer is still an open problem that must be re-examined in the context of modern EFT. It is also important to pursue the connection between QHD results for many-body systems and the recent calculations of few-nucleon systems within the EFT framework, including covariant chiral perturbation theory. A major challenge is to develop and apply systematic and consistent "power counting" schemes that lead to more general conserving approximations and to study renormalization-group methods that could determine the analytic structure of the ground-state energy functional [51]. In addition, we must learn how to extrapolate beyond the breakdown scale of the EFT description.

Ultimately we must answer the question: What is the best way to connect nuclear phenomenology to QCD? Quark models have been advocated, but they continue to lack a systematic framework and a direct connection to QCD. The EFT perspective makes explicit quark degrees of freedom nonessential. Nevertheless, finding an efficient, tractable, nonperturbative way to match the QCD lagrangian to the long-range, strong-coupling, effective field theory of QHD is a major goal for the future.

Acknowledgments

We thank B. C. Clark, H.-W. Hammer, D. B. Lichtenberg, and J. D. Walecka for useful discussions and comments. This work was supported in part by the National Science

Foundation under Grant No. PHY–9800964 and by the U.S. Department of Energy under Contract No. DE–FG02–87ER40365.

References

[1] L. I. Schiff, *Phys. Rev.* 84 (1951) 1, 10.

[2] J. D. Walecka, *Ann. Phys.* (N.Y.) 83 (1974) 491.

[3] B. D. Serot and J. D. Walecka, *Adv. Nucl. Phys.* 16 (1986) 1.

[4] B. D. Serot and J. D. Walecka, Recent Progress in Many-Body Theories, vol. 3, T. L. Ainsworth, C. E. Campbell, B. E. Clements, and E. Krotscheck, eds. (Plenum, New York, 1992), p. 49.

[5] J. D. Walecka, Theoretical Nuclear and Subnuclear Physics (Oxford Univ. Press, New York, 1995).

[6] J. Boguta and A. R. Bodmer, *Nucl. Phys.* A292 (1977) 413.

[7] H. Müller and B. D. Serot, *Nucl. Phys.* A606 (1996) 508.

[8] C. J. Horowitz and B. D. Serot, *Nucl. Phys.* A368 (1981) 503.

[9] P.-G. Reinhard, M. Rufa, J. Maruhn, W. Greiner, and J. Friedrich, *Z. Phys.* A323 (1986) 13.

[10] M. Rufa, P.-G. Reinhard, J. A. Maruhn, W. Greiner, and M. R. Strayer, *Phys. Rev.* C 38 (1988) 390.

[11] P.-G. Reinhard, *Rep. Prog. Phys.* 52 (1989) 439.

[12] Y. K. Gambhir, P. Ring, and A. Thimet, *Ann. Phys.* (N.Y.) 198 (1990) 132.

[13] B. D. Serot, *Rep. Prog. Phys.* 55 (1992) 1855.

[14] B. D. Serot and J. D. Walecka, *Int. J. Mod. Phys.* E 6 (1997) 515.

[15] J. A. McNeil, J. R. Shepard, and S. J. Wallace, *Phys. Rev. Lett.* 50 (1983) 1439.

[16] J. R. Shepard, J. A. McNeil, and S. J. Wallace, *Phys. Rev. Lett.* 50 (1983) 1443.

[17] B. C. Clark, S. Hama, R. L. Mercer, L. Ray, and B. D. Serot, *Phys. Rev. Lett.* 50 (1983) 1644.

[18] B. C. Clark, S. Hama, R. L. Mercer, L. Ray, G. W. Hoffmann, and B. D. Serot, *Phys. Rev.* C 28 (1983) 1421.

[19] B. C. Clark, S. Hama, E. Sugarbaker, M. A. Franey, R. L. Mercer, L. Ray, G. W. Hoffmann, and B. D. Serot, *Phys. Rev.* C 30 (1984) 314.

[20] J. F. Dawson and R. J. Furnstahl, *Phys. Rev.* C 42 (1990) 2009.

[21] R. Brockmann and R. Machleidt, *Phys. Lett.* 149B (1984) 283.

[22] R. Brockmann and R. Machleidt, *Phys. Rev.* C 42 (1990) 1965.

[23] R. Machleidt, *Adv. Nucl. Phys.* 19 (1989) 189.

[24] F. de Jong and R. Malfliet, *Phys. Rev.* C 44 (1991) 998.

[25] C. J. Horowitz and B. D. Serot, *Phys. Lett.* 137B (1984) 287.

[26] C. J. Horowitz and B. D. Serot, *Nucl. Phys.* A464 (1987) 613; A473 (1987) 760 (E).

[27] R. J. Furnstahl, R. J. Perry, and B. D. Serot, *Phys. Rev.* C 40 (1989) 321.

[28] J. Milana, *Phys. Rev.* C 44 (1991) 527.

[29] M. P. Allendes and B. D. Serot, *Phys. Rev.* C 45 (1992) 2975.

[30] J. Schwinger, *Ann. Phys.* (N.Y.) 2 (1957) 407.

[31] M. Gell-Mann and M. Lévy, *Nuovo Cim.* 16 (1960) 705.

[32] R. J. Furnstahl and B. D. Serot, *Phys. Rev.* C 47 (1993) 2338.

[33] R. J. Furnstahl and B. D. Serot, *Phys. Lett.* B 316 (1993) 12.

[34] R. J. Furnstahl, B. D. Serot, and H.-B. Tang, *Nucl. Phys.* A598 (1996) 539.

[35] S. Weinberg, *Phys. Rev. Lett.* 18 (1967) 188.

[36] S. Weinberg, *Phys. Rev.* 166 (1968) 1568.

[37] R. J. Furnstahl, H.-B. Tang, and B. D. Serot, *Phys. Rev.* C 52 (1995) 1368.

[38] E. Jenkins and A. V. Manohar, *Phys. Lett.* B 255 (1991) 558.

[39] P. J. Ellis and H.-B. Tang, *Phys. Rev.* C 57 (1998) 3356.

[40] T. Becher and H. Leutwyler, *Eur. Phys. J.* C 9 (1999) 643.

[41] R. J. Furnstahl, B. D. Serot, and H.-B. Tang, *Nucl. Phys.* A615 (1997) 441; A640 (1998) 505 (E).

[42] H. Georgi and A. Manohar, *Nucl. Phys.* B234 (1984) 189.

[43] H. Georgi, *Phys. Lett.* B 298 (1993) 187.

[44] J. L. Friar, D. G. Madland, and B. W. Lynn, *Phys. Rev.* C 53 (1996) 3085.

[45] R. J. Furnstahl and B. D. Serot, *Proc.* 15th *Particles and Nuclear International Conference (PANIC99),* Uppsala, Sweden, *Nucl. Phys.* A663 and A664 (2000) 513c.

[46] R. J. Furnstahl and B. D. Serot, *Nucl. Phys.* A671 (2000) 447.

[47] W. Kohn and L. J. Sham, *Phys. Rev.* A 140 (1965) 1133.

[48] W. Kohn, *Rev. Mod. Phys.* 71 (1999) 1253.

[49] N. Argaman and G. Makov, *Am. J. Phys.* 68 (2000) 69.

[50] Y. Hu, Ph.D. thesis, Indiana University (2000).

[51] H.-W. Hammer and R. J. Furnstahl, [arXiv:nucl-th/0004043 v2], *Nucl. Phys.* A678 (2000) 277.

[52] S. J. Brodsky, Comm. *Nucl. Part. Phys.* 12 (1984) 213.

[53] J. W. Negele, Comm. *Nucl. Part. Phys.* 14 (1985) 303.

[54] G. E. Brown, W. Weise, G. Baym, and J. Speth, *Comm. Nucl. Part. Phys.* 17 (1987) 39.

[55] S. Weinberg, *Phys. Lett.* B 251 (1990) 288.

[56] S. Weinberg, *Nucl. Phys.* B363 (1991) 363.

[57] S. Weinberg, *Phys. Lett.* B 295 (1992) 114.

[58] J. J. Rusnak and R. J. Furnstahl, *Nucl. Phys.* A627 (1997) 495.

[59] S. Weinberg, The Quantum Theory of Fields, vol. I (Cambridge Univ. Press, Cambridge, UK, 1995).

[60] L. L. Foldy and S. A. Wouthuysen, *Phys. Rev.* 78 (1950) 29.

[61] R. J. Furnstahl and B. D. Serot, *Nucl. Phys.* A673 (2000) 298.

[62] R. V. Reid, *Jr., Ann. Phys.* (N.Y.) 50 (1968) 411.

[63] J. Carlson, Few-Body Systems Suppl. 10 (1999) 1.

[64] M. V. Hynes, A. Picklesimer, P. C. Tandy, and R. M. Thaler, *Phys. Rev.* C 31 (1985) 1438.

[65] J. D. Lumpe and L. Ray, *Phys. Rev.* C 35 (1987) 1040.

[66] L. Ray, *Phys. Rev. C* 41 (1990) 2816.

[67] W. R. Coker and L. Ray, *Phys. Rev.* C 42 (1990) 659.

[68] W. Lin and B. D. Serot, *Phys. Lett.* B 233 (1989) 23.

[69] W. Lin and B. D. Serot, *Nucl. Phys.* A512 (1990) 637.

[70] R. J. Furnstahl and B. D. Serot, *Nucl. Phys.* A468 (1987) 539.

[71] R. J. Furnstahl and B. D. Serot, *Phys. Rev.* C 43 (1991) 105.

[72] G. E. Brown and D. G. Ravenhall, *Proc. Roy. Soc. London,* A208 (1951) 552.

[73] B. C. Clark, S. Hama, J. A. McNeil, R. L. Mercer, L. Ray, B. D. Serot, D. A. Sparrow, and K. Stricker-Bauer, *Phys. Rev. Lett.* 53 (1984) 1423.

[74] C. J. Horowitz and B. D. Serot, *Nucl. Phys.* A399 (1983) 529.

[75] R. J. Furnstahl, B. D. Serot, and H.-B. Tang, *Nucl. Phys.* A618 (1997) 446.

[76] J. Piekarewicz, [arXiv:nucl-th/0003029], *Phys. Rev.* C 62 (2000) 051304.

[77] D. H. Youngblood, H. L. Clark, and Y.-W. Lui, *Phys. Rev. Lett.* 82 (1999) 691.

In: Neutron Stars
Editors: Hiroshi Uechi et al, pp. 93-109
ISBN: 978-1-62081-123-8

Chapter 4

NAMBU-JONA-LASINIO MODEL IN COMPACT STARS

Nobutoshi Yasutake*
Research Center for the Early Universe,
School of Science, University of Tokyo,
Bunkyo-ku, Tokyo 113-0033, Japan.

Abstract

In this chapter we present the outline of the Nambu-Jona-Lasinio (NJL) model and its roles in neutron stars. The NJL model is the most famous model to represent the chiral symmetry breaking. After reading this chapter, the readers will know how quarks get their masses. We hope that this chapter becomes the first step to understand this model for the people without technical knowledge of nuclear physics and astrophysics. We only assume the readers have basic knowledge of the quantum mechanics.

1. Introduction

The Nambu-Jona-Lasinio model is the most famous model to represent the chiral symmetry breaking [1, 2]. The chiral symmetry breaking is the one of the important features in the quantum chromodynamics (QCD). The idea by Nambu and Jona-Lasinio was originally motivated by the BCS theory for superconductivity [3]. However, this model was not concerned with quarks at that time, since the their idea was proposed before the quark model [4] and QCD [5, 6] were established. After the development of QCD, this model is interpreted as the quark model which describes the spontaneous breaking of the chiral symmetry and how mass-less quarks (or quarks with small mass) get their masses at low energy [7, 8, 9, 10, 11, 12]. After almost a half century from his suggestions, Nambu was awarded the Nobel prize for "the discovery of the mechanism of spontaneous broken symmetry in subatomic physics" in 2008.

In this chapter, we present the outline of the NJL model and its role in neutron stars. Although many papers have been already published about this model, we try to explain on

*E-mail address: yasutake@resceu.s.u-tokyo.ac.jp

this model, especially for the people who are not familiar with nuclear physics and astrophysics. For the readers who want to understand this model more strictly, we recommend three reviews [13, 14, 15].

In this chapter, we begin by comparing both the BCS theory and the Nambu and Jona-Lasinio model. As for the BCS theory, the values of $\langle e(x)_\uparrow e(x)_\downarrow\rangle$ are zero in the normal metal, where the spin up and down electron fields are denoted as $e(x)_\uparrow$ and $e(x)_\downarrow$. It becomes $\langle e(x)_\uparrow e(x)_\downarrow\rangle \neq 0$ in the superconducting states, and accordingly the energy gap Δ opens at the Fermi surface. The electron pair with opposite spins is known as the Cooper pair, and this phenomenon is known as one example of Bose-Einstein condensations. Although we do not discuss the details, Δ^2 is proportional to the energy difference between the normal Fermi liquid and the BCS ground state. Since there is no density of state around the Fermi surface with the width of Δ, if electrons are excited, they must jump this energy gap. The energy of each particle is shown as $E = \sqrt{\xi(k)^2 + \Delta(k)^2}$, where $\xi(k)$ and $\Delta(k)$ are the kinetic energy and the energy gap. Clearly, this form is similar to the energy of Dirac particle, $E = \sqrt{\xi(k)^2 + m^2}$, although m is the mass in this case.

In the NJL model, it was constructed by employing the similar logic in the BCS theory and applying to the quark four fermion interaction (sometimes including six fermion interactions). In this model, the quark-antiquark condensates (or quark-quark condensates) correspond to $\langle e(x)_\uparrow e(x)_\downarrow\rangle$ in the BCS theory; e.g. the origin of mass comes from the condensations as we will show later.

In Table1, we summarize the comparison. Here, ψ_R and ψ_L are the right-hand parts and the left-hand parts of quark fields defined by

$$\psi = \frac{1+\gamma_5}{2}\psi + \frac{1-\gamma_5}{2}\psi = \psi_R + \psi_L, \tag{1}$$

where γ_5 defined as $\gamma_5 = i\gamma_0\gamma_1\gamma_2\gamma_3$, and we adopt the definitions of γ_0 and γ_i as;

$$\gamma_0 = \begin{pmatrix} I & 0 \\ 0 & -I \end{pmatrix}, \quad \gamma_i = \begin{pmatrix} 0 & \tau_i \\ -\tau_i & 0 \end{pmatrix}, \tag{2}$$

using the Pauli matrix τ_i.

Table 1. Comparison of superconductivity and strong interaction.

	super conductivity	strong interaction
effective theory	BCS theory	NJL model
symmetry	gauge symmetry	chiral symmetry
effective degrees of freedom	$e(x)_\uparrow, e(x)_\downarrow$	ψ_R, ψ_L
product by Bose-Einstein condensations	energy gap	mass
original reference	[3]	[1, 2]

2. Two-flavor Nambu-Jona-Lasinio Model

2.1. Lagrangian Density

We start with the Lagrangian density of two-flavor NJL model;

$$\mathcal{L} = \bar{\psi}(i\not{\partial} - \hat{m}_0)\psi + g\Big[(\bar{\psi}\psi)^2 + (\bar{\psi}i\gamma_5\tau^k\psi)^2\Big], \quad (3)$$

where ψ, $m_0 = diag(m_u, m_d)$, g, and τ^k are the quark field, the current quark mass, the coupling constant, and the Pauli matrix respectively.

In the NJL model for two flavors, Lagrangian does not change following transformations;

$$SU_R(2) : \psi_R \to \exp[i\theta_R^k\tau^k/2]\psi_R, \quad (4)$$
$$SU_L(2) : \psi_L \to \exp[i\theta_L^k\tau^k/2]\psi_L, \quad (5)$$
$$U_R(1) : \psi_R \to \exp[i\phi_R]\psi_R, \quad (6)$$
$$U_L(1) : \psi_L \to \exp[i\phi_L]\psi_L. \quad (7)$$

These transformations are equivalent to the following transformations;

$$SU_A(2) : \psi \to \exp[i\gamma^5\theta^k\tau^k/2]\psi, \quad (8)$$
$$SU_V(2) : \psi \to \exp[i\phi^k\tau^k/2]\psi, \quad (9)$$
$$U_A(1) : \psi \to \exp[i\gamma^5\alpha]\psi, \quad (10)$$
$$U_V(1) : \psi \to \exp[i\beta]\psi. \quad (11)$$

This equivalence of $SU_R(2) \times SU_L(2) \times U_R(1) \times U_L(1) \sim SU_A(2) \times SU_V(2) \times U_A(1) \times U_V(1)$ is checked easily. For example, let us consider the chiral transformation $SU_R(2) \times SU_L(2)$.

If we take the value $\theta_L = \theta_R = \phi$ in Eq.(4) and (5), ψ changes as,

$$\psi = \psi_R + \psi_L \to \exp[i\phi^k\tau^k/2]\psi_R + \exp[i\phi^k\tau^k/2]\psi_L = \exp[i\phi^k\tau^k/2]\psi, \quad (12)$$

which is equivalent to $SU_V(2)$ transformation.

When we take the value $\theta_L = -\theta_R = -\theta$ in Eq.(4) and (5), ψ changes as,

$$\begin{aligned}\psi &= \psi_R + \psi_L \to \exp[-i\theta^k\tau^k/2]\psi_R + \exp[-i\theta^k\tau^k/2]\psi_L, \\ &= \left(\cos\frac{|\boldsymbol{\theta}|}{2} + i\tau^k\hat{\theta}^k\sin\frac{|\boldsymbol{\theta}|}{2}\right)\frac{1+\gamma_5}{2}\psi + \left(\cos\frac{|\boldsymbol{\theta}|}{2} - i\tau^k\hat{\theta}^k\sin\frac{|\boldsymbol{\theta}|}{2}\right)\frac{1-\gamma_5}{2}\psi, \\ &= \left(\cos\frac{|\boldsymbol{\theta}|}{2} + i\gamma_5\tau^k\hat{\theta}^k\sin\frac{|\boldsymbol{\theta}|}{2}\right)\psi \\ &= \exp[i\gamma^5\theta^k\tau^k/2]\psi \end{aligned} \quad (13)$$

which is equivalent to $SU_A(2)$ transformation. Here, we set $\hat{\theta}^k$ as $\hat{\theta}^k = \theta^k/|\boldsymbol{\theta}|$, and used the relation $(\hat{\theta}^j\tau^j)(\hat{\theta}^k\tau^k) = 1$ in the derivation.

These transformations of Eq.(8)-(11) impose the constraints on the Lagrangians, but many types of the Lagrangians are still permitted in principle. The Lagrangian shown as

Table 2. Symmetries in the NJL model of two flavors.

Symmetry	Transformation	Current	Name
$SU_A(2)$	$\psi \to \exp[i\gamma^5\theta^i\tau^i/2]\psi$	$J^k_{\mu 5} = \bar{\psi}\gamma_\mu\gamma_5\tau^k\psi$	chiral
$SU_V(2)$	$\psi \to \exp[i\phi^i\tau^i/2]\psi$	$J^k_\mu = \bar{\psi}\gamma_\mu\tau^k\psi$	isospin
$U_A(1)$	$\psi \to \exp[i\gamma^5\alpha]\psi$	$j_{\mu 5} = \bar{\psi}\gamma_\mu\gamma^5\psi$	axial vector
$U_V(1)$	$\psi \to \exp[i\beta]\psi$	$j_\mu = \bar{\psi}\gamma_\mu\psi$	vector

(3) is one simple type in the NJL model. The unitary transformations $U_V(1)$ and $SU_V(2)$ are well known symmetries that correspond to baryon and iso-spin conservation. The other two transformations $U_A(1)$ and $SU_A(2)$ are so-called axial and chiral symmetries. We summarize these symmetries in Table 2. Here, we only show the currents corresponding to symmetries. However, if we consider "*quantum anomaly*", $U_A(1)$ symmetry is not conserved even for the mass less case, although we do not discuss this topic in this chapter.

We also do not discuss the Fierz transformations here. Readers can check it in the Ref. [13, 15]. The point is that, after the Fierz transformation, we can get the Lagrangian which includes all kinds of independent 4-Fermi interactions. After the Fierz transformation, the Lagrangian can be represented as

$$\mathcal{L} = \bar{\psi}(i\partial\!\!\!/ - \hat{m}_0)\psi + \frac{g}{8N_c}\Big[2(\bar{\psi}\psi)^2 + 2(\bar{\psi}i\gamma_5\tau^k\psi)^2 - 2(\bar{\psi}\tau^k\psi)^2 - 2(\bar{\psi}i\gamma_5\psi)^2$$
$$-4(\bar{\psi}\gamma^\mu\psi)^2 - 4(\bar{\psi}i\gamma^\mu\gamma_5\psi)^2 + (\bar{\psi}\sigma_{\mu\nu}\psi)^2 - (\bar{\psi}i\sigma_{\mu\nu}\tau^k\psi)^2\Big], \quad (14)$$

where we ignore the color non-singlet terms. We note that, nowadays, there are other approaches from the color-current interactions based on QCD[15], though we do not adopt them here. If we follow them, the factor on each term in Eq.(14) changes.

In this chapter, to show the features of chiral symmetry easily, we just adopt the simple interaction as

$$\mathcal{L} = \bar{\psi}(i\partial\!\!\!/ - \hat{m}_0)\psi + G_s\Big[(\bar{\psi}\psi)^2 + (\bar{\psi}i\gamma_5\tau^k\psi)^2\Big] + G_v(\bar{\psi}\gamma^\mu\psi)^2, \quad (15)$$

where G_s and G_v are the coupling constants. In other words, we assume that all other terms vanish for the symmetry properties of vacuum, hence that thermal averages vanish.

Let us check the $SU_A(2)$ symmetry in this Lagrangian for example, where the transformation of Eq.(8) is called the chiral transformation.

The term $\bar{\psi}\psi$ changes as;

$$\begin{aligned}
\bar{\psi}\psi &\to \psi^\dagger e^{-i\gamma^5\tau^j\hat{\theta}^j\theta/2}\gamma_0 e^{i\gamma^5\tau^k\hat{\theta}^k\theta/2}\psi, \\
&= \psi^\dagger(\cos\frac{\theta}{2} - i\gamma^5\tau^j\hat{\theta}^j\sin\frac{\theta}{2})\gamma^0(\cos\frac{\theta}{2} + i\gamma^5\tau^k\hat{\theta}^k\sin\frac{\theta}{2})\psi, \quad (16) \\
&= \bar{\psi}(\cos\frac{\theta}{2} + i\gamma^5\tau^j\hat{\theta}^j\sin\frac{\theta}{2})(\cos\frac{\theta}{2} + i\gamma^5\tau^j\hat{\theta}^j\sin\frac{\theta}{2})\psi, \\
&= \bar{\psi}\psi\cos\theta + \bar{\psi}i\gamma^5\tau^j\hat{\theta}^j\psi\sin\theta,
\end{aligned}$$

while the the term $\bar{\psi}i\gamma^5\tau^k\psi$ changes as;

$$\begin{aligned}
\bar{\psi}i\gamma^5\tau^k\psi &\rightarrow \psi^\dagger e^{-i\gamma^5\tau^j\hat{\theta}^j\theta/2}\gamma_0 i\gamma^5\tau^k e^{i\gamma^5\tau^l\hat{\theta}^l\theta/2}\psi, \\
&= \psi^\dagger(\cos\frac{\theta}{2} - i\gamma^5\tau^j\hat{\theta}^j\sin\frac{\theta}{2})\gamma_0 i\gamma^5\tau^k(\cos\frac{\theta}{2} + i\gamma^5\tau^l\hat{\theta}^l\sin\frac{\theta}{2})\psi, \\
&= \bar{\psi}i\gamma^5(\cos\frac{\theta}{2} + i\gamma^5\tau^j\hat{\theta}^j\sin\frac{\theta}{2})\tau^k(\cos\frac{\theta}{2} + i\gamma^5\tau^l\hat{\theta}^l\sin\frac{\theta}{2})\psi, \\
&= \bar{\psi}i\gamma^5(\tau^k\cos^2\frac{\theta}{2} - (2\delta^{jk} - \tau^k\tau^j)\hat{\theta}^j\tau^l\hat{\theta}^l\sin^2\frac{\theta}{2} \\
&\quad + i\gamma^5\tau^k\tau^j\hat{\theta}^j\cos\frac{\theta}{2}\sin\frac{\theta}{2} + i\gamma^5(2\delta^{jk} - \tau^j\tau^k)\hat{\theta}^j\sin\frac{\theta}{2}\cos\frac{\theta}{2}\Big)\psi \\
&= \bar{\psi}i\gamma^5\left(\tau^k\cos^2\frac{\theta}{2} - (2\hat{\theta}^k\tau^j\hat{\theta}^j - \tau^k)\sin^2\frac{\theta}{2} + 2i\gamma^5\hat{\theta}^k\cos\frac{\theta}{2}\sin\frac{\theta}{2}\right)\psi, \\
&= \bar{\psi}i\gamma^5\left(\tau^k - \hat{\theta}^k\tau^j\hat{\theta}^j(1-\cos\theta) + i\gamma^5\hat{\theta}^k\sin\theta\right)\psi, \\
&= \bar{\psi}i\gamma^5\tau^k\psi - (\bar{\psi}i\gamma^5\tau^j\hat{\theta}^j\psi)\hat{\theta}^k(1-\cos\theta) - \bar{\psi}\psi\hat{\theta}^k\sin\theta,
\end{aligned}$$

where we used $(\hat{\theta}^j\tau^j)(\hat{\theta}^k\tau^k) = 1$ from the property of the Pauli matrix.

After a simple calculation, one can check that the term $(\bar{\psi}\psi)^2 + (\bar{\psi}i\gamma^5\tau^k\psi)^2$ satisfies the chiral symmetry. The term of $\bar{\psi}\gamma^\mu\psi$ also satisfies this symmetry, since $\bar{\psi}\gamma^\mu\psi \rightarrow \bar{\psi}\gamma^\mu e^{-i\gamma_5\tau^j\hat{\theta}^j\theta/2}e^{i\gamma_5\tau^k\hat{\theta}^k\theta/2}\psi = \bar{\psi}\gamma^\mu\psi$ by way of $\gamma_5\gamma^\mu = -\gamma^\mu\gamma_5$. The first term obviously satisfies this symmetry in the chiral limit, $m_0 = 0$, since

$$\bar{\psi}\,\partial\!\!\!/\psi \rightarrow \bar{\psi}e^{i\gamma^5\tau^j\hat{\theta}^j\theta/2}\,\partial\!\!\!/ e^{i\gamma^5\tau^k\hat{\theta}^k\theta/2}\psi = \bar{\psi}\,\partial\!\!\!/\psi, \tag{17}$$

$$\bar{\psi}\psi \rightarrow \text{Eq.}(16) \neq \bar{\psi}\psi. \tag{18}$$

Hence, in the case of the small mass, the Lagrangian is almost invariant under $SU_A(2)$.

Now, we adopt the mean-field approximation to see features of the chiral symmetry breaking in matter;

$$\bar{\psi}\psi = \langle\bar{\psi}\psi\rangle + (\bar{\psi}\psi)', \tag{19}$$

$$\bar{\psi}i\gamma^5\tau^k\psi = \langle\bar{\psi}i\gamma^5\psi\rangle + (\bar{\psi}i\gamma^5\tau^k\psi)', \tag{20}$$

$$\bar{\psi}\gamma^\mu\psi = \langle\bar{\psi}\gamma^\mu\psi\rangle + (\bar{\psi}\gamma^\mu\psi)', \tag{21}$$

where $\langle\rangle$ and $()'$ represent the mean field and small fluctuation, respectively. We assume the isospin symmetry, $m_u = m_d$ for simplicity.

Under this approximation, the Lagrangian density of Eq.(14) becomes

$$\mathcal{L}_{MFA} = \bar{\psi}\{i\,\partial\!\!\!/ - (m_0 + \Sigma_s) + \Sigma_v\gamma_0\}\psi - U. \tag{22}$$

where

$$\Sigma_s = -2G_s\sigma, \tag{23}$$

$$\Sigma_v = -2G_v\omega, \tag{24}$$

$$U = G_s\sigma^2 + G_v\omega^2, \tag{25}$$

and $\langle\bar{\psi}i\gamma^5\tau^k\psi\rangle$ and $\langle\bar{\psi}\gamma^k\psi\rangle$ are vanished because of the symmetry of vacuum. The forms of the mean fields are denoted by $\sigma = \langle\bar{\psi}\psi\rangle$ and $\omega = \langle\bar{\psi}\gamma^0\psi\rangle$. Here, we neglect the second order of fluctuations.

In Eq.(22) we can see that Σ_s has the same role with the mass. Therefore, it is known that the quark condensate, in this case $\langle\bar{\psi}\psi\rangle$, gives the origin of mass. Hence, we introduce effective mass as $M = m_0 + \Sigma_s$ in the following discussions. At low temperature and low density (low chemical potential), $\langle\bar{\psi}\psi\rangle$ becomes large, and the chiral symmetry is spontaneously broken. From Eq. (22), the chemical potential for the free particles with mass M is represented as $\mu^* = \mu + \Sigma_v$ as shown in the next section, when μ is introduced as an input parameter.

2.2. Thermodynamic Potential and Other Physical Quantities

Let us start with the Lagrangian (22). In this case, the Hamiltonian is expressed as;

$$\begin{aligned} \hat{H} &= \int dx^3 \mathcal{H}(x), \\ &= \int dx^3 \Big(\bar{\psi}(x)(-i\gamma\cdot\nabla + M - \Sigma_v\gamma^0)\psi(x) + U \Big). \end{aligned} \tag{26}$$

Using the creation and annihilation operators, the quark field ψ is expanded as

$$\psi(x) = \frac{1}{\sqrt{V}} \sum_{\mathbf{p}} \sum_{s=1,2} \sqrt{\frac{M}{E_{\mathbf{p}}}} [a_{\mathbf{p}}^{(s)} u(\mathbf{p};s) e^{-ipx} + b_{\mathbf{p}}^{(s)\dagger} v(\mathbf{p};s) e^{ipx}], \tag{27}$$

where V is the volume of the system. Here, $a_{\mathbf{p}}^{(s)\dagger}$ ($b_{\mathbf{p}}^{(s)\dagger}$) and $a_{\mathbf{p}}^{(s)}$ ($b_{\mathbf{p}}^{(s)}$) are the quark (anti-quark) creation operator and annihilation operator, and $u(\mathbf{p};s)$ and $v(\mathbf{p};s)$ are given by

$$u(\mathbf{p};s) = \frac{\not{p} + M}{\sqrt{2M(E_{\mathbf{p}}+M)}} \begin{pmatrix} \omega(s) \\ 0 \end{pmatrix}, \tag{28}$$

$$v(\mathbf{p};s) = \frac{-\not{p} + M}{\sqrt{2M(E_{\mathbf{p}}+M)}} \begin{pmatrix} 0 \\ \omega(s) \end{pmatrix}, \tag{29}$$

$$\omega(1) = \begin{pmatrix} 1 \\ 0 \end{pmatrix}, \quad \omega(2) = \begin{pmatrix} 0 \\ 1 \end{pmatrix}. \tag{30}$$

Therefore, the Hamiltonian is

$$\hat{H} = \sum_{\mathbf{p}} \sum_{s=1,2} E_{\mathbf{p}} [a_{\mathbf{p}}^{(s)\dagger} a_{\mathbf{p}}^{(s)} - b_{\mathbf{p}}^{(s)} b_{\mathbf{p}}^{(s)\dagger} - \Sigma_v \hat{N}] + VU, \tag{31}$$

where $E_{\mathbf{p}} = \sqrt{\mathbf{p}^2 + M^2}$, and $\hat{N}$ is the number operator defined as $\hat{N} = a_{\mathbf{p}}^{(s)\dagger} a_{\mathbf{p}}^{(s)} -$

$b_{\mathbf{p}}^{(s)\dagger}b_{\mathbf{p}}^{(s)}$. In the derivation of Eq.(31), we used the following relations;

$$\{a_{\mathbf{p}}^{(s)}, a_{\mathbf{q}}^{(s)\dagger}\} = \delta^3(\mathbf{p}-\mathbf{q}), \tag{32}$$

$$\{a_{\mathbf{p}}^{(s)}, a_{\mathbf{q}}^{(s)}\} = \{a_{\mathbf{p}}^{(s)\dagger}, a_{\mathbf{q}}^{(s)\dagger}\} = 0, \tag{33}$$

$$(\not{p} - M)u(\mathbf{p}; s) = 0, \quad (\not{p} + M)v(\mathbf{p}; s) = 0, \tag{34}$$

$$u^\dagger(\mathbf{p}; s)u(\mathbf{p}; s') = \frac{E_{\mathbf{p}}}{M}\delta_{s,s'} = v^\dagger(\mathbf{p}; s)v(\mathbf{p}; s'), \tag{35}$$

$$u^\dagger(\mathbf{p}; s)v(-\mathbf{p}; s') = 0 = v^\dagger(\mathbf{p}; s)u(-\mathbf{p}; s'). \tag{36}$$

The grand canonical partition function is defined as

$$Z = \mathrm{Tr} e^{-\beta(\hat{H}-\mu\hat{N})}, \tag{37}$$

where β and μ are the inverse of temperature and the quark chemical potential.

The thermodynamical potential is

$$\begin{aligned}
\Omega &= -\tfrac{1}{\beta}\ln Z \\
&= -\tfrac{1}{\beta}\ln\Big[e^{-\beta VU}\mathrm{Tr}\textstyle\prod_{\mathbf{p}}\exp\Big(-\beta\sum_{s=1,2}E_{\mathbf{p}}[a_{\mathbf{p}}^{(s)\dagger}a_{\mathbf{p}}^{(s)} - b_{\mathbf{p}}^{(s)}b_{\mathbf{p}}^{(s)\dagger}] \\
&\qquad +\mu^*(a_{\mathbf{p}}^{(s)\dagger}a_{\mathbf{p}}^{(s)} - b_{\mathbf{p}}^{(s)\dagger}b_{\mathbf{p}}^{(s)})\Big)\Big] \\
&= -\tfrac{1}{\beta}\ln\Big[e^{-\beta VU}\mathrm{Tr}\textstyle\prod_{\mathbf{p}}\exp\Big(-\beta\sum_{s=1,2}E_{\mathbf{p}}\delta_{\mathbf{p},\mathbf{p}}\Big) \\
&\qquad\times\exp\Big(-\beta\textstyle\sum_{s=1,2}\{(E_{\mathbf{p}}+\mu^*)a_{\mathbf{p}}^{(s)\dagger}a_{\mathbf{p}}^{(s)} + (E_{\mathbf{p}}-\mu^*)b_{\mathbf{p}}^{(s)\dagger}b_{\mathbf{p}}^{(s)}\}\Big)\Big] \\
&= -\tfrac{1}{\beta}N_fN_c\Big[2\beta\textstyle\sum_{\mathbf{p}}E_{\mathbf{p}} + VU \\
&\qquad +\ln\textstyle\prod_{\mathbf{p}}\langle n_{\mathbf{p}}|\exp\Big(-\beta\sum_{s=1,2}\{(E_{\mathbf{p}}+\mu^*)a_{\mathbf{p}}^{(s)\dagger}a_{\mathbf{p}}^{(s)} + (E_{\mathbf{p}}-\mu^*)b_{\mathbf{p}}^{(s)\dagger}b_{\mathbf{p}}^{(s)}\}|n_{\mathbf{p}}\rangle\Big] \\
&= -2N_fN_c\Big[\textstyle\sum_{\mathbf{p}}\{E_{\mathbf{p}} + \tfrac{1}{\beta}\Big(\ln(1+\exp^{-\beta(E_{\mathbf{p}}+\mu^*)}) + \ln(1+\exp^{-\beta(E_{\mathbf{p}}-\mu^*)})\Big)\}\Big] + VU,
\end{aligned}$$

where N_f and N_c are the number of flavor and color.

Finally replacing the sum over momentum by the integral,

$$\sum_{\mathbf{p}} \to V\int\frac{d^3\mathbf{p}}{(2\pi)^3}, \tag{38}$$

we get the thermodynamic potential,

$$\begin{aligned}
\Omega &= -2N_fN_cV\Big[\int\frac{d^3\boldsymbol{p}}{(2\pi)^3}\{E_p + \frac{1}{\beta}\Big(\ln(1+\exp^{-\beta(E_{\mathbf{p}}+\mu^*)}) \\
&\qquad +\ln(1+\exp^{-\beta(E_{\mathbf{p}}-\mu^*)})\Big)\}\Big] + VU.
\end{aligned} \tag{39}$$

From this thermodynamic potential we can get other physical quantities using

$$\Omega = -PV, \tag{40}$$

$$N_i = -\frac{\partial\Omega}{\partial\mu_i}, \tag{41}$$

$$S = -\frac{\partial\Omega}{\partial T}, \tag{42}$$

$$E = -PV + TS + \mu_i N_i, \tag{43}$$

where, P, N, S, and E are pressure, number of particle, entropy, and energy.

2.3. Numerical Results of Chiral Phase Transition

Now we present some numerical results as examples. Actually, we get the values of order parameters σ, ω from the stationary equations;

$$\frac{\partial}{\partial\sigma}\left(\frac{\Omega}{V}\right) = 0, \tag{44}$$

$$\frac{\partial}{\partial\omega}\left(\frac{\Omega}{V}\right) = 0. \tag{45}$$

In Figure. 1, we show the relation of the thermodynamic potential Ω and quark mass M, namely the quark condensate $\langle\bar{\psi}\psi\rangle$. Here, we assume the isospin symmetric case. We adopt the coupling constant as $G_s = 2.337 \times (4N_c + 1)/4N_c$, $G_v = 2.337/2N_c$ from Ref.[16], where we set $N_c = 3$. Since the NJL model is non-renormalizable, it is needed to introduce a cutoff Λ in the momentum integration. Here we set the cutoff as $\Lambda = 925$ MeV. We also adopt the current quark mass as $m_u = m_d = 5$ MeV as shown in the same reference[16]. We note again that, nowadays, there are the other parameter sets given from the color current interactions based on QCD[15], though we do not adopt them here .

Depending on temperature and/or chemical potential, the shapes of thermodynamic potential change. As temperature (chemical potential) increases, the shape of the potential changes from a Mexican hut type to a parabolic type. The physical quantities are given at the minimum point of Ω in Figure 1 for each chemical potential and temperature.

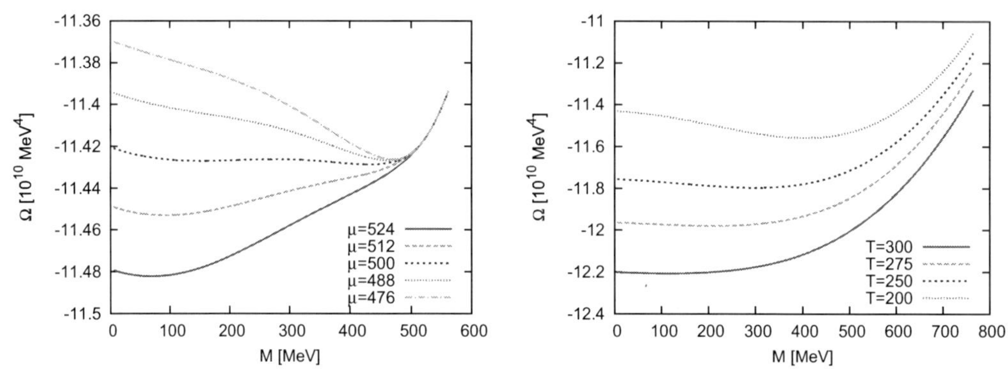

Figure 1. The thermodynamic potential Ω and the quark mass M ($\langle\bar{\psi}\psi\rangle$). Left panel shows the chemical potential dependency at $T = 0$ MeV. Right one the temperature dependency at $\mu = 0$ MeV. The units of μ in the left panel and T in the right panel are MeV.

Next, we show the order of the phase transition. The left panel of Figure 2 shows the relation between the chemical potentials and the mass which is the order parameter of the phase transition at each temperature. Clearly the phase transition at $T = 0$ MeV is the first order, and the one at $T = 90$ MeV is the cross over. In this case, the critical temperature exits at $T \sim 50$ MeV. In the right panel of Fig. 2, we show the phase diagram for the Lagrangian of Eq.(14). The color contour represents the mass in the unit, MeV. The derivatives of the mass corresponds the order of the phase transition. The critical point , which separates the chiral phase transition of the first order from the cross over, is around $\mu \sim 480$ MeV and $T \sim 50$ MeV.

However, depending on the models, there is the possibility that other types of the critical points may appear [17]. Moreover, if we take into account the interactions of the higher orders, above features will change [18]. In this sense, the NJL model is useful to know the chiral symmetry breaking qualitatively, but not adequate quantitatively.

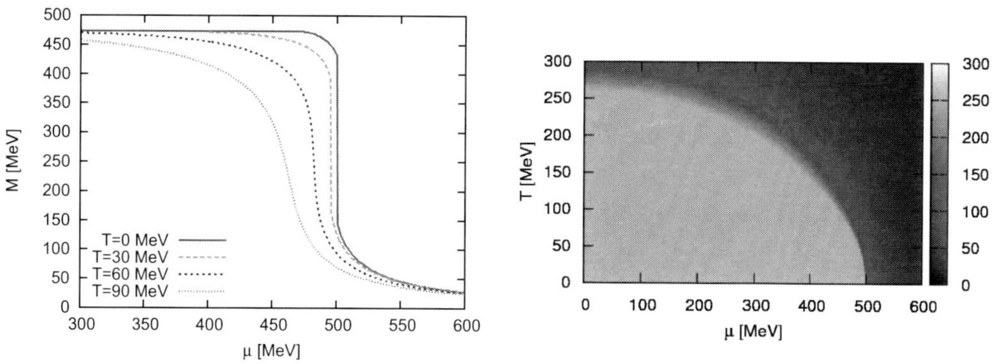

Figure 2. Left panel shows the chemical potential dependency of the order parameter M for each temperature. Right panel shows the color contour of M on $\mu - T$ plane, which is the phase diagram of the chiral phase diagram.

In Fig.3 we show the pressure and the density for this model. This is the equation of sate (EOS) which is the most important quantity to decide the structures of compact stars and the astrophysical phenomena related with these stars. The chiral phase transition occurs $n_B = 10n_0$ at zero temperature, for example. Here, n_B and n_0 are the baryon number density and the saturation density $n_0 = 0.17$ fm^{-3}, respectively. However, the density discontinuities become smooth at finite temperature because the order of phase transition becomes the cross over. If we adopt this EOS to calculate structures of compact stars, we must impose the charge neutrality condition instead of the isospin symmetric condition, which will be shown in the next section.

3. Three-flavor Nambu-Jona-Lasinio Model

3.1. Lagrangian Density and Thermodynamic Potential

In the last section, we have described the simple two-flavor system of the NJL model. However, actually, at high density and/or temperature the third flavor quark, strange quark, will appear.

The NJL of SU(3) have been first formulated by Kunihiro and Hatsuda [20], Bernard, Jaffe, and Meissner [21] including the t'Hooft term [22, 23]. In the case of the three flavor system, Lagrangian density of the NJL model is written as

$$\mathcal{L} = \bar{\psi}(i\partial\!\!\!/ - \hat{m}_0)\psi + \mathcal{L}_4 + \mathcal{L}_6, \tag{46}$$

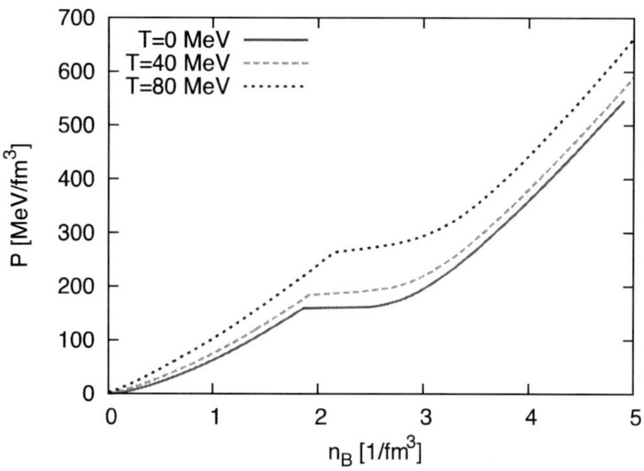

Figure 3. Density and pressure in two flavor NJL model under the isospin symmetric matter. The values for the parameters we adopt are shown in the text.

where

$$\mathcal{L}_4 = G_s \sum_{k=0}^{8} \left[(\bar{\psi}\lambda^k\psi)^2 + (i\bar{\psi}\gamma^5\lambda^k\psi)^2 \right] \tag{47}$$

$$\mathcal{L}_6 = -K\left[\det_{i,j}(\bar{\psi}_i(1+\gamma^5)\psi_j) + \det_{i,j}(\bar{\psi}_i(1-\gamma^5)\psi_j) \right], \tag{48}$$

$$\psi = \begin{pmatrix} \psi_u \\ \psi_d \\ \psi_s \end{pmatrix}, \quad \hat{m}_0 = \begin{pmatrix} m_u & 0 & 0 \\ 0 & m_d & 0 \\ 0 & 0 & m_s \end{pmatrix}, \tag{49}$$

where i, j and l denote the flavor of u, d and s, and λ^k is the Gell-Man Matrix. Here, we ignored the vector parts and the other higher order terms.

The symbol of ψ and $\hat{m}_0$ are quark field and current-mass matrix, respectively. $\mathcal{L}_4$ and $\mathcal{L}_6$ generate four-quark and six-quark interaction. The six-quark interaction, $\mathcal{L}_6$, comes from $U(1)_A$ anomaly, and sometimes called as the 't Hooft term . In the chiral limit, the Lagrangian density have $SU(3)_L \times SU(3)_R \times SU(3)_C \times U(1)_V$ symmetry. In following, we explicitly write the pseudo-scalar interaction, because it is related with the pseudo-scalar mesons but the vacuum expectation value becomes zero.

Inserting the mean fields,

$$\bar{\psi}_m\psi_n \rightarrow \langle\bar{\psi}_m\psi_n\rangle, \tag{50}$$

$$\bar{\psi}_m\gamma^0\psi_n \rightarrow \langle\bar{\psi}_m\gamma^0\psi_n\rangle, \tag{51}$$

to the Lagrangian density, we have

$$\mathcal{L}_{MFA} = \sum_{f=i,j,k} \bar{\psi}_f(i\partial\!\!\!/ - (m_f + \Sigma_{s,f}))\psi_f - U, \tag{52}$$

where

$$\Sigma_{s,i} = -4G_s\sigma_i + 2K\sigma_j\sigma_k, \tag{53}$$

$$U = \sum_{l=u,d,s}(2G_s\sigma_l^2) - 4K\sigma_u\sigma_d\sigma_s, \tag{54}$$

with $i \neq j \neq k$. The forms of the mean fields are denoted by $\sigma_i = \langle\bar{\psi}_i\psi_i\rangle$.

Thus the thermodynamic potential renders

$$\begin{aligned}\Omega &= -\frac{1}{\beta}\ln Z \\ &= -2N_c V\sum_{f=u,d,s}\int\frac{d^3\mathrm{p}}{(2\pi)^3}\Big[E_f(\mathrm{p}) + \frac{1}{\beta}\ln(1+e^{-\beta E_f^-(\mathrm{p})}) + \frac{1}{\beta}\ln(1+e^{-\beta E_f^+(\mathrm{p})})\Big] \\ &\quad + UV.\end{aligned} \tag{55}$$

where $E_f(\mathrm{p}) = \sqrt{\mathbf{p}^2 + M_f^2}$, $E_f^\pm = E_f(\mathrm{p}) \pm \mu_f$,and $M_f = m_f + \Sigma_{s,f}$.

Here, we use the three-dimensional momentum cutoff, and the parameter sets are obtained in Ref. [19]. The cutoff Λ is 0.6023 GeV, and the coupling constant of four-quark interaction is defined as $G_s\Lambda^2 = 1.835$, that of six-quark interaction is defined as $K\Lambda^5 = 12.36$. The current quark mass of u- and d-quark is fixed to be 5.5 MeV, that of s-quark is fixed to be 140.7 MeV. From this thermodynamic potential, we can get other physical quantities.

The stationary condition for σ is

$$\begin{aligned}\frac{1}{V}\frac{\partial\Omega}{\partial\sigma_{i=u,d,s}} &= 4(G_s\sigma_i - K\sigma_j\sigma_k) + \sum_{f=i,j,k}\Big(4G_s\delta_{i,f} - 2K(\sigma_j\delta_{k,f} + \sigma_k\delta_{j,f})\Big)\times \\ &\quad 6\int\frac{d^3p}{(2\pi)^3}\Big\{\frac{M_f}{E_f(\mathbf{p})}\{1 - n_{q_i} - n_{\bar{q}_i}\}\Big\} = 0,\end{aligned} \tag{56}$$

where

$$n_{q_i} = \frac{1}{\exp(\beta E_i^-) + \exp(-\beta E_i^-)}, \quad n_{\bar{q}_i} = \frac{1}{\exp(\beta E_i^+) + \exp(-\beta E_i^+)}. \tag{57}$$

3.2. Numerical Results

3.2.1. Thermodynamic Potential and Quark-Antiquark Condensations

From the stationary condition Eq.(56), we get the value of each order parameter $\langle\bar{\psi}_i\psi_i\rangle$ by searching the minimum point of thermodynamic potential. In Figure 4, we show the thermodynamic potential Ω and $\langle\bar{\psi}_i\psi_i\rangle$ at zero temperature for each chemical potential. The panels represent the values of Ω in the unit of MeV. We impose $\langle\bar{\psi}_u\psi_u\rangle = \langle\bar{\psi}_d\psi_d\rangle$ and all lepton chemical potentials are zero in these figures.

The minimum points depend on the chemical potential, which determines each condensate, namely each mass. At high chemical potential, all the condensates of $\langle\bar{\psi}_i\psi_i\rangle$ become close to zero, which implies the chiral symmetry restoration.

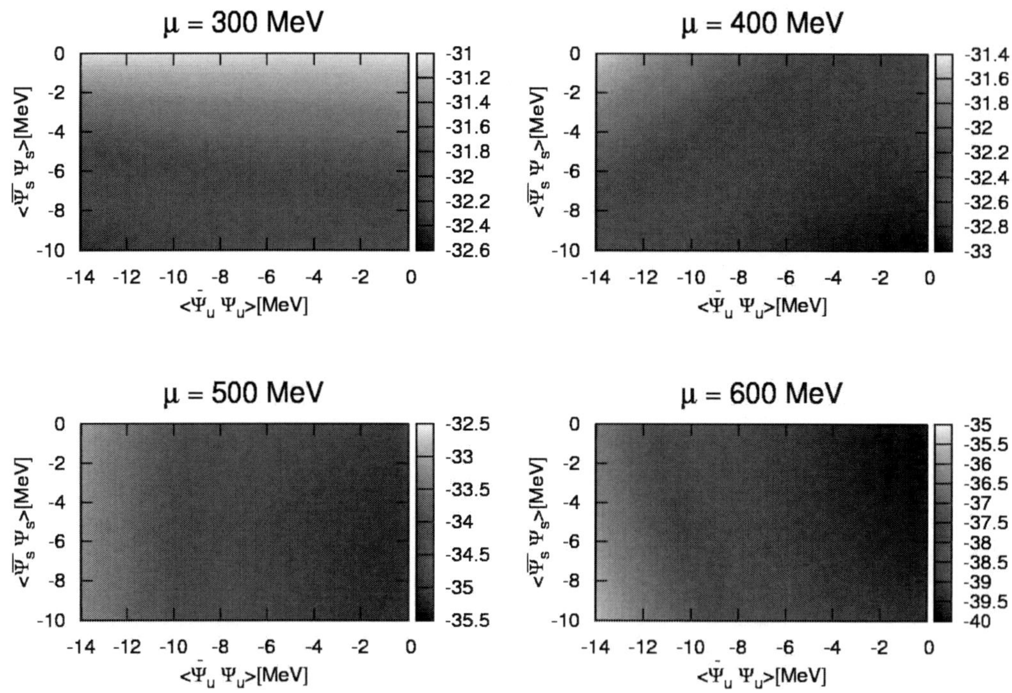

Figure 4. The thermodynamic potential Ω and $\langle\bar{\psi}_i\psi_i\rangle$ at zero temperature for each chemical potential. Note that we impose $\langle\bar{\psi}_u\psi_u\rangle = \langle\bar{\psi}_d\psi_d\rangle$ and all lepton potentials are zero to show these figures.

3.2.2. NJL Model and Compact Stars

The possible existence of quark matter in compact stars has been proposed [24, 25]. It should be noted that the uncertainty of EOS is always a big problem in the study of compact star structures. The recent observations imply the existence of compact objects with large mass [26, 27, 28, 29].

The effects of quark matter on various astrophysical phenomena have been studied extensively. For example, cooling effects on compact star evolutions have been studied in Ref. [30, 31, 32, 33]. Other examples include the effects of quark matter on gravitational wave radiation [34, 35, 36], neutrino emissions [37, 38], rotational frequencies [39], and the maximum energy release during the transition from neutron stars to quark stars [40, 41], etc..

Inside of the compact stars, we must impose the beta equilibrium and charge neutrality. Therefore, we rewrite the chemical potentials of u-,d- and s-quark as

$$\mu_s = \mu_d = \mu_u + (\mu_e - \mu_{\nu_e}), \tag{58}$$

and impose the condition of $\partial\Omega/\partial\mu_e = 0$ for chemical potential of electron μ_e. Under this condition, the relation of charge neutrality, $2n_u/3-(n_d+n_s)/3-n_e = 0$, is satisfied. Here, we neglect the quark-quark condensations, namely the effects of color superconductivity. Therefore, we do not consider the color neutrality.

Considering these conditions, we show the equation of state and the mass for each temperature and chemical potential in Fig 5. We set the realistic values of temperature and chemical potential shown in astrophysical phenomena; zero temperature and neutrino free for the inside of cold neutron stars (NSs), $T = 40$ MeV and lepton fraction $Y_L = 0.4$ for supernovae (SNe)[42], and $T = 40$ MeV and neutrino free for binary neutron star mergers (BSs)[43]. Clearly, the EOS for SN matter is harder than the ones for the other two matters shown in the left panel in Figure 5. Since neutrinos are trapped in SN matter, the electron fraction is enhanced for the chemical equilibrium (see $\mu_e - \mu_{\nu_e}$ in Eq. (58)). Hence, the chiral symmetry restoration of s-quark are suppressed shown in the right panel of Figure. 5. As the result, the EOS for SN matter becomes hard.

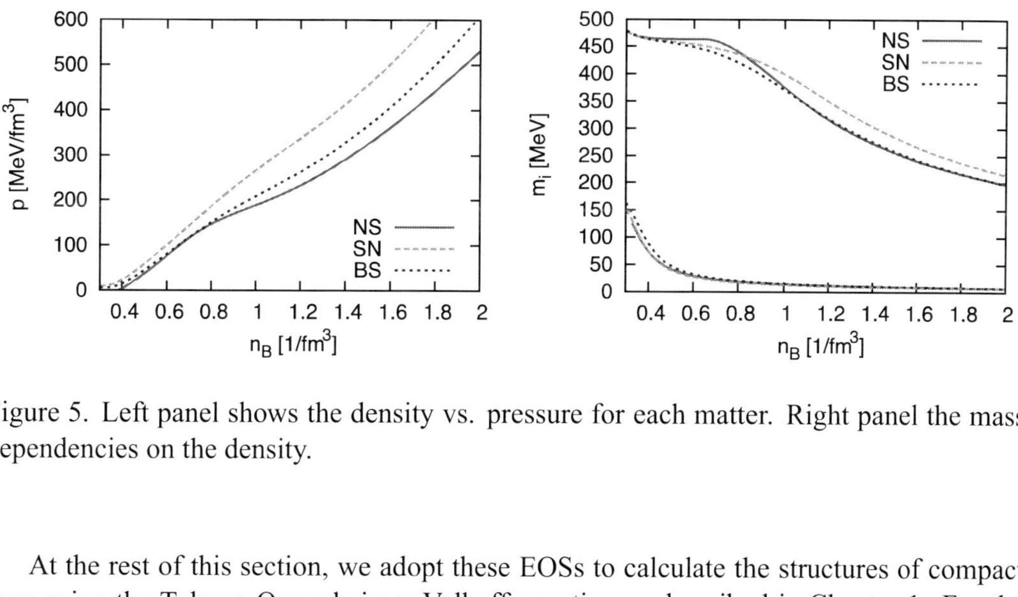

Figure 5. Left panel shows the density vs. pressure for each matter. Right panel the mass dependencies on the density.

At the rest of this section, we adopt these EOSs to calculate the structures of compact stars using the Tolman-Oppenheimer-Volkoff equation as described in Chapter 1. For the crusts, we adopt the EOS by Shen et al.[44] imposing the beta equilibrium condition without neutrinos. In Figure 6, we show the mass-radius relation of the compact star for each matter. The maximum masses are respectively 1.23 M_s for NS matter, 1.37 M_s for SN matter, and 1.26 M_s for BS matter. The similar results are also shown in Ref. [45].

However, the observations for NS-NS mergers show 1.44 M_s as the mass of the compact star, PSR B1913+16. Moreover, the other observations have suggested larger masses [46, 47]. Hence, the maximum masses we calculated above must be unrealistic. Originally the NJL models are not suitable to show the realistic hadron matter, because this model is the effective theory to show the chiral phase transitions in the frameworks of the quark model. Hence, it needs the other realistic EOS models of hadron matter to compare with the observations. In this context, some researchers have adopted more realistic EOSs for hadron phase [48, 49], though there are uncertainties of "*finite size effects*". Generally the properties of the mixed phase should strongly depend on electromagnetic interaction and surface tension etc., and these effects, sometimes called "*the finite-size effects*", which are discussed in Chapter 9.

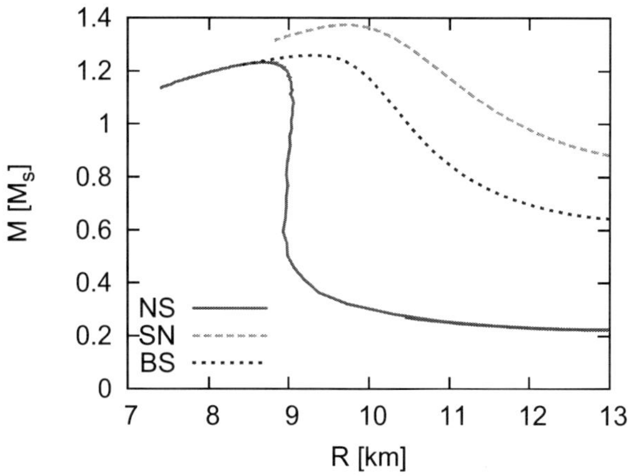

Figure 6. Mass-radius relation of compact stars for each matter.

4. Summary and Concluding Remarks

In this chapter, we described the Nambu-Jona-Lasinio model which is the most famous model to show the chiral symmetry restoration, effectively. For simplicity, we neglect the higher order interactions. We also do not take into account the quark-quark condensation, $\langle\psi\psi\rangle$. Hence, this chapter does not include the topic such as color superconductivity. If we consider them, above features may change [50].

Moreover, we note that the simple NJL model, such as in this chapter, does not also contain the effects of gluons. Nowadays some researchers suggested the Polyakov-loop Nambu-Jona-Lasinio (PNJL) model that includes the effects of gluons introducing the concepts of imaginary chemical potentials [51, 52]. The PNJL model has the same symmetry of the QCD Lagrangian. Therefore the PNJL models are compared by the lattice QCD calculations at zero chemical potential regions. Moreover, the imaginary chemical potential is recently considered in the PNJL model to investigate the chemical potential effects for thermodynamics because the imaginary chemical potential has deep relation with the real chemical potential; for example see Ref.[53, 54] and references therein.

We note that there are the uncertainties in the phase transition itself which comes from the finite-size effects, and so on. The readers can see these topics later (Chapter 9).

The NJL model is the effective theory to show the chiral phase transition. Hence, it is not unrealistic, especially for hadron matter. However, it is interesting to see how the chiral symmetry restoration affects the structures of compact stars or other astrophysical phenomena such as supernovae. Since there is a serious problem in the lattice QCD simulations, namely the sign problem, at high-density region, astrophysical phenomena should provide the possible windows to observe the aspects of QCD at that region, for the present.

Acknowledgement

We are grateful to K. Kashiwa, M. Nakashima, Y. Miyamoto T. Maruyama, T. Tatsumi, and T. Hatsuda. The discussions and collaborations with them are very useful for this work.

References

[1] Y. Nambu, G. Jona-Lasinio, *Phys. Rev.*, 122 (1961) 345.

[2] Y. Nambu, G. Jona-Lasinio, *Phys. Rev.*, 124 (1961) 246.

[3] J. Bardeen, C. N. Cooper, J. R. Schreiffer, *Phys.Rev.*, 108 (1957) 1175.

[4] M. Gell-Mann, *Phys. Lett.*, 8 (1964) 214.

[5] H. Politzer, *Phys.Rev. Lett.*, 30 (1973) 1346.

[6] D. J. Gross, F. Wilczek, *Phys. Rev. Lett.*, 30 (1973) 1340.

[7] T. Kunihiro, T. Hatsuda, *Prog. Theor. Phys.*, 71 (1984) 1332.

[8] T. Hatsuda, T. Kunihiro, *Prog. Theor. Phys.*, 74 (1985) 1175.

[9] D. Bailin, J. Cleymans, M. D. Scadron, *Phys. Rev.* D31 (1985) 164.

[10] V. Bernard, *Phys. Rev.* D34 (1986) 1601.

[11] V. Bernard, Ulf-G. Meissner, I. Zhed, *Phys. Rev.* D36 (1987) 819.

[12] V. Bernard, R. L. Jaffe, Ulf-G. Meissner, *Phys. Lett.* B198 (1987) 92.

[13] S. P. Klevansky, *Rev. of Mod. Phys.*, 64 (1992) 3.

[14] T. Hatsuda, T. Kunihiro, *Phys. Rep.*, 247 (1994) 221.

[15] M. Buballa, *Phys. Rep.*, 407 (2005) 205.

[16] M. Asakawa, K. Yazaki, *Nuc. Phys.*, A504 (1989) 668.

[17] T. Hatsuda, M. Tachibana, N. Yamamoto, and G. Baym, *Phys. Rev. Lett.*, 97 (2006) 122001.

[18] K. Kashiwa, H. Kouno, T. Sakaguchi, M. Matsuzaki, M. Yahiro, *Phys. Lett.*, B647 (2006) 446.

[19] P. Rehberg, S. P. Klevansky, and J. Hufner, *Phys. Rev.* C 53 (1996) 410 .

[20] T. Kunihiro and T. Hatsuda, *Phys. Lett.*, B206 (1988) 385.

[21] V. Bernard, R. L. Jaffe, Ulf-G. Meissner, *Nucl. Phys.*, B308 (1988) 753.

[22] G. 't Hooft, *Phys. Rev. Lett.* 37 (1976) 8.

[23] G. 't Hooft, *Phys. Rev.* D14 (1976) 3432.

[24] N. Itoh, Prog. *Theor. Phys.* 44, (1970) 291.

[25] E. Witten, *Phys. Rev.* D 30 (1984) 272.

[26] S. M. Ransom, et al., *Science* 307, (2005) 892.

[27] P. C. Freire, S. M. Ransom, S. Bégin, and I. H. Stairs, *Astrophys. J.* 675, (2008) 670.

[28] P. C. Freire, A. Wolszczan, M. van den Berg, and J. W. T. Hessels, *Astrophys. J.* 679, (2008) 1433.

[29] J. P. W. Verbiest, M. Bailes, W. van Straten, G. B. Hobbs, R. T. Edwards, R. N. Manchester, N. D. R. Bhat, J. M. Sarkissian, B. A. Jacoby, and S. R. Kulkarni, *Astrophys. J.* 679, (2008) 675.

[30] D. Page, M. Prakash, J. M. Lattimer, *Phys. Rev. Lett.* 85, (2000) 2048.

[31] D. Blaschke, T. Klahn and D. Voskresensky, *Astrophys. J.* 533, (2000) 406.

[32] D. Blaschke, H. Grigorian, and D. Voskresensky, *Astron. Astrophys.* 368, (2001) 561.

[33] H. Grigorian, D. Blaschke, and D. Voskresensky, *Phys. Rev.* C71, (2005) 045801.

[34] L. M. Lin, K. S. Cheng, M. C. Chu, W. M Suen, *Astrophys. J.*, 639, (2006) 382.

[35] N. Yasutake, K. Kotake, M. Hashimoto, and S. Yamada, *Phys. Rev.* D75, (2007) 084012.

[36] E.B. Abdikamalov, H. Dimmelmeier, L. Rezzolla, and J. C. Miller, arXiv:astro-ph/0806.1700.

[37] K. Nakazato, K. Sumiyoshi, and S. Yamada, *Phys. Rev.* D77, (2008) 103006.

[38] I. Sagert, M. Hampel, G. Pagliara, J. Schaffner-Bielich, T. Fischer, A. Mezzacappa, F.K. Thielmann, and M. Liebendörfer, arXiv:astro-ph/0809.4225 *[Phys. Rev. Lett. (to be published)*].

[39] G. F. Burgio, H. J. Schulze, and F. Weber, *Astron. Astrophys.* 408, (2003) 675.

[40] N. Yasutake, M. Hashimoto, and Y. Eriguchi, *Prog. Theor. Phys.* 113, (2005) 953.

[41] J. L. Zdunik, M. Bejger, P. Haensel, and E. Gourgoulhon, *Astron. Astrophys.* 465, (2007) 533.

[42] Y. Suwa, K. Kotake, T. Takiwaki, S. C. Whitehouse, M. Liebendörfer, and K. Sato, Publ. Astron. Soc. Japan 62, (2010) L49-L53. and the references.

[43] Y. Sekiguchi, K. Kiuchi, K. Kyutoku, and M. Shibata, arXiv:1105.2125.

[44] H. Shen, H. Toki, K. Oyamatsu, and K. Sumiyoshi, *Nucl. Phys.* A637, (1998) 435.

[45] D. P. Menezes, C. Provideñcia, and D. B. Melrose, *J. Phys.* G. 32 (2006) 1081-1095.

[46] Ožel, *Nature,* 441, (2006) 1115-1117.

[47] P. B. Demorest, T. Pennucci, S. M. Ransom, M. S. E. Roberts, and J. W. T. Hessels, *Nature*, 467 (2010) 1081-1083.

[48] G. F. Burgio, S. Plumari, *Phys. Rev.* D77 (2008) 085022.

[49] N. Yasutake, K. Kashiwa, *Phys. Rev.* D79 (2009) 043012.

[50] M. Alford et al., *Rev. Mod. Phys.* 80, (2008) 1455, and the references.

[51] K. Fukushima, *Phys. Rev.* D 77, (2008) 114028.

[52] K. Fukushima and H.J. Warringa, *Phys. Rev.* D 78, (2008) 039902.

[53] K. Kashiwa, M. Yahiro, H. Kouno, M. Matsuzaki, Y. Sakai., *J. Phys. G. Nucl. Part. Phys.* 36, (2009) 105001

[54] H. Kouno, Y. Sakai, T. Sasaki, K. Kashiwa, M. Yahiro, *Phys. Rev.* D 82, (2010) 076003.

In: Neutron Stars
Editors: Hiroshi Uechi et al, pp. 111-134

ISBN: 978-1-62081-123-8

Chapter 5

THE RELATIVISTIC (σ,π,ω) CHIRAL PARTICLE APPROACH

Schun T. Uechi and Hiroshi Uechi
Research Center for Nuclear Physics,
Osaka University, Ibaraki, Osaka 567-0047 Japan
Department of Distributions and Communication Sciences,
Osaka Gakuin University
2-36-1 Kishibe-minami, Suita, Osaka 564-8511 Japan.

Abstract

Density-dependent relations among saturation properties of symmetric nuclear matter and hyperonic matter, the coupling ratios (strengths) of hyperon matter, and properties of hadronic stars are discussed by applying the conserving chiral nonlinear (σ, π, ω) hadronic mean-field theory. The chiral nonlinear (σ, π, ω) mean-field theory is an extension of the conserving nonlinear (nonchiral) σ-ω hadronic mean-field theory which is a thermodynamically consistent, relativistic, and Lorentz-covariant mean-field theory of hadrons. In the extended chiral (σ, π, ω) mean-field model, all the masses of hadrons are produced by the spontaneous breaking of chiral symmetry. By comparing both nonchiral and chiral mean-field approximations, the effects of the chiral symmetry breaking mechanism on the mass of σ-meson, coefficients of nonlinear interactions, coupling ratios of hyperons to nucleons and Fermi-liquid properties are investigated in nuclear matter, hyperonic matter, and neutron stars.

1. Introduction

A renormalizable quantum field theory based on hadronic degrees of freedom provides us with an intuitively and physically accessible approach from finite nuclei to infinite nuclear matter, *e.g.*, high density nuclear and hyperonic matter such as neutron (hadronic) stars [1–8]. The linear neutral scalar and vector (σ,ω), nonlinear (σ,ω), and nonlinear (σ,ω,ρ) mean-field models are actively studied and applied to finite and infinite hadronic many-body systems. Though hadronic pictures of mean-field models render nuclear and astronomical phenomena readily understandable, they are mainly composed of strongly interacting particles. Those strong interactions make the hadronic approaches and extensions much more complicated. One may investigate the hadronic system by starting from quantum chromodynamics (QCD), but because of strong interactions, QCD calculations become complicated to apply directly to the nuclear energy domain. Consequently, one is led to introduce certain effective hadronic models to simulate strong interactions of hadrons [9].

The hadronic mean-field models must be constructed to reproduce the binding energy at the saturation of symmetric nuclear matter (assumed to be -15.75 MeV at $\rho_0 = 0.148$ fm^{-3} or $k_F = 1.30$ fm^{-1} in the current calculation), which is one of the fundamental requirements for nuclear physics as explained in Chap. 1. The pressure must vanish at saturation ($p = 0$), and simultaneously, the self-consistent single particle energy, $E(k_F)$, must be obtained by the functional derivative of energy density with respect to baryon density, $\delta\mathcal{E}/\delta\rho_B = E(k_F)$, as a dynamical constraint for any employed approximation. The energy density and pressure must maintain a thermodynamic relation, such as $\mathcal{E} + p = \mu\rho_B$ (at $T = 0$), to be a self-consistent approximation for nuclear matter. In terms of dynamical quantities, the self-consistent requirement can be stated that Green function, self-energy and energy density must maintain conditions of conserving approximations, termed *thermodynamic consistency*. Thermodynamic consistency is explicitly expressed as the requirement that functional derivatives of energy density with respect to self-energies must vanish, $\delta\mathcal{E}/\delta\Sigma = 0$ [10], which becomes equivalent to Landau's hypothesis of quasiparticles and the fundamental requirement of density functional theory [10–12].

Although the linear and nonlinear (σ,ω,ρ) mean-field models appropriately simulate properties of symmetric nuclear matter and neutron stars, they have many free parameters, masses and nonlinear coupling constants, coming from meson fields and nonlinear interactions. The upper bounds of values of nonlinear coefficients are confined by maintaining conditions of thermodynamic consistency to an employed approximation [5] and by reproducing empirical data. The values of nonlinear coefficients are bounded for thermodynamically consistent approximations. Self-consistent renormalization to observables restrict values of nonlinear conefficients, which is discussed as a manifestation of *naturalness* for self-consistent approximations. However, it is beyond the linear and nonlinear, nonchiral mean-field models to answer why coupling constants for nonlinear interactions are needed and restricted with such strengths. The chiral mean-field model reveals important features and strengths of nonlinear interactions [13–16].

The chiral (σ, π, ω) mean-field approximation provides the following:

1. Generations of hadron masses by the spontaneous chiral symmetry breaking mechanism produce coefficients of nonlinear meson interactions. This indicates that the fundamental requirement of nuclear matter saturation is directly related to experimental values of hadron masses ($M_N, m_\pi, m_\omega, m_\rho, \cdots$) and coupling constants ($g_\sigma, g_\pi, g_\omega, g_\rho, \cdots$). In the mean-field (Hartree) approximation, pion contributions vanish, and σ-meson compensates for attractive contributions expected to be given by pions at saturation density. Hence, the saturation property restricts the mass of the sigma meson, m_σ and g_σ.

2. The coupling constants for hyperons are important for studying phase transitions from β-equilibrium (n, p, e) asymmetric nuclear matter to (n, p, H_1, e) hyperon matter, binding energy of pure-hyperon matter and masses of hadronic stars. It is found that the Λ-hyperon coupling ratio to nucleon, $r^{\omega}_{\Lambda N} = g_{\omega\Lambda}/g_{\omega N}$, is expected to be $r^{\omega}_{\Lambda N} \sim 1.0$ by the requirement of thermodynamic consistency [6, 10], whereas the SU(6) quark model for hadrons demands $r^{\omega}_{\Lambda N} \sim 2/3$, or $1/3$ [17, 18]. Although the discrepancy of the coupling

ratios seems small, the differences of $r^{\omega}_{\Lambda N}$ result in significant discrepancies in the effective masses of hadrons, onset densities of nucleon-hyperon phase transitions, saturation properties of hyperons, and masses of hadron stars [6]. If the current chiral (σ, π, ω) mean-field model is applied to phase transition to β-equilibrium lambda matter (n, p, Λ, e), the coupling ratio: $r^{\sigma}_{\Lambda N} = g_{\sigma\Lambda}/g_{\sigma N} = M_{\Lambda}/M_{N} \approx 1.187$ can be examined to be reasonable to produce experimental values of nuclear matter and neutron stars (see, sec. 3). It is consistent with the analysis of the conserving, nonchiral (σ, ω, ρ) mean-field approximation. However, we would like to remark that it is not our purpose to critisize some effective theories and approximate calculations for causing the defects or differences, but it is to exhibit that the discrepancies between effective hadronic and quark theories could be a profound problem for clear comprehension of hadronic and quark approaches to nuclear physics. It may be a complicated task more than one expects to reconcile certain consequences derived from effective hadronic and quark theories.

The current extended chiral (σ, π, ω) mean-field model starts from a Lagrangian without hadron masses and generates all the hadron masses by way of spontaneous chiral symmetry breaking. This is different from other chiral mean-field models, which introduce the isoscalar-vector particle, ω, externally in order to produce the repulsive interaction and saturation mechanism. The current chiral (σ, π, ω) mean-field model produces masses of σ, π and ω particles by the chiral symmetry breaking mechanism. The chiral symmetric Lagrangian, spontaneous chiral symmetry breaking, and binding energy are discussed in Sec. 2. Fermi-liquid properties of nuclear matter, such as incompressibility and symmetry energy, K and a_4, and numerical results are shown in Sec. 3.

Vacuum fluctuation corrections to the chiral (σ, π, ω) mean-field approximation, applications to β-equilibrium (n, p, e) asymmetric nuclear matter and properties of hadron (neutron) stars are discussed in Sec. 4. The phase transition from symmetric nuclear matter to β-equilibrium hyperon matter, (n, p, H_1, e), and important results regarding coupling ratios given by the spontaneous chiral symmetry breaking are also discussed. Concluding remarks are in Sec. 5.

2. An Extended Chiral (σ,π,ω) Nonlinear Mean-Field Approximation

The conventional chiral mean-field models for hadrons assume that the Lagrangian with interaction potential, $V(\sigma^2 + \boldsymbol{\pi}^2)$, should be invariant under the chiral transformation and constrain only σ and $\boldsymbol{\pi}$ mesons as a *chiral partner*. Moreover, a massive isoscalar vector field ω_μ is input externally to supply repulsive nuclear-nuclear interactions, as in QHD-I [1,2]. The conventional chiral mean-field models for hadrons show that, when the chiral symmetry breaking parameter vanishes, the masses m_σ and m_π also vanish: $m_\sigma \rightarrow 0$, $m_\pi \rightarrow 0$, whereas $m_\omega \not\rightarrow 0$.

We introduce an extended chiral symmetric mean-field Lagrangian for hadrons with the interaction potential $V(\sigma^2 + \boldsymbol{\pi}^2 - a\omega_\mu^2)$. The Lagrangian is invariant under the chiral transformation and produces all hadron masses and nonlinear mean-field interactions by way of the spontaneous chiral symmetry breaking. The parameter a is constant, which will be identified as $m_\omega^2/m_\pi^2 \sim 31.65$ in the nuclear domain, after chiral symmetry breaking. Therefore, the current extended chiral mean-field model generates ω-meson as *chiral particles* such that all the meson masses are required to vanish simultaneously: $m_\sigma \to 0$, $m_\pi \to 0$, and $m_\omega \to 0$ when the chiral breaking parameter vanishes, $\varepsilon \to 0$. In other words, we assume that all the hadron masses $(M_N, m_\sigma, m_\pi, m_\omega)$ and nonlinear interactions should be generated by the Lagrangian with interaction potential $V(\sigma^2 + \boldsymbol{\pi}^2 - a\omega_\mu^2)$ under the chiral symmetry breaking mechanism.

The current extended chiral mean-filed lagrangisn that produces all the hadron masses $(M_N, m_\sigma, m_\pi, m_\omega)$ with the chiral symmetry breaking is based on a relativistic chiral $(\sigma, \boldsymbol{\pi}, \omega)$ model discussed by Walecka, Serot and others [13–16]. The extended [chiral $(\sigma, \boldsymbol{\pi}, \omega)$ Lagrangian is generated from,

$$
\begin{aligned}
\mathcal{L} =& \bar{\psi}\left[\gamma_\mu(i\partial^\mu - g_\omega\omega^\mu) + g(\sigma + i\gamma_5\boldsymbol{\tau}\cdot\boldsymbol{\pi})\right]\psi \\
&+ \frac{1}{2}(\partial_\mu\sigma\partial^\mu\sigma + \partial_\mu\boldsymbol{\pi}\cdot\partial^\mu\boldsymbol{\pi}) - \frac{1}{4}F_{\mu\nu}F^{\mu\nu} - V(\sigma^2 + \boldsymbol{\pi}^2 - a\omega_\mu^2) - \delta_{csb},
\end{aligned}
\tag{1}
$$

where $\delta_{csb} = \varepsilon\sigma$ is the chiral symmetry breaking term. The nucleon is $\psi = \begin{pmatrix}\psi_p \\ \psi_n\end{pmatrix}$, and $\sigma, \boldsymbol{\pi}, \omega_\mu$ are neutral scalar meson, pseudo-scalar isovector pion and neutral isoscalar omega meson fields, respectively. The field strength tensor $F_{\mu\nu} = \partial_\mu\omega_\nu - \partial_\nu\omega_\mu$ is for the vector-isoscalar ω-meson. Note that there are no baryon and meson masses in the Lagrangian (1). Baryons and mesons are coupled as $g_\omega\bar{\psi}\gamma_\mu\omega^\mu\psi$, and $g\bar{\psi}(\sigma + i\gamma_5\boldsymbol{\tau}\cdot\boldsymbol{\pi})\psi$. The coupling constant, g, is the pion-nucleon (and σ-nucleon) coupling constant to be required from invariance under the chiral transformation ($g_\sigma = g_\pi = g$ is assumed).

When the chiral symmetry is restored, the lagrangian (1) satisfies SU(2)×SU(2)×U(1) global gauge symmetry, SU(2) isospin current conservation and axial current conservation. We introduce the chiral-invariant potential of the following form:

$$
V(\sigma^2 + \boldsymbol{\pi}^2 - a\omega_\mu^2) = \frac{\lambda}{4}\left[\sigma^2 + \boldsymbol{\pi}^2 - a\omega_\mu^2\right]^2 , \tag{2}
$$

where $\lambda \neq 0$ and $a > 0$ are constants determined in the ground state after the spontaneous chiral symmetry breaking. Hence, the free parameters of the current chiral mean-field model are g, g_ω and λ. Note that $(\sigma, \boldsymbol{\pi}, \omega)$ mesons make the Lagrangian chiral invariant all together, not by way of the chiral-invariance generated by the $(\sigma, \boldsymbol{\pi})$-chiral partner. In this sense, we call $(\sigma, \boldsymbol{\pi}, \omega)$ mesons *chiral particles*.

The current chiral Lagrangian is invariant under the following gauge transformations [13]:

$$
\begin{aligned}
\delta\psi &= \frac{i}{2}\boldsymbol{\epsilon}\cdot\boldsymbol{\tau}\gamma_5\psi , \\
\delta\boldsymbol{\pi} &= -\boldsymbol{\epsilon}\sigma , \\
\delta\sigma &= \boldsymbol{\epsilon}\cdot\boldsymbol{\pi} ,
\end{aligned}
\tag{3}
$$

where ϵ is assumed to be an infinitesimal value, and the ω meson is invariant under the gauge transformation: $\delta\omega_\mu = 0$. After chiral symmetry breaking , the interaction potential is given in the new ground state as,

$$\begin{aligned} V &= \frac{\lambda}{4}\left[(\sigma^2+\boldsymbol{\pi}^2-a\omega_\mu^2)-v^2\right]^2+\delta_{csb} \\ &= \frac{\lambda}{4}\left[(\sigma^2+\boldsymbol{\pi}^2-a\omega_\mu^2)-v^2\right]^2+\varepsilon\sigma\ , \end{aligned} \tag{4}$$

where λ, v, a and ε are constants determined at the ground state.

The mesons are excited from the new ground state as follows,

$$\begin{aligned} \sigma &\to \langle\sigma\rangle+\phi, \\ \boldsymbol{\pi} &\to \langle\boldsymbol{\pi}\rangle+\boldsymbol{\pi}, \\ \omega_\mu &\to \langle\omega_\mu\rangle+\omega_\mu, \end{aligned} \tag{5}$$

where $\langle\sigma\rangle$, $\langle\boldsymbol{\pi}\rangle$ and $\langle\omega_\mu\rangle$ are values for the meson fields in the vacuum defined by minimization of (4) with respect to σ, $\boldsymbol{\pi}$, and ω_μ. The interaction potential V has the following form at the ground state in the new vacuum,

$$V=\frac{\lambda}{4}\left[\left(\langle\sigma\rangle^2+\langle\boldsymbol{\pi}\rangle^2-a\langle\omega_\mu\rangle^2\right)-v^2\right]^2+\varepsilon\langle\sigma\rangle\ , \tag{6}$$

and the minimization conditions give

$$\begin{aligned} \frac{\partial V}{\partial\langle\sigma\rangle} &= \lambda\langle\sigma\rangle\left[(\langle\sigma\rangle^2+\langle\boldsymbol{\pi}\rangle^2-a\langle\omega_\mu\rangle^2)-v^2\right]+\varepsilon=0, \\ \frac{\partial V}{\partial\langle\boldsymbol{\pi}\rangle} &= \lambda\langle\boldsymbol{\pi}\rangle\left[(\langle\sigma\rangle^2+\langle\boldsymbol{\pi}\rangle^2-a\langle\omega_\mu\rangle^2)-v^2\right]=0, \\ \frac{\partial V}{\partial\langle\omega_\mu\rangle} &= \lambda a\langle\omega_\mu\rangle\left[(\langle\sigma\rangle^2+\langle\boldsymbol{\pi}\rangle^2-a\langle\omega_\mu\rangle^2)-v^2\right]=0\ . \end{aligned} \tag{7}$$

The conditions, $\lambda\neq 0$ and $\varepsilon\neq 0$, lead to $\langle\sigma\rangle\equiv\sigma_0\neq 0$, $[\langle\sigma\rangle^2+\langle\boldsymbol{\pi}\rangle^2-a\langle\omega_\mu\rangle^2-v^2]\neq 0$, and

$$\langle\boldsymbol{\pi}\rangle=0, \quad \langle\omega_\mu\rangle=0\ . \tag{8}$$

The ground state value, σ_0, is then defined as,

$$\langle\sigma\rangle\equiv\sigma_0=-\frac{M}{g}. \tag{9}$$

By expanding the interaction potential $V(\sigma^2+\boldsymbol{\pi}^2-a\omega_\mu^2)$, the terms in (6) are collected as follows:

(1) Constant terms are

$$V_0=\frac{\lambda}{4}(\sigma_0^2-v^2)^2+\varepsilon\sigma_0. \tag{10}$$

(2) The terms that are linear in ϕ are

$$V_1 = \left\{\lambda\sigma_0(\sigma_0^2 - v^2) + \varepsilon\right\}\phi = 0\,. \tag{11}$$

This expression vanishes because of the minimization conditions, (7) and (8).

(3) The terms that are quadratic in $\boldsymbol{\pi}$ are

$$V_2 = -\frac{1}{2}\frac{\varepsilon}{\sigma_0}\boldsymbol{\pi}^2 = \frac{1}{2}\frac{g\varepsilon}{M}\boldsymbol{\pi}^2 \equiv \frac{1}{2}\mu_2^2\boldsymbol{\pi}^2. \tag{12}$$

(4) The terms that are quadratic in ω_μ are derived in the same way as,

$$V_3 = -\frac{1}{2}\frac{g\varepsilon}{M}a\omega_\mu^2 \equiv -\frac{1}{2}\mu_3^2\omega_\mu^2\,. \tag{13}$$

(5) The terms that are quadratic in ϕ are

$$V_4 = \frac{\lambda}{2}(\sigma_0^2 - v^2)\phi^2 + \lambda\sigma_0^2\phi^2 \equiv \frac{1}{2}\mu_1^2\phi^2\,, \tag{14}$$

and λ is given by

$$\lambda \equiv \frac{1}{2}\left(\frac{g}{M}\right)^2(\mu_1^2 - \mu_2^2)\,. \tag{15}$$

(6) The remaining cubic and quartic interactions of the meson fields $(\phi, \boldsymbol{\pi}, \omega)$ are then given by

$$\begin{aligned} V_5 &= \frac{\lambda}{4}\left[4\sigma_0\phi(\phi^2 + \boldsymbol{\pi}^2 - a\omega_\mu^2) + (\phi^2 + \boldsymbol{\pi}^2 - a\omega_\mu^2)^2\right] \\ &= \frac{1}{2}(\mu_1^2 - \mu_2^2)\left[\left(\frac{g}{2M}\right)^2(\phi^2 + \boldsymbol{\pi}^2 - a\omega_\mu^2)^2 - 2\left(\frac{g}{2M}\right)\phi(\phi^2 + \boldsymbol{\pi}^2 - a\omega_\mu^2)\right]\,. \end{aligned} \tag{16}$$

A collection of these terms then yields the interaction potential V written as,

$$\begin{aligned} V =& \frac{1}{2}\mu_1^2\phi^2 + \frac{1}{2}\mu_2^2\boldsymbol{\pi}^2 - \frac{1}{2}\mu_3^2\omega_\mu^2 \\ &+ \frac{1}{2}(\mu_1^2 - \mu_2^2)\left[\left(\frac{g}{2M}\right)^2(\phi^2 + \boldsymbol{\pi}^2 - a\omega_\mu^2)^2 - 2\left(\frac{g}{2M}\right)\phi(\phi^2 + \boldsymbol{\pi}^2 - a\omega_\mu^2)\right]\,. \end{aligned} \tag{17}$$

The Lagrangian density (1) with the generation of hadron masses by spontaneous symmetry breaking finally takes the following form:

$$\begin{aligned} \mathcal{L}_{csb} =& \bar{\psi}\left[\gamma_\mu(i\partial^\mu - g_\omega\omega^\mu) - \{M - g(\phi + i\gamma_5\boldsymbol{\tau}\cdot\boldsymbol{\pi})\}\right]\psi \\ &+ \frac{1}{2}(\partial_\mu\phi\partial^\mu\phi - \mu_1^2\phi^2) + \frac{1}{2}(\partial_\mu\boldsymbol{\pi}\cdot\partial^\mu\boldsymbol{\pi} - \mu_2^2\boldsymbol{\pi}^2) - \frac{1}{4}F_{\mu\nu}F^{\mu\nu} + \frac{1}{2}\mu_3^2\omega_\mu^2 \\ &- \frac{1}{2}(\mu_1^2 - \mu_2^2)\left[\left(\frac{g}{2M}\right)^2(\phi^2 + \boldsymbol{\pi}^2 - a\omega_\mu^2)^2 - 2\left(\frac{g}{2M}\right)\phi(\phi^2 + \boldsymbol{\pi}^2 - a\omega_\mu^2)\right] \\ &+ \text{constant}\,. \end{aligned} \tag{18}$$

The parameters are identified to be: $\mu_1 = m_\sigma, \mu_2 = m_\pi, \mu_3 = m_\omega$, and $a = m_\omega^2/m_\pi^2 \sim 31.65$ in the nuclear ground state after the spontaneous chiral symmetry breaking.

3. Chiral Invariance and Currents

If A and B are two fixed noncommuting operators and ξ is a parameter, then we have

$$e^{\xi A} B e^{\pm \xi A} = B + \xi [A, B]_{\pm} + \frac{\xi^2}{2!}[A, [A, B]]_{\pm} + \frac{\xi^3}{3!}[A, [A, [A, B]]]_{\pm} + \cdots , \tag{19}$$

where the square bracket denotes $[A, B]_{\pm} = AB \pm BA$. The following formulas are useful for checking invarience of a lagrangian,

$$\begin{aligned}
&(\boldsymbol{\tau} \cdot \boldsymbol{a})(\boldsymbol{\tau} \cdot \boldsymbol{b}) = \mathrm{a} \cdot \mathrm{b} + i\boldsymbol{\tau} \cdot (\boldsymbol{a} \times \boldsymbol{b}) , \\
&(\boldsymbol{\tau} \cdot \boldsymbol{a})(\boldsymbol{\tau} \cdot \boldsymbol{b}) + (\boldsymbol{\tau} \cdot \boldsymbol{b})(\boldsymbol{\tau} \cdot \boldsymbol{a}) = 2\boldsymbol{a} \cdot \boldsymbol{b} , \\
&(\boldsymbol{\tau} \cdot \boldsymbol{a})(\boldsymbol{\tau} \cdot \boldsymbol{b}) - (\boldsymbol{\tau} \cdot \boldsymbol{b})(\boldsymbol{\tau} \cdot \boldsymbol{a}) = 2i\boldsymbol{\tau} \cdot (\boldsymbol{a} \times \boldsymbol{b}) , \\
&\gamma_5 \equiv i\gamma_0\gamma_1\gamma_2\gamma_3 , \qquad \{\gamma_5, \gamma_\mu\} = 0 , \qquad e^{i\gamma_5}\gamma_\mu = \gamma_\mu e^{-i\gamma_5} .
\end{aligned} \tag{20}$$

The operator relation and formulas are used to examine invarience of an lagrangian. It is known from Noether's theorem that an invarience (symmetry) of a lagrangian leads directly to a conserved current of the following form,

$$\delta\mathcal{L} = \partial_\lambda \Big[\frac{\partial \mathcal{L}}{\partial(\partial_\lambda q_1)} \delta q_1 + \cdots \Big] = 0 , \tag{21}$$

where $q_1, q_2, \cdots$ are independent fields.

The lagrangian (18) is invarient under the SU(2) isospin transformation defined by

$$U_I(\boldsymbol{\epsilon}) = e^{\frac{i}{2}\boldsymbol{\epsilon} \cdot \boldsymbol{\tau}} , \tag{22}$$

which is proved by applying (22) and the operator relation (19) and formulas (20) on the lagrangian (18). However, the SU(2) (global) chiral transformation defined by

$$U_\chi(\boldsymbol{\epsilon}) = e^{\frac{i}{2}\gamma_5 \boldsymbol{\epsilon} \cdot \boldsymbol{\tau}} , \tag{23}$$

is invarient in the limit $\epsilon \to 0$, which is denoted as a partial invarience.

The SU(2) (global) isospin-symmetry invariance of (18) generates the conserved current:

$$j_\mu = \bar{\psi}\gamma_\mu \frac{\boldsymbol{\tau}}{2}\psi + \boldsymbol{\pi} \times \partial_\mu \boldsymbol{\pi} , \tag{24}$$

that can be proved to be

$$\partial_\mu j^\mu = 0 , \tag{25}$$

from the Lagrangian (18) and equations of motion for baryons and mesons. The SU(2) (global) chiral symmetry breaking of (18) results in the partially conserved axial-vector current (PCAC) :

$$A_\mu = \bar{\psi}\gamma_\mu\gamma_5 \frac{\boldsymbol{\tau}}{2}\psi + \phi\partial_\mu \boldsymbol{\pi} - \boldsymbol{\pi}\partial_\mu \phi , \tag{26}$$

which is shown to satisfy the PCAC,

$$\partial_\mu A^\mu = \varepsilon \boldsymbol{\pi} \,, \tag{27}$$

with the use of equations of motion. The symmetry breaking parameter, ε, is expressed in the interaction potential (6).

4. Effective Masses of Mesons and Baryons

The chiral $(\sigma, \boldsymbol{\pi}, \omega)$ mean-field approximation is defined by replacing meson quantum fields with classical fields: $\hat{\phi} \rightarrow \phi_0, \hat{\omega}_\mu = (\hat{\omega}_0, \hat{\boldsymbol{\omega}}) \rightarrow (\omega_0, \boldsymbol{\omega})$. They are constants independent of x_μ. The spatial part of the vector field $\langle \boldsymbol{\omega} \rangle$ should vanish by the requirement of rotational invariance of static and homogeneous nuclear matter [1]. In addition, π-meson contributions vanish in the (mean-field) Hartree approximation. Thus, the chiral mean-field Lagrangian is given by

$$\begin{aligned}\mathcal{L}_{csb} =& \bar{\psi}\left[i\gamma_\mu\partial^\mu - g_\omega\gamma_0\omega_0 - (M - g\phi_0)\right]\psi \\ & -\frac{1}{2}m_\sigma^2\phi_0^2 + \frac{1}{2}m_\omega^2\omega_0^2 \\ & -\frac{1}{2}(m_\sigma^2 - m_\pi^2)\left[\left(\frac{g}{2M}\right)^2(\phi_0^2 - a\omega_0^2)^2 - 2\left(\frac{g}{2M}\right)\phi_0(\phi_0^2 - a\omega_0^2)\right].\end{aligned} \tag{28}$$

The equations of motion for the scalar and vector mesons are given by

$$m_\sigma^2\phi_0 - \frac{g}{2M}(m_\sigma^2 - m_\pi^2)\left(3\phi_0 + 2a\omega_0^2 - 2\frac{g}{2M}\phi_0^2\right)\phi_0 = g\rho_\sigma^*, \tag{29}$$

$$m_\omega^2\omega_0 - \frac{g}{2M}(m_\sigma^2 - m_\pi^2)a\left\{2\phi_0 + \frac{g}{2M}\left(2a\omega_0^2 - 2\phi_0^2\right)\right\}\omega_0 = g_\omega\rho_B, \tag{30}$$

where ρ_σ^* is the scalar source, and ρ_B is the baryon density: $\rho_B = \sum_B k_{F_B}^3/3\pi^2$, where k_{F_B} is a baryon Fermi-momentum. The energy density and pressure can be derived from the energy momentum tensor [2, 5]:

$$\begin{aligned}\mathcal{E} =& \sum_{B=n,p}\frac{2}{(2\pi)^3}\int^{k_{F_B}} d^3k E_B(k) + \frac{1}{2}m_\sigma^2\phi_0^2 - \frac{g}{2M}(m_\sigma^2 - m_\pi^2)\left(\phi_0^3 - \frac{1}{2}\frac{g}{2M}\phi_0^4\right) \\ & -\frac{1}{2}m_\omega^2\omega_0^2 + \frac{g}{2M}(m_\sigma^2 - m_\pi^2)a\left(\phi_0\omega_0^2 + \frac{1}{2}\frac{g}{2M}a\omega_0^4 - \frac{g}{2M}\phi_0^2\omega_0^2\right),\end{aligned} \tag{31}$$

$$\begin{aligned}p =& \sum_{B=n,p}\frac{1}{3}\frac{2}{(2\pi)^3}\int^{k_{F_B}} d^3k\frac{k^2}{E_B^*(k)} - \frac{1}{2}m_\sigma^2\phi_0^2 + \frac{g}{2M}(m_\sigma^2 - m_\pi^2)\left(\phi_0^3 - \frac{1}{2}\frac{g}{2M}\phi_0^4\right) \\ & +\frac{1}{2}m_\omega^2\omega_0^2 - \frac{g}{2M}(m_\sigma^2 - m_\pi^2)a\left(\phi_0\omega_0^2 + \frac{1}{2}\frac{g}{2M}a\omega_0^4 - \frac{g}{2M}\phi_0^2\omega_0^2\right),\end{aligned} \tag{32}$$

where $E_B(k) = E^*_B(k) + \Sigma^0_\omega = \sqrt{k^2 + M^{*2}_B} - g_\omega \omega_0$. The scalar source ρ^*_σ is derived from the functional derivative with respect to ϕ_0 [5]:

$$\rho^*_\sigma = \sum_B \frac{1}{\pi^2} \int^{k_{F_B}} q^2 dq \frac{M^*}{E^*(q)} - \frac{1}{2M}(m^2_\sigma - m^2_\pi) a \omega^2_0 \, . \tag{33}$$

The self-consistent effective masses of hadrons are determined by satisfying conditions of thermodynamic consistency:

$$\begin{aligned} M^*_N &= M - g\phi_0, \\ m^{*2}_\sigma &= m^2_\sigma - \frac{g}{2M}(m^2_\sigma - m^2_\pi)\left\{3\phi_0 - 2\frac{g}{2M}\left(\phi^2_0 - a\omega^2_0\right)\right\}, \\ m^{*2}_\omega &= m^2_\omega - \frac{g}{2M}(m^2_\sigma - m^2_\pi)\left\{2a\phi_0 - \frac{g}{2M}(2a\phi^2_0 - 2a^2\omega^2_0)\right\}, \end{aligned} \tag{34}$$

and self-consistent scalar and vector self-energies are given by:

$$\Sigma^s = -\frac{g^2}{m^{*2}_\sigma}\rho^*_\sigma \, , \qquad \Sigma^\mu_\omega = -\frac{g^2_\omega}{m^{*2}_\omega}\rho_B \delta_{\mu,0} \, , \tag{35}$$

where m^*_σ and m^*_ω are the effective masses of σ and ω mesons. One should note that the effective masses (34) and self-energies (35) are derived from $\delta\mathcal{E}/\delta\rho_B = E(k_F)$ of energy density (31) and equations of motion for mesons. Hence, renormalized effective masses and self-energies are consistent with conditions of thermodynamic consistency [10] and the density functional theory [11, 12].

The self-consistent self-energies (35) and single particle energy $E(k) = E^*_B(k) + \Sigma^0_\omega$ are essential to understand the effect of coupling constant on the equation of state. Though the single particle energy is a complicated function of coupling constants, it becomes formally simple when nonlinear interactions are renormalized by the condition of thermodynamic consistency in the current chiral mean-field approximation. From the eq. (35), the single particle energy behaves, $\sim g^2_{\omega N}\rho_B/m^{*2}_\omega$, at high densities. Therefore, the coupling ratio $r^\omega_{HN} = g_{\omega H}/g_{\omega N} \lesssim 2/3$ required by SU(6) effective quark model [17, 18] produces smaller single particle energy, $\lesssim (4/9)g^2_{\omega N}\rho_B/m^{*2}_\omega$, in a hyperonic high density matter. In other words, the single particle energy $E(k)$ or chemical potential $\mu = E(k_F)$ becomes small when the coupling ratios $r_{HN} \lesssim 2/3$ are employed, resulting in a softer equation of state which becomes difficult to generate observed masses of neutron stars. In addition, the chemical potential $\mu = E(k_F)$ of baryons will be redistributed among baryons and hyperons to maintain phase-transition and charge-conservation conditions in high-density hadronic matter, which will reduce the magnitude of chemical potentials [6].

The 3-dimensional image of the interaction potential after spontaneous symmetry breaking defined by

$$V \equiv \mathcal{E} - \sum_{B=n,p} \frac{2}{(2\pi)^3} \int^{k_{F_B}} d^3k E_B(k), \tag{36}$$

is shown in Fig. 1. In the current chiral mean-field approximation, the interaction potential is self-consistently constructed by σ and ω mesons. Sigma mesons produce attractive

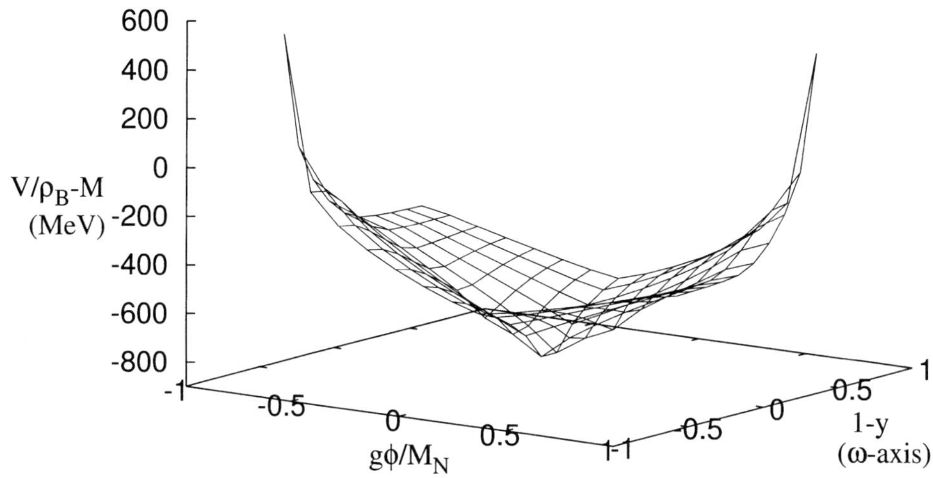

Figure 1. This is the 3-dimensional image of interaction potential V defined by classical meson fields as $V \equiv \mathcal{E} - \sum_{B=n,p} \frac{2}{(2\pi)^3} \int^{k_{F_B}} d^3k E_B(k)$. The field ϕ produces attractive interaction at low densities. The ω-axis is written by the variable $1 - y$, where $y = (m_\omega^2/g_\omega \rho_B)\omega_0$, and ω-field produces repulsive interaction at high densities. The origin is set in the center of ϕ-ω plain. It shows that the interaction potential is bound when $m_\sigma < m_\pi$.

interactions at low densities, whereas omega mesons mainly generate repulsive contributions at high densities. In the chiral (σ, π) Hartree approximation, the pion field will vanish completely, and the meson interaction potential in the new ground state becomes only the function of $V(\sigma)$, which has the Mexican hat symmetry. In the current paper, the pion field vanishes but the omega field obtains mass. Hence, the current interaction potential after SSB is expressed by $V(\sigma, \omega)$. It seems different from the well-known Mexican-hat type potential ; however, the potential is bound and produces the Mexican-hat symmetry when the limits, $g_\omega, a \to 0$, are taken.

The energy density and pressure satisfy $\mathcal{E} + p = \mu\rho_B$ and $\mu = E(k_F)$ at all densities. The binding energies of symmetric nuclear matter are shown in Fig. 2. The linear (σ, ω, ρ) Hartree approximation denoted as MFT-II [1] is listed for comparison in order to see the effect of chiral nonlinear corrections.

5. Fermi Liquid Properties at Nuclear Matter Saturation

The chiral (σ, π, ω) mean-field model exhibits remarkable properties when it is compared to the nonchiral, nonlinear (σ, ω, ρ) mean-field model. The nonchiral mean-field model is applied to (n, p) symmetric, (n, p, e) asymmetric, (n, p, H, e) hyperonic matter, and neutron stars [6]. Although the nonchiral model reasonably simulates properties of nuclear and neutron matter, it has many free nonlinear parameters which cannot be determined uniquely. The upper and lower bound values of the coupling constants and effective masses of hadrons are constrained by empirical data and self-consistent conditions of approxima-

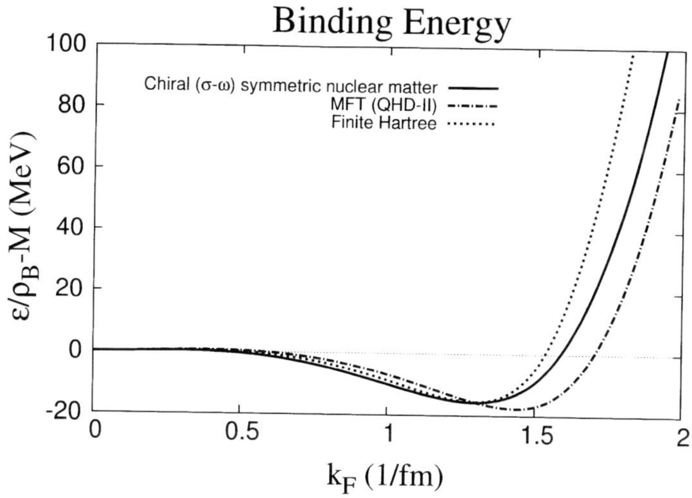

Figure 2. The binding energies of isospin symmetric (n, p) nuclear matter (solid line), MFT-II [1](dash-dotted line) and Finite Hartree [1] (dotted line) are shown. Note that $\mathcal{E}/\rho_B = E(k_F)$ is exactly satisfied at the saturation density, $\rho_B = 0.148$ fm^{-3} ($\rho_B = 0.193$ fm^{-3}) for MFT-II.

tions. The nonlinear nonchiral mean-field approximations determine nonlinear coupling constants by empirical data and self-consistent conditions, but could not clearly explain why values of nonlinear coupling constants are bound in a characteristic way [6]. The chiral symmetry approach will sharply restrict nonlinear parameters by the chiral invariance of model Lagrangians, and it clarifies relations among nonlinear coupling constants, hadron masses and observables in the chiral (σ, π, ω) mean-field model.

All the hadron masses and nonlinear coefficients are related to the properties of symmetric nuclear matter, such as binding energy, K and a_4, because the chiral breaking mechanism determines nonlinear interactions in terms of hadron masses and coupling constants, g and g_ω, respectively. Consequently, the mass of σ-meson, m_σ, is related to the binding energy of symmetric nuclear matter ($\mathcal{E}/\rho_B - M = -15.75$ MeV, at $k_F = 1.30$ fm^{-1}) and is adjusted self-consistently. Incompressibility is calculated by

$$K = 9\rho_B \frac{\partial^2 \mathcal{E}}{\partial \rho_B^2} = 9\rho_B \left(\frac{\partial \mu}{\partial \rho_B} \right), \tag{37}$$

where μ is the chemical potential and is equal to the Fermi energy, $\mu = E(k_F)$, because the current chiral mean-field approximation is thermodynamically consistent and Landau's hypothesis for quasiparticles is maintained exactly. The symmetry energy is calculated by

$$a_4 = \frac{1}{2}\rho_B \left[\left[\frac{\partial^2 \mathcal{E}}{\partial \rho_3^2} \right]_{\rho_B} \right]_{\rho_3 = 0}, \tag{38}$$

where ρ_3 is the difference between the proton and neutron density: $\rho_3 = \rho_p - \rho_n = (k_{F_p}^3 - k_{F_n}^3)/3\pi^2$ at a fixed baryon density, $\rho_B = \rho_p + \rho_n = 2k_F^3/3\pi^2$.

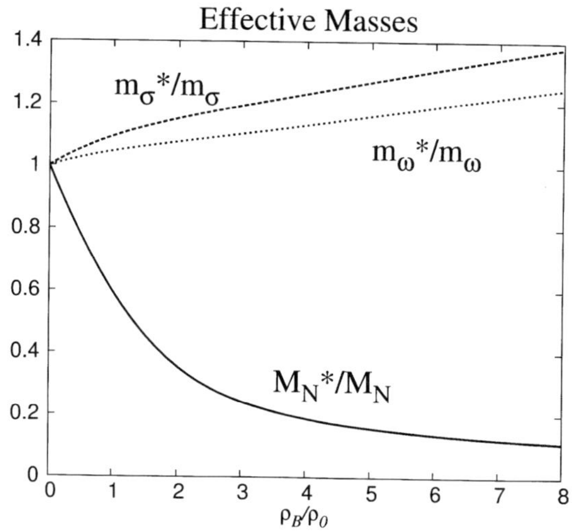

Figure 3. Effective masses of nucleons, M_N^*/M_N, and mesons, m_σ^*/m_σ and m_ω^*/m_ω. The qualitative behavior of the effective masses is consistent with those derived from nonlinear, nonchiral (σ, ω, ρ) mean-field approximations.

The coupling constants and effective masses of hadrons and the Fermi-liquid properties of symmetric nuclear matter are listed in Table 1. The effective masses of mesons are shown in Fig. 3: $M_N^*/M_N \sim 0.60$, $m_\sigma^*/m_\sigma \sim 1.09$, $m_\omega^*/m_\omega \sim 1.04$, at saturation density. The effective mass of a nucleon, $M_N^*/M_N \sim 0.60$, would be considered to produce a hard EOS and large masses of neutron stars in nonchiral mean-field approximations, but the chiral mean-field approximation produces a softer EOS.

The incompressibility and symmetry energy are shown in Fig. 4 and Fig. 5, respectively. They are $K = 371$ MeV and $a_4 = 17.4$ MeV, at saturation density. These observables are expected to be $K \sim 300$ MeV and $a_4 \sim 30$ MeV in the nonchiral, nonlinear (σ, ω, ρ) mean-field approximation [6]. One can notice that ρ-meson contribution would be important when

Table 1. Coupling constants and Fermi-liquid properties of nuclear matter. In the current chiral mean-field approximation, the masses of π and ω mesons are identified as $\mu_2 = m_\pi = 139.0$ MeV and $\mu_3 = m_\omega = 783.0$ MeV in the nuclear domain. Hence, the adjustable parameters are only g, g_ω, and m_σ. The effective masses, K and a_4, are values at saturation of nuclear matter: $\rho_B = 0.148$ fm^{-3} (confer, Table 2).

g	g_ω	m_σ		
2.4095	13.4232	120.0		
M_N^*/M_N	m_σ^*/m_σ	m_ω^*/m_ω	K (MeV)	a_4 (MeV)
0.60	1.09	1.04	371	17.4
M_{max}	$\mathcal{E}_c$	I	R (km)	
2.60	1.58	418	12.8	

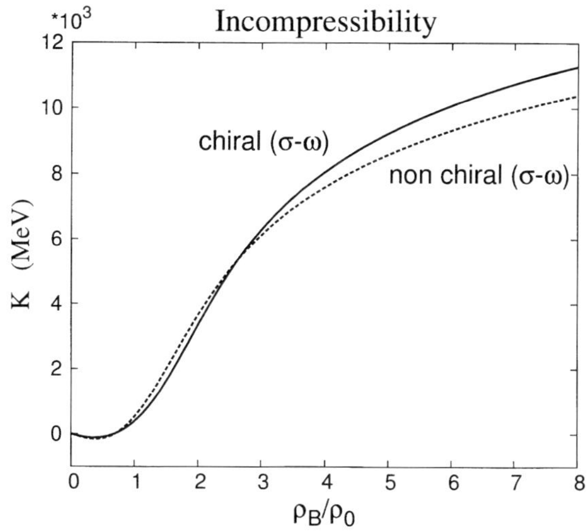

Figure 4. Incompressibilities in the nonchiral and chiral (σ, ω) mean-field approximations. The effect of chiral symmetry on incompressibilities is not significant around saturation but is important at high densities.

a_4 in the nonchiral (σ, ω, ρ) is compared to that of chiral (σ, ω) in Fig. 5. Hence, in order to examine calculations quantitatively, the chiral (σ, π, ω) model must be extended to the chiral $(\sigma, \pi, \omega, \rho)$ model [19], which is expected to clarify the chiral hadronic models.

The mass of σ-meson is important because all observables, EOS, and masses of neutron stars, depend only on the three adjustable parameters: m_σ and coupling constants, g and g_ω. Therefore, the binding energy of symmetric nuclear matter ($\mathcal{E}/\rho_B - M = -15.75$ MeV, at $k_F = 1.30$ fm^{-1}) and the maximum mass of neutron stars ($M_{\max} \simeq 2.50$ M$_\odot$) determine the mass of σ-mesons to be $m_\sigma \approx 120.0$ MeV and $g = 2.4095$. The mass of σ-meson seems to be very small compared to the masses employed in other mean-field models. However, the dimensionless parameter, $C \equiv g^2(M_N^2/m_\sigma^2)$, is similar to the values derived from nonchiral linear and nonlinear mean-field approximations [1, 5]. Binding energies are compared in the Fig. 2, and incompressibility, symmetry energy (Figs. 4 and 5) and the maximum mass of neutron stars (Fig. 9 and Tables 1, 2) show reasonable results in the level of relativistic Hartree (σ, ω) mean-field approximation.

In (Hartree) mean-field approximations, contributions of π-mesons vanish in infinite matter due to spin-saturation, and hence, σ-meson compensates for π-meson contributions in order to produce the saturation mechanism of symmetric nuclear matter. The σ-meson produces attractive interactions at low densities with the mass: $m_\sigma \approx 120.0$ MeV, which is close to the pion mass. Moreover, $m_\sigma \lesssim m_\pi$ is required to obtain solutions that are consistent with those of conserving nonchiral mean-field approximations. If one assumes $m_\sigma > m_\pi$, solutions are restricted to low densities. However, the chiral mean-field approximation is not appropriate in this case because the interaction potential V shown in Fig. 1 becomes unbound and decreases at high densities. The behaviour of σ-meson should be investigated further when higher-order interactions are included, because π-meson interactions appear explicitly from Hartree-Fock approximation and σ-meson contributions may

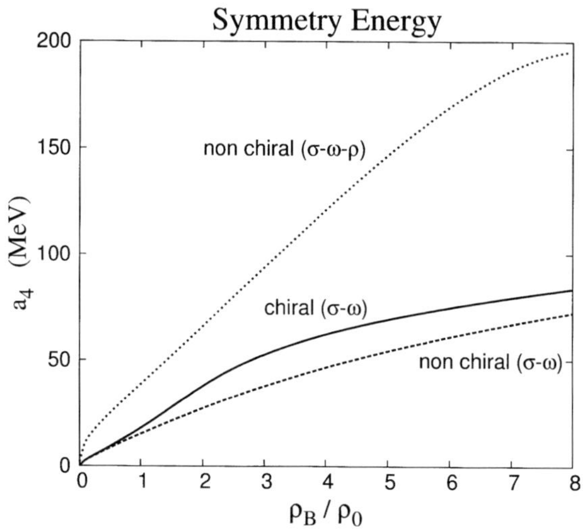

Figure 5. Symmetry energies in nonchiral (σ, ω, ρ) isospin asymmetric matter, and nonchiral, chiral (σ, ω) isospin symmetric matter. The ρ-meson contribution is more important for a_4.

be replaced by π-meson contributions. This conjecture is important to check and currently in investigation.

6. Vacuum Fluctuation Corrections and Neutron Star Properties

The full relativistic chiral Hartree approximation, including vacuum fluctuation corrections (VFC), is derived in this section and applied to the properties of neutron stars. The divergent integrals coming from the occupied negative energies (Dirac vacuum) will be rendered finite by including appropriate counterterms in the current chiral Lagrangian. By applying the method discussed in the linear σ-ω mean-field approximation [1] to the nonlinear σ-ω-ρ mean-field approximation [5], the baryon and meson propagators, self-energies are defined, and appropriate counterterms that render divergent integrals finite are introduced.

The baryon propagator in the mean-field (Hartree) approximation is assumed to be [1]:

$$\begin{aligned} G_B^H(k) &= (\gamma_\alpha k^\alpha + M_B^*)\left\{\frac{1}{k^2 - M_B^{*2} + i\epsilon} + \frac{i\pi}{E_B^*(k)}\delta(k^0 - E_B(k))\theta(k_{F_B} - |\boldsymbol{k}|)\right\} \\ &= G_B^F(k) + G_B^D(k)\,, \end{aligned} \tag{39}$$

where $G_B^F(k)$, $(B = n, p, \Lambda, \cdots)$, is the propagator for negative energy Dirac-sea and $G_B^D(k)$ is for density-dependent Fermi-sea particles. It can be readily shown that the energy density, pressure and self-energies in Sec. 2 are computed by assuming $G^H(k) = G_B^D(k)$. Hence, we recalculate (35) by including $G^F(k)$, which requires renormalization of infinities into physical parameters of the model. By employing the full propagator (39) in the chiral nonlinear σ-ω Hartree approximation, the vector meson self-energy in eq. (35) becomes

$$\begin{aligned}\Sigma^{\mu}_{\omega} &= -i\frac{g_\omega}{m^{*2}_{\omega}}\sum_B \int \frac{d^4k}{(2\pi)^4}\mathrm{Tr}\left[ig_\omega\gamma_\mu G^H_B(k)\right] \\ &= 4i\frac{g_\omega}{m^{*2}_{\omega}}\sum_B g_\omega \int \frac{d^4k}{(2\pi)^4}\frac{k^\mu}{k^2 - M^{*2} + i\epsilon} - \frac{g^2_\omega}{m^{*2}_\omega}\rho_\omega\delta_{\mu,0}.\end{aligned} \tag{40}$$

The first term of vector self-energy (40) is a divergent integral evaluated using the technique of dimensional regularization as follows,

$$\Sigma^{\mu}_{\omega} = 4i\frac{g_\omega}{m^{*2}_{\omega}}\sum_B g_\omega \int \frac{d^nk}{(2\pi)^4}\frac{k^\mu}{k^2 - M^{*2}_B + i\epsilon} - \frac{g^2_\omega}{m^{*2}_\omega}\rho_\omega\delta_{\mu,0}, \tag{41}$$

where the first term of integration is performed in n dimensions, and the final result of any calculation will be obtained by taking the physical limit $n \to 4$. The integral (41) vanishes by symmetric integration, which indicates that counterterm corrections (CTC) for the chiral mean-field (Hartree) approximation are produced only by way of ϕ fields.

The counterterms that make the scalar self-energy finite are evaluated by expanding the full propagator of G^H in a power series in the *renormalized* scalar self-energy Σ^σ. Using the Dyson equation, G^H is formally expanded as,

$$\begin{aligned}G^H(k) &= G^0(k) + G^0(k)\Sigma^s G^H(k) \\ &= \sum_{m=0}^{\infty}\left[G^0(k)\right]^{m+1}\left[\Sigma^s\right]^m.\end{aligned} \tag{42}$$

Insertion of this expression into the scalar self-energy produces,

$$\Sigma^s_H = i\frac{g_\sigma}{m^{*2}_\sigma}\sum_B\int\frac{d^nq}{(2\pi)^4}\mathrm{Tr}\left[\sum_{m=0}^{\infty}\frac{1}{m!}[\Sigma^s]^m\frac{\partial^m G^0(q)}{\partial M^m}\right] - \frac{g^2}{m^{*2}_\sigma}\rho^*_\sigma + \Sigma^s_{\mathrm{CTC}}. \tag{43}$$

It is clearly shown that the terms of $m = 0, 1, 2, 3$ in (43) have divergence when the power counting of q is performed in the physical dimension $n = 4$. These divergences can be removed by including the counterterm contribution in the Lagrangian density:

$$\mathcal{L}_{\mathrm{CTC}} = \alpha_1\phi + \frac{1}{2!}\alpha_2\phi^2 + \frac{1}{3!}\alpha_3\phi^3 + \frac{1}{4!}\alpha_4\phi^4. \tag{44}$$

The coefficients of $\alpha_1, \alpha_2, \alpha_3, \alpha_4$ are evaluated explicitly by dimensional regularization [1]. They are given by

$$\begin{aligned}\alpha_1 &= \frac{g}{4\pi^2}M^3_B\left\{\Gamma(1-n/2) + 2\ln M_B + O(n-4)\right\}, \\ \alpha_2 &= -\frac{g^2}{4\pi^2}3M^2_B\left\{\Gamma(1-n/2) + 2\ln M_B + \frac{2}{3} + O(n-4)\right\}, \\ \alpha_3 &= \frac{g^3}{4\pi^2}M_B\left\{6\Gamma(1-n/2) + 12\ln M_B + 10 + O(n-4)\right\}, \\ \alpha_4 &= -\frac{g^4}{4\pi^2}\left\{6\Gamma(1-n/2) + 12\ln M_B + 22 + O(n-4)\right\}.\end{aligned} \tag{45}$$

The Lagrangian density, $\mathcal{L}_{\rm CTC}$, is related to the self-energy $\Sigma^{s}_{\rm CTC}$ by the functional derivative as,

$$\Sigma^{s}_{\rm CTC} = -\frac{g}{m_{\sigma}^{*2}}\frac{\delta\mathcal{L}_{\rm CTC}}{\delta\phi_0}\,, \tag{46}$$

and the full self-energy is finally calculated as,

$$\begin{aligned}\Sigma^{\sigma}_{H} =& i\frac{g}{m_{\sigma}^{*2}}\sum_{B}\frac{g}{2\pi^2}\Big[M_B^{*3}\ln\frac{M_B^*}{M_B} - M_B^2(M_B^* - M_B) \\ &- \frac{5}{2}M_B(M_B^* - M_B)^2 - \frac{11}{6}(M_B^* - M_B)^3\Big] - \frac{g^2}{m_{\sigma}^{*2}}\rho_{\sigma}^{*}.\end{aligned} \tag{47}$$

The full energy density is calculated using the energy-momentum tensor and (44), (45) as,

$$\mathcal{E}_{\rm Dirac}(M_B^*) = \frac{2\pi^{n/2}}{(2\pi)^4}\Gamma(-n/2)M_B^{*n}. \tag{48}$$

The vacuum expectation value of the energy density defined in the limit $k_F \to 0$ is given by,

$$\mathcal{E}_{\rm Dirac}(M_B) = \frac{2\pi^{n/2}}{(2\pi)^4}\Gamma(-n/2)M_B^{n}. \tag{49}$$

The finite vacuum fluctuation correction to energy density is determined from (44) as $(-\langle\psi|\mathcal{L}_{\rm CTC}|\psi\rangle)$, and is calculated as follows:

$$\begin{aligned}\Delta\mathcal{E}_{\rm VFC} =& \mathcal{E}_{\rm Dirac}(M_B^*) - \mathcal{E}_{\rm Dirac}(M_B) - \alpha_1\phi - \frac{1}{2!}\alpha_2\phi^2 - \frac{1}{3!}\alpha_3\phi^3 - \frac{1}{4!}\alpha_4\phi^4 \\ =& -\frac{1}{8\pi^2}\sum_{B}\Big[M_B^{*4}\ln\left(\frac{M_B^*}{M_B}\right) + M_B^3(M_B - M_B^*) - \frac{7}{2}M_B^2(M_B - M_B^*)^2 \\ &+ \frac{13}{3}M_B(M_B - M_B^*)^3 - \frac{25}{12}(M_B - M_B^*)^4\Big].\end{aligned} \tag{50}$$

Pressure is given by $\Delta p_{\rm VFC} = -\Delta\mathcal{E}_{\rm VFC}$, which is obtained by an energy-momentum tensor as: $p = \frac{1}{3}\langle T^{ii}\rangle$, ($i$ is summed, $i = x, y, z$). The VFC gives repulsive contributions for all densities. The model parameters, m_σ, g and g_ω, must be adjusted and fixed to reproduce saturation of nuclear matter, where pressure $p = 0$ and $\mathcal{E}/\rho_B = E(k_F)$ must be satisfied.

The effective masses of baryons and mesons, including VFC, are shown in Fig. 6. At saturation density, they are $M_N^*/M_N \sim 0.74$, $m_\sigma^*/m_\sigma \sim 1.06$, $m_\omega^*/m_\omega \sim 1.03$. Meson effective masses are almost unity around saturation. The baryon effective mass increases slightly at saturation, which produces a softer EOS at high densities and decreases the masses of neutron stars. The scalar source is decreased a little by VFC, and accordingly, other fields are similarly decreased by self-consistent relations required by thermodynamic consistency. The coupling constants and effective masses of hadrons and the Fermi-liquid properties of symmetric nuclear matter including VFC are listed in Table 2.

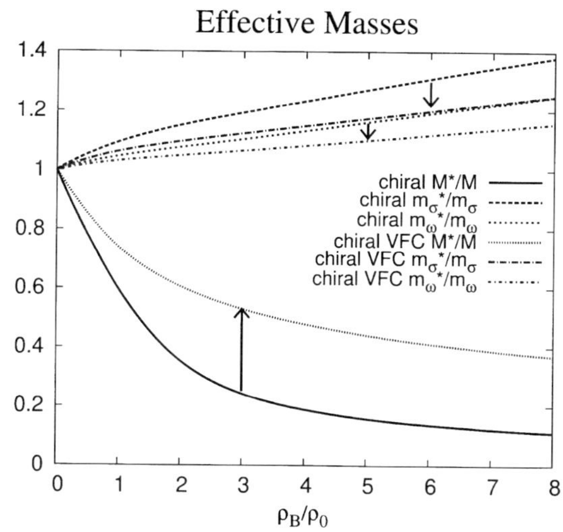

Figure 6. The vacuum fluctuation corrections to effective masses in the chiral model.

The incompressibility and symmetry energy with VFC are shown in Fig. 7 and Fig. 8, respectively. These Fermi-liquid properties are almost similar at saturation density, but incompressibility, K, is softened at high densities. This character shows that the effect of VFC is noticeable at high densities but is not so important at low densities. The symmetry energy, including VFC, gives similar results as discussed in Sec. 3. One can see from Fig. 8 that the dominant contribution to a_4 should be expected from ρ-mesons. The Fock-exchange corrections produce important contributions to a_4 and K [20]. Because the chiral symmetry breaking model confines coupling constants strictly, it is important to extend chiral models with ρ-meson to the conserving chiral Hartree-Fock and Brueckner HF approximations.

Table 2. Coupling constants and Fermi-liquid properties of nuclear matter with VFC. This result indicates that ρ-meson is necessary to obtain reasonable results for properties of Fermi-liquid and neutron stars. Analysis with the chiral $(\sigma, \pi, \omega, \rho)$ model [19] is needed to extract quantitative results. M_{max} is the maximum mass in the solar mass unit $(\mathrm{M}_{\odot})$, and $\mathcal{E}_c(10^{15}\mathrm{g/cm}^3)$ is the central energy density. I is the inertial mass $(\mathrm{M}_{\odot}\mathrm{km}^2)$ and $R(\mathrm{km})$ is the radius of a (n, p, e) asymmetric neutron star.

g	g_ω	m_σ		
1.972	10.2235	120.0		
M_N^*/M_N	m_σ^*/m_σ	m_ω^*/m_ω	K (MeV)	a_4 (MeV)
0.74	1.06	1.03	383	14.8
M_{max}	$\mathcal{E}_c$	I	R (km)	
2.19	1.88	249	11.6	

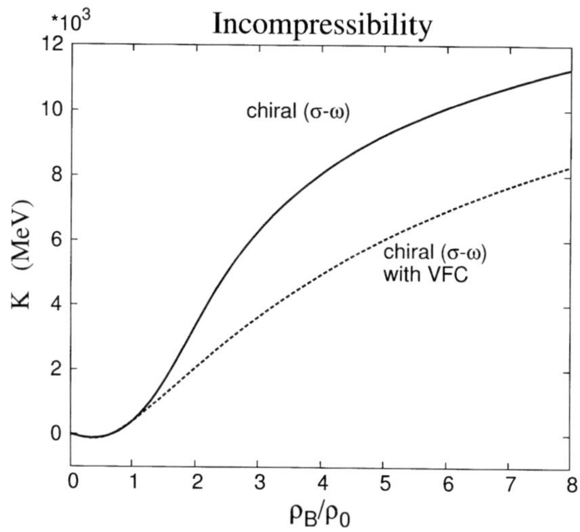

Figure 7. The vacuum fluctuation corrections to incompressibility in the chiral model.

7. Hyperon Coupling Constants and Onset Densities

As explained in chap. I that B. Day, K. Brueckner and others intensely examined binding energy and saturation density of symmetric nuclear matter, the binding energy and density at saturation of hyperon matter should be studied in terms of thermodynamic consistency, or the density functional theory. The binding energy and saturation properties of hyperon matter are important for constructing self-consistent calculations, renormalization of effective masses and coupling constants of hyperons.

The phase transition from β-equilibrium (n, p, e) to (n, p, Λ, e) or (n, p, Σ^-, e) matter is discussed in the article [6]. The hyperon-onset densities depend explicitly on nucleon-hyperon coupling ratios , $r^{\omega}_{HN} = g_{\omega H}/g_{\omega N}$ and $r^{\sigma}_{HN} = g_{\sigma H}/g_{\sigma N}$ ($H = \Lambda$ or Σ^-). They are given by

$$r^{\omega}_{HN} = \frac{m^{*2}_{\omega}}{g_{\omega N} g^*_{\omega N} \rho_{\omega}} \left(\frac{g_{\sigma H}}{g^*_{\sigma N}} (M_N - M^*_N) + \alpha_H \right) = \frac{m^{*2}_{\omega}}{g_{\omega N} g^*_{\omega N} \rho_{\omega}} \left(M_H - M^*_H + \alpha_H \right), \quad (51)$$

where $\rho_{\omega} = \rho_p + \rho_n$, and $g^*_{\omega N}$ is a density-dependent coupling constant ; α_H is the lowest binding energy of a hyperon. The coupling ratios are required to be $r^{\omega}_{\Lambda N} \gtrsim 1.0$ and $r^{\omega}_{\Sigma^- N} \gtrsim 1.0$ in the nonchiral, nonlinear (σ, ω, ρ) mean-field approximation in order to obtain optimum empirical values of symmetric nuclear matter and neutron stars. If chiral symmetry breaking is applied to phase transitions from (n, p, e) to (n, p, Λ, e) or (n, p, Σ^-, e) matter, it supports the results that coupling ratios should be $r^{\omega}_{HN} \gtrsim 1.0$, which is explained as follows. In (σ, π, ω) chiral symmetry breaking models, σ-mesons generate the mass of nucleons in the new ground state: $\sigma \to \sigma_0 + \phi$ and $\sigma_0 = -M_N/g_{\sigma N}$. Let us include baryons $(n, p, \Lambda, \Sigma^-, \cdots)$ in the Lagrangian (1). The σ-hyperon coupling constants are $g_{\sigma n}, g_{\sigma p}, g_{\sigma\Lambda}, g_{\sigma\Sigma^-}, \cdots$, respectively. Suppose that the ground state expectation value σ_0 is equipartitioned to baryons in the new ground state after chiral symmetry breaking. Then,

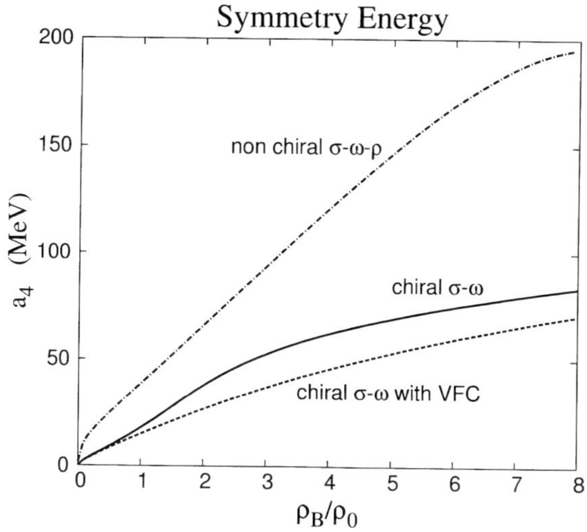

Figure 8. The vacuum fluctuation corrections to symmetry energy. The nonchiral (σ, ω, ρ) calculation is listed for comparison.

one obtains $-M_n/g_{\sigma n} = -M_p/g_{\sigma p} = -M_\Lambda/g_{\sigma\Lambda} = -M_{\Sigma^-}/g_{\sigma\Sigma^-}$, and it results in,

$$r^\sigma_{pn} = \frac{g_{\sigma p}}{g_{\sigma n}} = \frac{M_p}{M_n}, \quad r^\sigma_{\Lambda n} = \frac{g_{\sigma\Lambda}}{g_{\sigma n}} = \frac{M_\Lambda}{M_n}, \quad r^\sigma_{\Sigma^- n} = \frac{g_{\sigma\Sigma^-}}{g_{\sigma n}} = \frac{M_{\Sigma^-}}{M_n}. \tag{52}$$

The hyperon coupling ratios in the ground state of nuclear matter are $r^\sigma_H \gtrsim 1$.

These values agree with those concluded independently in the calculation of the nonchiral, nonlinear (σ, ω, ρ) conserving mean-field approximation. The hyperon coupling ratios, $r^\sigma_H \gtrsim 1$, are derived from the saturation condition of binding energy of pure-hyperon matter. Let us suppose that binding energy of pure-hyperon, for example, pure-lambda matter is self-bound as is symmetric nuclear matter. Then, one can produce saturation by computing energy density of pure-lambda matter by employing (σ, ω) mean-field approximation. However, as it is proved in the paper [6], the coupling ratios are constrained by the eq. (51). With the constraints, the coupling ratios that produce saturation of hyperon matter are shown to be $r^\sigma_H \gtrsim 1$. If the coupling ratios are $r^\sigma_H \lesssim 2/3$ or $1/3$ as suggested by the SU(6) effective quark model [17, 18], it is not possible to produce saturation of binding energy of pure-hyperon matter, which is a fundamental error as a self-consistent theory of nuclear matter [21]. This is one of conclusions from the hadronic (σ, ω, ρ) mean-field approximation [6]; the current chiral model and results (52) agree and support the conclusion of coupling ratios.

8. Applications to Neutron Stars

The masses of neutron stars are calculated using the Tollman-Oppenfeimer-Volkoff (TOV) equation [22], energy density and pressure (EOS) obtained in Sec. 4 and Sec. 6. For a given central pressure p_c, the energy density $\mathcal{E}_c$ is calculated from EOS, and then, TOV equation is used to calculate pressure $p(r)$ at $r = r_c + dr$. Again, EOS is used to calculate energy

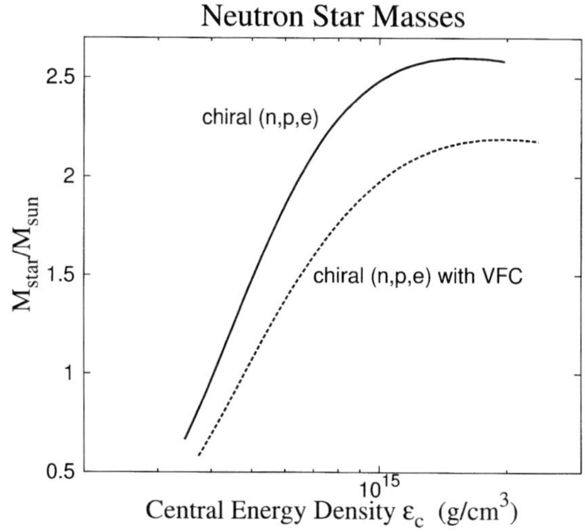

Figure 9. The masses of neutron stars in the chiral mean-field approximation, with or without VFC.

density $\mathcal{E}(r)$ until pressure becomes zero ($p(R) = 0$, R is the star radius), which means that the calculation has reached the surface of a star. The mass of a neutron star is calculated from energy density $\mathcal{E}(r)$ as,

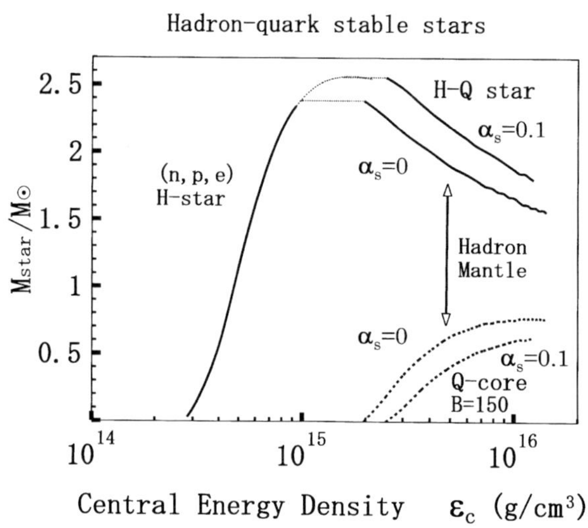

Figure 10. Hadron-quark stars based on isospin asymmetric β-equilibrium (n, p, e) matter and MIT-bag model. Note that the quark-core of H-Q stars is stable at high densities. The maximum masses of H-Q stars are $M_{max}^{H-Q} = 2.39\ \mathrm{M}_\odot$ ($\alpha_c = 0$) and $M_{max}^{H-Q} = 2.49\ \mathrm{M}_\odot$ ($\alpha_c = 0.1$).

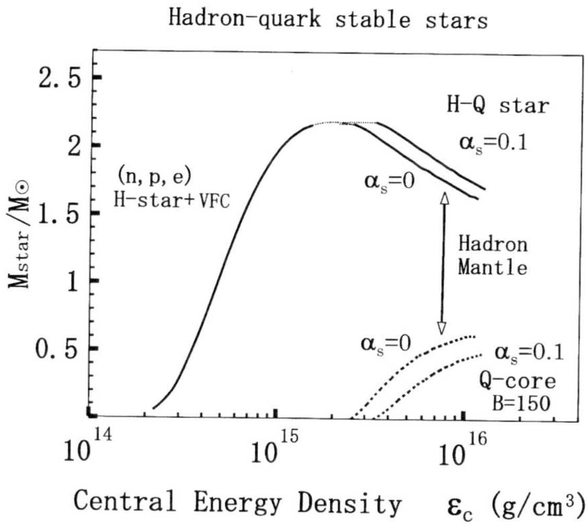

Figure 11. Hadron-quark stars based on isospin asymmetric β-equilibrium (n, p, e)+VFC and MIT-bag model. The masses of H-Q stars are small but stable at high densities. The maximum masses of H-Q stars are $M_{max}^{H-Q} = 2.18$ M$_\odot$ ($\alpha_c = 0$) and $M_{max}^{H-Q} = 2.19$ M$_\odot$ ($\alpha_c = 0.1$).

$$M(R) = 4\pi \int_0^R \mathcal{E}(r) r^2 dr \, . \tag{53}$$

The space-time geometry of general relativity and derivation of TOV equation are also discussed in detail in chap. I. The numerical results of maximum masses of neutron stars are shown in Fig. 9 as a function of a central energy density, $\mathcal{E}_c$. Vacuum fluctuation correction softens EOS and reduces the maximum mass of neutron stars by about 20 %. It should be noticed that the conserving nonlinear, nonchiral σ, ω mean-field approximation [5] produces similar results for K, a_4, $M_{\rm star}$, when $m_\sigma = 120.0$ MeV and $g_\sigma = 1.9 \sim 2.0$ are assumed. Hence, the chiral symmetry breaking mechanism provides a consistent method for understanding solutions to nonchiral, nonlinear mean-field models.

The current chiral model is applied to examine the possibility of hadron-quark 2-phase stars by way of (n, p, e) β-equilibrium matter. The β-equilibrium matter is defined by adding the electron part to the chiral lagrangian and assuming the lepton and hadron interaction only with charge neutrality [6]. The results are shown in Fig. 10 and Fig. 11, and the hadron-quark 2-phase stars are more stable than pure hadron or quark stars, which will be discussed in detail again in chapter 6.

9. Conclusions

In the current extended chiral mean-field model, all the masses of baryons and mesons are produced through spontaneous chiral symmetry breaking of nonlinear interaction potentials. Adjustable free parameters are limited to m_σ, g and g_ω after the hadron masses are identified and fixed in the nuclear domain, *e.g.* $M_N = 939.0$, $m_\pi = 139.0$ and $m_\omega = 783.0$ MeV.

Constraints on the chiral mean-field approximation are properties at saturation ($\mathcal{E}/\rho_B - M = -15.75$ MeV, at $k_F = 1.30$ fm^{-1}) and the maximum mass of isospin-asymmetric neutron stars ($M_{\max}(n, p, e) \lesssim 2.50\,\mathrm{M}_{\odot}$). The mass of σ-mesons is determined to maintain the constraints and is given by $m_\sigma \simeq 120$ MeV ($g_\sigma = 1.9 \sim 2.0$), which is also necessary so that the interaction potential V is positive and bounded at high densities, and one should note that the dimensionless parameter, $C^2 \equiv g^2(M_N^2/m_\sigma^2)$, is similar to linear and nonlinear mean-field approximations applied to finite and infinite nuclear systems [1, 5]. The chiral mean-field approximation indicates that a scalar particle close to the mass of π-mesons should be needed to produce saturation of nuclear matter.

The effective masses of nucleons and mesons, $M_N^*, m_\sigma^*, m_\omega^*$, are similar to those derived from conserving nonchiral, nonlinear (σ, ω) mean-field approximations. The effective mass of a nucleon M_N^*/M_N monotonically decreases, but the effective masses of mesons are $1.0 \lesssim m_\sigma^*/m_\sigma, m_\omega^*/m_\omega$, at or around saturation density. The vacuum fluctuation corrections exhibit repulsive effects for all densities, but after adjusting coupling constants to reproduce properties of saturation and neutron stars, the effect of VFC was less significant at saturation than that at high densities. The effect of nonlinear interactions is more important than that of VFC in the Hartree approximation. A similar conclusion is also obtained in the nonchiral, nonlinear (σ, ω, ρ) mean-field approximation. As shown in Fig. 5, ρ-mesons give noticeable contributions, so the chiral nonlinear $(\sigma, \boldsymbol{\pi}, \omega)$ mean-field approximation should be extended and examined by including ρ-mesons.

The nonchiral, nonlinear (σ, ω, ρ) mean-field approximations have many adjustable nonlinear coupling constants. The nonlinear coupling constants have upper bounds restricted by self-consistent conditions to approximations and properties of saturation and neutron stars [5], which is expected as a manifestation of *naturalness* of nonlinear coefficients [14]. The current chiral mean-field approximation determines all the nonlinear constants in terms of three adjustable parameters: m_σ, g and g_ω. The masses of mesons, m_π and m_ω, are identified and fixed by experimental values, after chiral symmetry breaking. The nonlinear constants expressed by m_σ, g and g_ω support the properties of naturalness and the bounded values of nonlinear constants given by nonchiral, nonlinear (σ, ω, ρ) mean-field approximations. Self-consistent and optimum solutions to the nonchiral, nonlinear (σ, ω) mean-field approximation with $m_\sigma = 120.0$ MeV and $g = 1.90 \sim 2.40$ become similar to those of the chiral (σ, ω) mean-field approximation, suggesting that chiral symmetry serves to restrict solutions to nonlinear mean-field approximations.

Because chiral symmetry breaking relates nonlinear coefficients to hadron masses, the chiral mean-field approximation suggests that nucleon-proton and nucleon-hyperon coupling ratios be given by ratios of hadron masses, such that $r_{\Lambda N}^\sigma = M_\Lambda/M_N \approx r_{\Lambda N}^\omega$ and $r_{\Sigma N}^\sigma = M_\Sigma/M_N \approx r_{\Sigma N}^\omega$. It is remarkable that the values of the coupling ratios are consistent with those obtained by the conditions at hyperon-onset density, which is determined by the requirement of thermodynamic consistency at the saturation of hyperon matter [6]. The coupling ratios produce reasonable density-dependent properties of nuclear matter and neutron stars in the calculation of conserving nonchiral, nonlinear (σ, ω, ρ) mean-field approximations. On the contrary, the coupling ratios given by the SU(6) quark model for vector coupling constants [17, 18] are expected to be $r_{\Lambda N}^\omega = 2/3$ and $r_{\Sigma N}^\omega = 2/3$, but the ratios do not generate consistent results for properties of nuclear and neutron matter. Chiral symmetry breaking is useful in understanding the relations among nonlinear coupling constants for nonlinear mean-field models.

The effect of VFC is not prominent at saturation density compared with that of nonlinear interactions, but VFC softens EOS at high densities, which would be softened further when hyperons are generated. This fact is also consistent with the results derived from the nonchiral, nonlinear (σ, ω, ρ) mean-field approximation [7]. Because chiral symmetry breaking clarifies relations among nonlinear interactions, it is important to understand how hyperon-onset densities, binding energy and saturation properties of hyperon matter, and masses of hadron and hadron-quark stars would be modified by chiral models of hadrons. The chiral symmetry breaking mechanism helps us understand the physical meanings of chiral symmetry in masses and coupling constants of hadrons. Quantitative analysis, in terms of chiral symmetry breaking, may help us understand if σ-meson is a real particle state or a virtual state characteristic of the Hartree (mean-field) approximation. The problems of high-energy hadron scatterings and properties of infinite matter, such as hadron-quark stars, suggest that hadronization from QCD and phase transitions from bound state hadrons to quark matter could be an important topics in the near future. However, quantitative analysis in terms of both quantumhadrodynamics (QHD) and QCD is necessary. The current self-consistent chiral mean-field model and other chiral models should be extended to be more sophisticated approximations, such as conserving HF and BHF approximations, in order to obtain quantitative results.

References

[1] B. D. Serot and J. D. Walecka, *Advances in Nuclear Physics*, edited by J. W. Negele and E. Vogt (Plenum, New York, 1986), Vol. 16.

[2] B. D. Serot and J. D. Walecka, *Int. J. Mod. Phys.*, Vol. E6, (1997) 515.

[3] B. D. Serot and H. Uechi, *Ann. Phys.* 179 (1989) 272.

[4] N. K. Glendenning, *Compact Stars*, Springer-Verlag New York, 2000.

[5] H. Uechi, *Nucl. Phys.* A780 (2006) 247; H. Uechi, *Nucl. Phys.* A 799 (2008) 181.

[6] S. T. Uechi and H. Uechi, *Adv. in High Ene. Phys.*, (2009) 640919; H. Uechi and S. T. Uechi, The Open Nucl. and Part. *Phys. J.*, V2, (2009) 47.

[7] Schun T. Uechi and Hiroshi Uechi, arXiv:1003.4815v1 [nucl-th] 25 Mar 2010.

[8] Schun T. Uechi and Hiroshi Uechi, *AIP Conference Proceedings,* Eds., W. Bentz, et.al., V1355, (2010) pp285-292.

[9] B. D. Serot, *Rep. Prog. Phys.* 55 (1992) 1855.

[10] H. Uechi, *Prog. Theor. Phys.* 111 (2004) 525.

[11] W. Kohn and L. J. Sham, *Phys. Rev.* 140 (1965) A1133.

[12] W. Kohn, *Rev. Mod. Phys.*, 71 (1999) 1253.

[13] J. D. Walecka, *Theoretical Nuclear and Subnuclear Physics*, Oxford University Press, 1995.

[14] R. J. Furnstahl, B. D. Serot, H-B. Tang, *Nucl. Phys.* A 598 (1996) 539; R. J. Furnstahl, B. D. Serot, H-B Tang, *Nucl. Phys.* A 615 (1997) 441.

[15] W.-Z. Jiang, B.-A. Li, L.-W. Chen, *Phys. Lett.* B 653 (2007) 184.

[16] B. D. Serot, *Annals of Phys.*, 322 (2007) 2811.

[17] J. Schaffner and I. N. Mishustin, *Phys. Rev.* C53 (1996) 1416.

[18] G.-yun Shao, and Y.-xin Liu, *Phys. Rev.* C79 (2009) 025804.

[19] B. D. Serot and J. D. Walecka, *Acta Phys. Pol.* B23 (1992) 655.

[20] H. Uechi, Phys. Rev. C41 (1990) 744; *Nucl. Phys.* A696 (2001) 511.

[21] B. D. Day, *Reviews of Modern Phys.* 50 (1978) 495.

[22] C. W. Misner, K. S. Thorne and J. W. Wheeler, *Gravitation*, W. H. Freeman and Company, New York, 1973.

In: Neutron Stars
Editors: Hiroshi Uechi et al, pp. 135-170
ISBN: 978-1-62081-123-8

Chapter 6

HADRON AND HADRON-QUARK COMPACT STARS

Hiroshi Uechi[1,*] ***and Schun T. Uechi***[2,†]
[1]Department of Distributions and Communication Sciences,
Osaka Gakuin University
2-36-1 Kishibe-minami, Suita, Osaka 564-8511 Japan
[2]Research Center for Nuclear Physics,
Osaka University,
Ibaraki, Osaka 567-0047 Japan.

Abstract

The conserving nonlinear, nonchiral σ-ω-ρ hadronic mean-field approximation is applied to saturation properties of symmetric nuclear and hyperonic matter, properties of hadron and hadron-quark neutron stars. Nonlinear interactions are renormalized self-consistently as effective coupling constants, effective masses, and sources of equations of motion by maintaining thermodynamic consistency to the mean-field approximation. The conditions of thermodynamic consistency to an approximation are essential for the analysis of density-dependent correlations. We found that neutron stars with hadron-quark stratum structure become more stable at high densities compared to pure hadronic and quark stars.

1. Self-Consistency of Hadronic Many-Body Theory

1.1. The Evolution of Linear σ-ω Mean-Field Approximation

Symmetric nuclear matter is self-bound, whose binding energy exhibits a characteristic concave curve at its saturation density. The binding energy at saturation is taken as -15.75 MeV at $k_F = 1.30$ fm^{-1} ($\rho_0 = 0.148$ fm^{-3}) with the symmetry energy $a_4 = 30.0$ MeV. At saturation (the minimum point binding energy), the pressure of nuclear matter vanishes ($p = 0$). It is used as a constraint in examining self-consistency to nuclear many-body approximations. This self-consistency is fundamental for many-body approximations in terms of Landau's Fermi-liquid theory, Kadanoff-Baym's theory of conserving approximations,

*E-mail address: uechi@ogu.ac.jp
†E-mail address: suechi@rcnp.osaka-u.ac.jpv

and density functional theory. These theoretical background and evolutions are explained in chap. 1. They are fundamentally important to examine consequences of space-time and gauge symmetries of lagrangians and calculations, symmetries of a class of Feynman diagrams and self-consistency in many-body theories.

The relativistic, field-theoretical, and linear σ-ω mean-field approximations that maintain the conditions of conserving approximations was applied to finite nuclei, nuclear matter, and neutron stars [1, 2]. These nuclear mean-field approximations reveal an important relation among self-consistent single particle energy, energy density, and particle density, which is denoted as *thermodynamic consistency* [3, 4]. The conserving nonlinear σ-ω-ρ mean-field approximation, which maintains thermodynamic consistency, was applied to investigate density-dependent correlations among properties of nuclear, neutron, and hyperonic matter [5–7].

The linear σ-ω model was extended to a variety of nonlinear σ, σ-ω, σ-ω-ρ mean-field models [8, 9]. The nonlinear mean-field approximations are defined by replacing all quantum fields with classical fields, and these approximations improve the results of linear σ-ω mean-field approximation. However, one may anticipate that calculations could be sensitive to which interactions and approximations one should choose, because nuclear interactions are strong and coupling constants are large. One of the important consequences from the nonlinear models suggests that nonlinear interactions behave as if they were perturbative corrections to observables when interactions are self-consistently renormalized. This was first discussed as a hypothesis of *naturalness* to nonlinear interactions, which assumes that the coefficients of all higher-order nonlinear terms in a lagrangian are of order unity in appropriate dimensionless units [10, 11], and it was explicitly shown in the analysis of self-consistent effective masses of mesons and coupling constants that all employed nonlinear corrections are small in the level of nonlinear Hartree approximations [5–7]. The self-consistent renormalization in terms of Kadanoff-Baym's theory of conserving approximations [12] as well as the density functional theory is essential to extract renormalized effective masses of mesons and coupling constants from employed nonlinear mean-field approximations.

The conserving nonlinear, nonchiral σ-ω-ρ mean-field approximation yields density-dependent effective masses and coupling constants. The density-dependence leads to significant effects on incompressibility, K, and symmetry energy, a_4. These Fermi liquid properties monotonically increase about the saturation density, but they are piecewise continuously softened at each hyperon onset density. The piecewise continuous behavior at phase transition comes from the first-order phase transitions from β-equilibrium matter, (n, p, e), to hyperonic β-equilibrium matter, $(n, p, \Lambda, ..., e)$, which is assumed to exist in neutron stars after supernova explosion [13, 14]. The energy density is discontinuous at phase transition, since the time-scale for the evolution from asymmetric (n, p, e) beta-equilibrium matter to symmetric nuclear matter is considered to be an order of millions of years. They may be connected piecewise-continuously by Maxwell construction method, maintaining conservation laws such as baryon number conservation, charge conservation, and phase-equilibrium conditions. Discontinuities of energy density and pressure are small. All physical quantities are self-consistently calculated by maintaining conservation laws, thermodynamic relations and conditions of thermodynamic consistency [15, 16]. Noticeable physical properties obtained from the current self-consistent approximation are the softening of EOS, relations

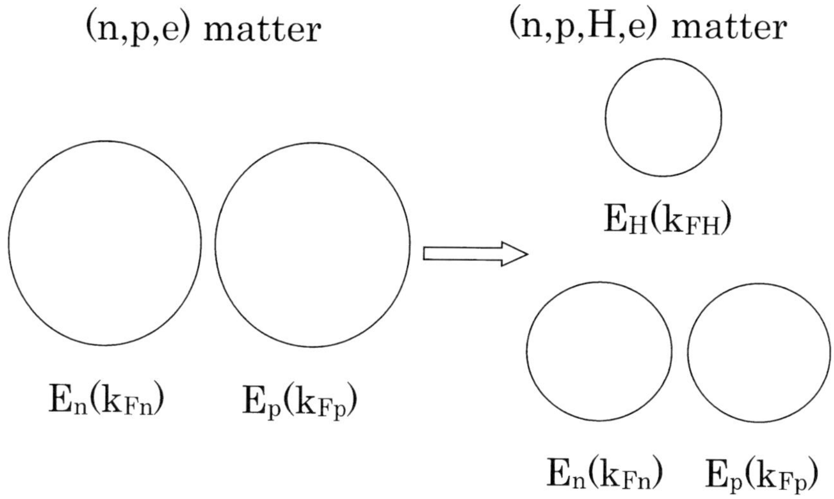

Figure 1. The schematic picture of Fermi energies in a hyperon generation from (n, p, e) to (n, p, H, e) phase transition. The nucleon Fermi energies are redistributed in the phase transition from (n, p, e) to (n, p, H, e) matter, and Fermi energies become relatively small, resulting in a softer EOS.

among effective masses, hyperon coupling constants and binding energy at each onset density. Effective masses and hyperon coupling constants are important observables to check density-dependence of K and a_4.

Hyperon-onset densities at phase transitions are given by equilibrium conditions of the chemical potentials:

$$\mu_H = \mu_n - q_H \mu_e \,, \tag{1}$$

where μ_H, μ_n, and μ_e are hyperon, neutron, and electron chemical potentials; q_H is the hyperon charge with units of e. The conditions of (1) are given by minimizing the energy density $\mathcal{E}(n, p, H, e)$, with respect to baryons and leptons being restricted by baryon-number conservation and charge neutrality. The electron and muon densities slowly increase in a low density region and decrease rapidly at high densities. The muon can be generated but restricted in a density region much narrower than that of the electron because of $\mu_{\mu^-} = \mu_{e^-}$ and $m_{\mu^-} \gg m_{e^-}$; hence, the effect of muon is smaller than that of the electron. The chemical potential μ is equal to the Fermi energy $E(k_F)$ (k_F is the Fermi momentum), which is an important property to check self-consistency to an approximation.

At a hyperon onset density from (n, p, e) to (n, p, H, e) matter, the EOS suddenly becomes softer. This is because the nucleon Fermi energies, $E_n(k_{Fn}), E_p(k_{Fp})$ in the phase (n, p, e), will be redistributed to the Fermi energies, $E_n(k_{Fn}), E_p(k_{Fp}), E_H(k_{F_H})$ in the phase (n, p, H, e). Consequently, the Fermi energies become relatively small in the phase (n, p, H, e), compared to those of (n, p, e). The redistribution and slow increase of Fermi energies occur whenever a hyperon is generated, resulting in a softer EOS. The schematic picture of redistribution of Fermi energies are shown in Fig. 1. The softer EOS and discontinuous changes in physical quantities, such as the effective masses of hadrons, incompressibility, symmetry energy and energy density, can be numerically checked [6, 7].

As the hyperon-onset will confine the Fermi-energies of baryons, as explained above, single and double hyperon generations exhibit different density-dependent phenomena. For example, the Λ-hyperon onset density, such as in (n, p, e)-(n, p, Λ, e), is $\rho_\Lambda/\rho_0 \sim 2.2$. However, it gives $\rho_\Lambda/\rho_0 \sim 4.2$ when Λ is generated in (n, p, e)-(n, p, Σ^-, e)-$(n, p, \Sigma^-, \Lambda)$. The same phenomena are observed with other hyperons, and generally the onset-density of a hyperon is pushed up to a higher density. This is denoted as the *push-up phenomena* of hyperon onset-densities in many-fold hyperon generations [6]. Because of this push-up phenomena of the hyperon onset density, we found that hyperon generations are suppressed at high densities. Hyperons relevant to determine the maximum mass of neutron stars are Σ^- and Λ. Similar results are discussed in nonrelativistic Brueckner-Hartree-Fock [17,18]. The hyperon onset densities are related to hyperon single particle energies by way of phase equilibrium conditions. Hence, the Hugenholtz-Van Hove theorem and the thermodynamic consistency at saturation of the hyperon binding energy are essential to define a self-consistent approximation of hyperonic matter.

The hyperon onset densities are related to hyperon single particle energies by way of phase equilibrium conditions. Moreover, hyperon single particle energy, E_H, energy density, $\mathcal{E}_H$, and particle density, ρ_H, must satisfy $\delta\mathcal{E}_H/\delta\rho_H = E_H$, which is called the Hugenholtz-Van Hove (HV) theorem extended to the hyperon binding energy. The HV theorem was theoretically extended by Kadanoff and Baym as the theory of conserving approximations. By employing the theory of conserving approximations, Serot and Walecka's mean-field (Hartree and relativistic Hartree) approximations were examined as thermodynamically consistent approximations [3,4]. The relation between the theory of conserving approximations and the density functional theory was discussed and thermodynamic consistency was explicitly expressed by the functional derivative of energy density with respect to particle distribution [16], which is briefly shown in the Chap. 1. The effective masses of mesons and effective coupling constants are naturally produced by the functional derivative of energy density with respect to mean fields of mesons, or alternatively, by the requirement of the density functional theory. Thermodynamic consistency is essential to obtain a self-consistent single particle energy for a hyperon.

1.2. Binding Energies of Hyperons

Density-dependent effective masses and effective coupling constants, as well as saturation properties for nucleons and hyperons, are discussed in the nonlinear σ-ω-ρ mean-field approximation [6,7]. The results show that:

(1) Coupling constants of hyperons are related to those of nucleons by effective masses, effective coupling constants, and the binding energy at the onset-density of respective hyperons. Hence, this shows that binding energies of symmetric nuclear matter and hyperon matter are self-consistently related to each other.

(2) Self-consistency suppresses hyperon generations at high densities, denoted as *the push-up phenomena* of hyperon onset densities. The suppression of hyperon generations also was discussed in different calculations [17,18].

(3) Coupling ratios of hyperons are expected as, $g_{\sigma H}/g_{\sigma N}, g_{\omega H}/g_{\omega N} \sim 1$, to be consistent with the conditions of thermodynamic consistency, as well as the empirical values of nuclear matter and neutron stars.

We have included vacuum fluctuation corrections (VFC) into the nonlinear σ-ω-ρ approximation and examined properties of (n, p, e), (n, p, H_1, e), (n, p, H_1, H_2, e) hyperonic matter, which is constrained by transitions to strange quark matter. We will show that hadron-quark (H-Q) stars are more stable than pure hadronic and hadron-quark stars.

The conserving nonlinear σ-ω-ρ mean-field approximation [5–7] and quark-based effective nuclear models [19] were applied to finite nuclei, symmetric nuclear, isospin-asymmetric, and high-density matter. We compared hyperon coupling constants required by the nonlinear σ-ω-ρ mean-field model with those required by the SU(6) quark model for vector coupling constants [20, 21]. The hyperon coupling constants required by hadronic and SU(6) quark models exhibit quite different results for effective masses, binding energies of hyperons [6, 7], and properties of hadron-quark (H-Q) stars. The coupling ratios required by the SU(6) quark model cannot reproduce hyperon saturation properties, which will be discussed in terms of the conditions of thermodynamic consistency in Sec. 4. The analysis of discrepancies in predictions using hadronic and quark-based models would provide insight into the construction of self-consistent nuclear many-body problems [2–7]. The effective masses of hyperons depend on coupling ratios ($g_{\sigma H}/g_{\sigma N} \equiv r^{\sigma}_{HN}$, $g_{\omega H}/g_{\omega N} \equiv r^{\omega}_{HN}$, $g_{\rho H}/g_{\rho N} \equiv r^{\rho}_{HN}$) and decrease analogously to the effective masses of nucleons at high densities. This shows strong density-dependency of hadronic interactions. However, the effective masses of hyperons exhibit weak density-dependent results with $r^{\omega}_{HN} \leq 2/3$ required by the SU(6) quark model for vector coupling constants. In general, hadrons exhibit strong density-dependent interactions and correlations among properties of nuclear matter, hyperonic matter, and neutron stars. In the hadronic mean-field approximation, $r^{\omega}_{HN} \sim 1.0$ is preferred to be consistent with properties of nuclear matter and the maximum mass of isospin asymmetric neutron stars, $M_{max}(n, p, e) = 2.00 \sim 2.50\ M_{\odot}$. The value $r^{\omega}_{HN} = 2/3$ results in different effective masses and binding energies of hyperons. This discrepancy originates from density-dependent interactions of hadrons [7].

The density-dependent many-body effects produced by the conserving nonlinear σ-ω-ρ mean-field approximation should be compared to chiral hadronic mean-field approximations [22–25]. The effective quark-based chiral lagrangian approach suggests that the appropriate in-medium scaling law, $m^*_{\sigma}/m_{\sigma} \approx m^*_N/m_N \approx m^*_{\omega}/m_{\omega} \approx f^*_{\pi}/f_{\pi} \approx m^*_{\rho}/m_{\rho}$, should be expected [26]. Although it is ambiguous as to whether $m^*_N/m_N \approx m^*_{\omega}/m_{\omega}$ means $m^*_{\omega}/m_{\omega} \lesssim 1.0$ or $m^*_{\omega}/m_{\omega} \gtrsim 1.0$ at saturation density, the effective masses of mesons, m^*_{σ}/m_{σ}, m^*_{ω}/m_{ω}, m^*_{ρ}/m_{ρ} are certainly decreasing above saturation density, because m^*_N/m_N is expected to be decreasing model-independently. This is significantly different from nonlinear mean-field approximations, as the self-consistency of the hadronic mean-field approximation demands $m^*_{\sigma}/m_{\sigma} \approx m^*_{\omega}/m_{\omega} \gtrsim 1.0$ [5, 27] and $m^*_{\rho}/m_{\rho} \approx m^*_{\pi}/m_{\pi} \lesssim 1.0$ [28] at high densities. In other words, because the effective masses and coupling constants are self-consistently related to each other, if some effective masses decrease, the others have to counterbalance the variations. Therefore, all correlated effective masses and coupling constants must decrease or increase simultaneously to compensate for variations among others. At saturation density, the effective masses of mesons are density-dependent and their ratios are approximately equal: $m^*_{\sigma}/m_{\sigma} \approx m^*_{\omega}/m_{\omega} \approx m^*_{\rho}/m_{\rho} \approx m^*_{\pi}/m_{\pi} \approx 1.0$. However, mesons exhibit a different behavior above saturation densities, as explained. This discrepancy was discussed in terms of the thermodynamic consistency in hadronic mean-field approximations [5–7], and should be investigated in other hadronic models to extract consistent model-independent results. The discussions so far indicate that thermodynamic consistency is fundamental for the analysis of nuclear physics. Although it is difficult to construct conserving approximations, conditions of conserving approximations must be maintained in order to obtain consistent results from models of hadrons.

1.3. Hadron and Hadron-Quark Stars

We applied the nonlinear σ-ω-ρ mean-field approximation and MIT-bag model to hadron-quark (H-Q) stars [29]. The MIT-bag model attempts to include asymptotic freedom and confinement in a very intuitive way by a constant energy density (the bag constant B) which keeps quarks and gluons confined. The ground state calculations of nuclear matter are linked to those of quark matter defined by MIT-bag model in order to obtain possible and appropriate results, constraints to hadron-quark phase transition, properties of hadron and hadron-quark stars. Our numerical calculations exhibit new results, demonstrating that hadron-(strange) quark stars can be more stable in a wide range of central energy density than pure hadronic and strange quark stars, respectively. Our calculations also suggest a correlation between the bag constant and QCD strong coupling constant, (B, α_s) [30,31], to the central density and maximum mass of hadron-(strange) quark stars, ($\mathcal{E}_c$, M_{max}). Results obtained in the current calculations should be examined using astronomical data to assess whether or not H-Q stars are possible, at the same time if values of the bag constant and strong coupling constant (B, α_s) could be compatible with astronomical data for neutron stars.

The H-Q phase transition is assumed to be the first-order and defined by Maxwell construction. As nuclear matter approximations are usually constructed at zero temperature ($T = 0$), low temperature isotherms for nuclear matter exhibit constant pressure, discontinuous phase transitions [1]. As temperature increases, corresponding isotherms become smooth and change continuously, resulting in continuous phase transition. This phenomenon comes from the finite temperature effect that changes a sharp Fermi-surface to a *smeared* Fermi-surface. The *smeared* Fermi-surface is theoretically expected in excited, high energy and density matter during the phase transition from hadron to quark matter. However, mean-field approximations of hadrons are typically constructed in the ground state at $T = 0$, and quark matter defined by MIT-bag model is also a crude realization of the confinement and ground state at $T = 0$. Hence, it is not possible to have a continuous phase transition. The piecewise continuous phase transition defined by Maxwell construction is assumed physically appropriate in a thermodynamical or phenomenological sense, because of unknown dynamics for the phase transition from hadron to quark matter.

We employed the concept of excitation of Fermi surface (an artificial excitation of Fermi surface by employing a method analogous to a finite temperature effect) to determine phase transition from hadron to quark matter. The numerical calculations exhibit that hadrons produce strong bound state (pressure, $P_H < 0$) at a high density in order to confine themselves, which becomes compatible with confinement ($P_Q < 0$) in the sense of MIT-bag model. The numerical results also indicate that the energy density and pressure during hadron-quark phase transition are used for confinement of hadrons and deconfinement of quarks when quark-gluon plasma appears, and consequently, the phase transition processes do not contribute to producing the mass of hadron-quark stars. Therefore, the simple concept of hadron-quark mixing matter [14] could not be possible, which indicates that hadron and quark phases be clearly divided, resulting in the hadron-mantle and quark-core stratum structure. The hadron-mantle and quark-core stratum structure, the method of phase transition with the artificial excitation of Fermi surface, and determination of critical values (μ_c, P_c, ρ_c) by Maxwell construction are discussed in detail in Sec. 6.

Self-consistent relations among the saturation properties of nucleons and hyperons are briefly reviewed in Sec. 2, and the relation among hyperon effective masses, coupling constants, hyperon-onset, and saturation densities are explained in Sec. 3. Thermodynamic consistency at saturation density for hyperonic matter is discussed in Sec. 4. The pure-hadron matter and vacuum fluctuation corrections, pure-quark matter defined by MIT-bag model, and the hyperon coupling ratios suggested by properties of pure-hadron and pure-quark stars are discussed in Sec. 5. The hadron-quark stars with hadron-mantle and quark-core stratum structure are discussed in Sec. 6, and conclusions are given in Sec. 7.

2. Self-Consistent Effective Masses and Coupling Constants

The nonlinear, nonchiral σ-ω-ρ hadronic lagrangian which yields density-dependent effective masses and effective coupling constants is given by [5]:

$$\begin{aligned}
\mathcal{L} = & \sum_B \bar{\psi}_B[\gamma_\mu(i\partial^\mu - g^*_{\omega B}V^\mu - \frac{g^*_{\rho B}}{2}\boldsymbol{\tau}\cdot\mathbf{R}^\mu) - (M_B - g^*_{\sigma B}\phi)]\psi_B \\
& + \frac{1}{2}(\partial_\mu\phi\partial^\mu\phi - m_\sigma^2\phi^2) - \frac{g_{\sigma 3}}{3!}\phi^3 - \frac{g_{\sigma 4}}{4!}\phi^4 \\
& - \frac{1}{4}F_{\mu\nu}F^{\mu\nu} + \frac{1}{2}m_\omega^2 V_\mu V^\mu + \frac{g_{\omega 4}}{4!}(V_\mu V^\mu)^2 + \frac{g_{\sigma\omega}}{4}\phi^2 V_\mu V^\mu \\
& - \frac{1}{4}\mathbf{L}_{\mu\nu}\cdot\mathbf{L}^{\mu\nu} + \frac{1}{2}m_\rho^2\mathbf{R}_\mu\cdot\mathbf{R}^\mu + \frac{g_{\rho 4}}{4!}(\mathbf{R}_\mu\cdot\mathbf{R}^\mu)^2 + \frac{g_{\sigma\rho}}{4}\phi^2\mathbf{R}_\mu\cdot\mathbf{R}^\mu \\
& + \frac{g_{\omega\rho}}{4}V_\mu V^\mu\mathbf{R}_\mu\cdot\mathbf{R}^\mu \\
& + \sum_l \bar{\psi}_l(i\gamma_\mu\partial^\mu - m_l)\psi_l + \delta\mathcal{L}
\end{aligned} \tag{2}$$

where ψ_B $(B = n, p, \Lambda, \Sigma, \cdots)$ and ψ_l $(l = e^-, \mu^-)$ denote the fields of baryons and leptons, respectively. The meson-fields operators are: ϕ for the σ-field, V for the vector-isoscalar ω-meson, $V_\mu V^\mu = V_0^2 - \mathbf{V}^2$, $(\mu = 0, 1, 2, 3)$, and R_μ for ρ-meson. The vector field strengths, $F_{\mu\nu}$ and $\mathbf{L}_{\mu\nu}$, are defined as $F_{\mu\nu} = \partial_\mu V_\nu - \partial_\nu V_\mu$ and $\mathbf{L}_{\mu\nu} = \partial_\mu\mathbf{R}_\nu - \partial_\nu\mathbf{R}_\mu - g_\rho\mathbf{R}_\mu\times\mathbf{R}_\nu$.

All dynamics generated by baryon fields are mediated by mean fields of mesons, which will be self-consistently defined in an employed approximation. The effective coupling constants, g^*_{iB} $(i = \sigma, \omega, \rho)$, denote renormalized, density-dependent coupling constants that are defined by the self-consistent mean-field of the σ-meson. The nonlinear σ-ω-ρ mean-field model maintains the structure of Serot and Walecka's linear σ-ω mean-field approximation [1], Lorentz-invariance and renormalizability, the Hugenholtz-Van Hove theorem [32], conditions of conserving approximations [15, 16], the virial theorem [3, 4] and Landau's hypothesis of quasiparticles [33–35]. Nonlinear mean-field approximations are equivalent to the Hartree approximation when nonlinear interactions are properly renormalized [5], and the concepts of *effective masses* and *effective coupling constants* are naturally generated by nonlinear mean-field interactions. Self-consistent relations among the single particle energy, effective masses, and coupling constants will restrict empirical values of low-density nuclear matter and high-density hadronic matter. The admissible values of effective coupling constants and masses are confined to certain values owing to strong density-dependent correlations among physical quantities of nuclear matter and neutron stars [5–7].

Meson-fields operators are replaced by mean-fields denoted by ϕ_0, V_0, and R_0. The equations of motion for baryons are given by:

$$\Big[(i\gamma_\mu\partial^\mu - g^*_{\omega B}\gamma_0 V_0 - \frac{g^*_{\rho B}}{2}\gamma_0\tau_3 R_0) - (M_B - g^*_{\sigma B}\phi_0)\Big]\psi_B = 0\,, \tag{3}$$

where $g^*_{\sigma B}$, $g^*_{\omega B}$, and $g^*_{\rho B}$ are *effective coupling constants* for σ, ω, and ρ mesons, respectively. It should be noted that effective coupling constants cannot be simply introduced as experimental, external inputs into an employed approximation, because density-dependent coupling constants will modify equations of motion and effective masses of mesons. We will assume that only nucleon-meson coupling constants are density-dependent through σN (σ-meson and nucleons) nonlinear interactions, as we are interested in the density-dependent correlations among properties of symmetric nuclear matter. The density-dependent nucleon-meson coupling constants that maintain thermodynamic consistency are:

$$\begin{aligned} g^*_{\sigma N} &= g_{\sigma N} + g_{\sigma\sigma N}\phi_0/2m_\sigma\,, \\ g^*_{\omega N} &= g_{\omega N} + g_{\sigma\omega N}\phi_0/m_\sigma\,, \\ g^*_{\rho N}/2 &= g_{\rho N}/2 + g_{\sigma\rho N}\phi_0/m_\sigma\,. \end{aligned} \tag{4}$$

The equations of motion for mesons are given by:

$$\begin{aligned} m_\sigma^2\phi_0 + \frac{g_{\sigma 3}}{2!}\phi_0^2 + \frac{g_{\sigma 4}}{3!}\phi_0^3 - \frac{g_{\sigma\omega}}{2}V_0^2\phi_0 - \frac{g_{\sigma\rho}}{2}R_0^2\phi_0 - \frac{g_{\sigma\sigma N}}{2m_\sigma}\rho_s\phi_0 & \\ = g^*_{\sigma N}\rho_s - \frac{g_{\sigma\omega N}}{m_\sigma}V_0\rho_\omega - \frac{g_{\sigma\rho N}}{m_\sigma}R_0\rho_3\,, & \\ m_\omega^2 V_0 + \frac{g_{\omega 4}}{3!}V_0^3 + \frac{g_{\sigma\omega}}{2}\phi_0^2 V_0 + \frac{g_{\omega\rho}}{2}R_0^2 V_0 = g^*_{\omega N}\rho_\omega\,, & \\ m_\rho^2 R_0 + \frac{g_{\rho 4}}{3!}R_0^3 + \frac{g_{\sigma\rho}}{2}\phi_0^2 R_0 + \frac{g_{\omega\rho}}{2}V_0^2 R_0 = \frac{g^*_{\rho N}}{2}\rho_3\,, & \end{aligned} \tag{5}$$

where $\rho_s = \sum_B \rho_{sB}$ is the total scalar source, ρ_ω is the isoscalar density, and $\rho_3 = (k^3_{F_p} - k^3_{F_n})/3\pi^2$ is the isovector density. The density-dependent coupling constants will modify the equation of motion for the σ-meson, which acquires a mass term $(\frac{g_{\sigma\sigma N}}{2m_\sigma}\rho_s\phi_0)$ and new source terms $(-g_{\sigma\omega N}V_0\rho_\omega/m_\sigma - g_{\sigma\rho N}R_0\rho_3/m_\sigma)$ from the density-dependency of effective coupling constants.

The introduction of a nonlinear $\sigma\sigma N$-vertex interaction leads to the effective mass of the nucleon:

$$M^*_N = M_N - g^*_{\sigma N}\phi_0 = M_N - g_{\sigma N}\phi_0 - (g_{\sigma\sigma N}/2m_\sigma)\phi_0^2\,. \tag{6}$$

Effective masses of nucleons and hyperons are related to each other as:

$$M_H - M_H^* = \frac{g_{\sigma H}}{g_{\sigma N}^*}(M_N - M_N^*)\,. \tag{7}$$

where nonlinear interactions in the hyperon coupling constants are not assumed for simplicity; effective masses of hyperons are defined by $M_H^* = M_H - g_{\sigma H}\phi_0$. The total scalar source is obtained by the requirement of self-consistency:

$$\Sigma^s = \Sigma_N^s + \Sigma_H^s = -\frac{g_{\sigma N}^{*2}}{m_\sigma^{*2}}(\rho_{sN}^* + \rho_{sH})\,, \tag{8}$$

where ρ_{sN}^* is the modified scalar density defined by $g_{\sigma N}^*\rho_{sN}^* = g_{\sigma N}^*\rho_{sN} - g_{\sigma\omega N}V_0\rho_B/m_\sigma - g_{\sigma\rho N}R_0\rho_3/m_\sigma$. The scalar sources are respectively given by

$$\rho_{sN} = \frac{1}{\pi^2}\int_0^{k_{F_N}} dq q^2 \frac{M_N^*}{E_N^*(q)}\,, \qquad \text{and} \qquad \rho_{sH} = \frac{g_{\sigma H}/g_{\sigma N}^*}{\pi^2}\int_0^{k_{F_H}} dq q^2 \frac{M_H^*}{E_H^*(q)}\,, \tag{9}$$

where N is used to denote the proton and neutron, $N = (p, n)$. Hyperons are denoted as $H = \Lambda, \Sigma^-, \Sigma^0, \Sigma^+, \cdots$. The ω-meson and ρ-meson contributions to self-energies are given by:

$$\Sigma_\omega^\mu = -\frac{g_{\omega N}^{*2}}{m_\omega^{*2}}\rho_\omega\delta_{\mu,0} \quad \text{and} \quad \Sigma_{\rho\binom{p}{n}}^\mu = \mp\frac{g_{\rho N}^{*2}}{4m_\rho^{*2}}\rho_3\delta_{\mu,0}\,, \tag{10}$$

where the isoscalar density, ρ_ω, is given by

$$\rho_\omega = \rho_p + \rho_n + \sum_H r_{HN}^\omega \rho_H, \tag{11}$$

and $r_{HN}^\omega = g_{\omega H}/g_{\omega N}^*$ is the density-dependent ratio of hyperon-nucleon coupling. The self-energies, $\Sigma_{\rho p}^\mu$ and $\Sigma_{\rho n}^\mu$, are briefly denoted as $\Sigma_{\rho\binom{p}{n}}^\mu$. The isovector density is denoted as $\rho_3 = (k_{F_p}^3 - k_{F_n}^3)/3\pi^2$, where the Fermi momentum k_{F_p} is for the proton and k_{F_n} is for the neutron [5, 6]. The baryon-isovector density, ρ_{3B}, and the ratios of sigma-nucleon coupling constants on the ρ-meson are also defined; for example, $\rho_{3B} = \rho_3 + r_{\Sigma N}^\rho \rho_{3\Sigma}$, where $r_{\Sigma N}^\rho = g_{\rho\Sigma}/g_{\rho N}^*$ and $\rho_{3\Sigma} = \rho_{\Sigma^+} - \rho_{\Sigma^-}$.

Thermodynamically consistent effective masses of mesons compatible with effective coupling constants (4) are required to be [5]:

$$\begin{aligned}
m_\sigma^{*2} &= m_\sigma^2\left(1 + \frac{g_{\sigma 3}}{2m_\sigma^2}\phi_0 + \frac{g_{\sigma 4}}{3!m_\sigma^2}\phi_0^2 - \frac{g_{\sigma\omega}}{2m_\sigma^2}V_0^2 - \frac{g_{\sigma\rho}}{2m_\sigma^2}R_0^2 - \frac{g_{\sigma\sigma N}}{2m_\sigma^3}\rho_{sN}\right)\,,\\
m_\omega^{*2} &= m_\omega^2\left(1 + \frac{g_{\omega 4}}{3!m_\omega^2}V_0^2 + \frac{g_{\sigma\omega}}{2m_\omega^2}\phi_0^2 + \frac{g_{\omega\rho}}{2m_\omega^2}R_0^2\right)\,,\\
m_\rho^{*2} &= m_\rho^2\left(1 + \frac{g_{\rho 4}}{3!m_\rho^2}R_0^2 + \frac{g_{\sigma\rho}}{2m_\rho^2}\phi_0^2 + \frac{g_{\omega\rho}}{2m_\rho^2}V_0^2\right)\,.
\end{aligned} \tag{12}$$

As the effective masses of mesons and coupling constants depend on the mean fields of mesons, they are density-dependent through meson fields. Note that the effective mass of the σ-meson depends on the (n, p) scalar source of nucleons, ρ_{sN}. Modifications to the equations of motion, propagators, and self-energies produced by density-dependent effective coupling constants and masses have to be carefully checked. All nonlinear correction terms with coupling constants and mean-fields in (12) to bare masses of mesons, m_σ^2, m_ω^2, m_ρ^2, are much less than unity. The nonlinear coupling constants in the nonlinear σ-ω-ρ mean-field approximation [6, 7] seem to be large at first glance, but they maintain naturalness [10]. Nonlinear corrections are less than unity in dimensionless forms of gF_i/m_i^2 or gF_i^2/m_i^2 ($F_i = \phi_0, V_0, R_0$) in (12) for the current nonlinear σ-ω-ρ mean-field approximation. It is essential that nonlinear coupling constants, effective masses, and mean-fields are thermodynamically consistent and adjusted to maintain conditions at saturation and maximum masses of neutron stars at high densities.

The energy density and pressure of isospin-symmetric, asymmetric, and charge-neutral hadronic matter are calculated by way of energy-momentum tensor as:

$$\begin{aligned}
\mathcal{E}_{NHA} &= \\
&\sum_B \frac{1}{\pi^2}\int_0^{k_{F_B}} dk k^2 E_B(k) + \frac{m_\sigma^2}{2}\phi_0^2 + \frac{g_{\sigma 3}}{3!}\phi_0^3 + \frac{g_{\sigma 4}}{4!}\phi_0^4 - \frac{m_\omega^2}{2}V_0^2 - \frac{g_{\omega 4}}{4!}V_0^4 - \frac{g_{\sigma\omega}}{4}\phi_0^2 V_0^2 \\
&- \Big(\frac{m_\rho^2}{2} + \frac{g_{\rho 4}}{4!}R_0^2 + \frac{g_{\sigma\rho}}{4}\phi_0^2 + \frac{g_{\omega\rho}}{4}V_0^2\Big)R_0^2 + \sum_{l=e^-,\mu^-}\frac{1}{\pi^2}\int_0^{k_{F_l}} dk k^2 E_l(k)\,,
\end{aligned} \tag{13}$$

$$\begin{aligned}
p_{NHA} &= \\
&\frac{1}{3\pi^2}\sum_B \int_0^{k_{F_B}} dk \frac{k^4}{E_B^*(k)} - \frac{m_\sigma^2}{2}\phi_0^2 - \frac{g_{\sigma 3}}{3!}\phi_0^3 - \frac{g_{\sigma 4}}{4!}\phi_0^4 + \frac{m_\omega^2}{2}V_0^2 + \frac{g_{\omega 4}}{4!}V_0^4 + \frac{g_{\sigma\omega}}{4}\phi_0^2 V_0^2 \\
&+ \Big(\frac{m_\rho^2}{2} + \frac{g_{\rho 4}}{4!}R_0^2 + \frac{g_{\sigma\rho}}{4}\phi_0^2 + \frac{g_{\omega\rho}}{4}V_0^2\Big)R_0^2 + \sum_{l=e^-,\mu^-}\frac{1}{3\pi^2}\int_0^{k_{F_l}} dk \frac{k^4}{E_l^*(k)}\,,
\end{aligned} \tag{14}$$

where k_{F_B} is the Fermi-momentum. $E_B(k)$ and $E_l(k)$ are single particle energies for baryons and leptons, respectively. One can check that thermodynamic relations, such as $\mathcal{E}_{NHA} + p_{NHA} = \rho_B E_n(k_{F_n})$, and the chemical potential, $\mu = \partial\mathcal{E}_{NHA}/\partial\rho_B = E_n(k_{F_n}) = E^*(k_{F_n}) - \Sigma^0(k_{F_n})$, are exactly satisfied with a given baryon density, $\rho_B = 2k_F^3/3\pi^2$.

To start self-consistent hadronic matter calculations, coupling constants, $g_{\sigma N}$, $g_{\omega N}$, $g_{\rho N}$, and another nine nonlinear coupling constants should be supplied. However, admissible values of nonlinear coupling constants are confined to satisfy properties of symmetric nuclear matter at saturation (-15.75 MeV at $k_F = 1.30$ fm^{-1}, $a_4 = 30.0$ MeV), simultaneously searching for the minimum value of incompressibility and reproducing the maximum mass of isospin-asymmetric neutron stars ($M_{\max}(n,p,e) = 2.50\ M_\odot$). In addition, hadronic phase transitions from (n,p,e) to $(n,p,H_1,H_2,\cdots,e)$ should be carefully incorporated into the EOS, which will be self-consistently determined by phase equilibrium conditions, baryon and charge conservation laws. The binding energies of (n,p) symmetric matter and β-equilibrium isospin asymmetric (n,p,e)-(n,p,Σ^-,e), as well as (n,p,e)-(n,p,Λ,e), matter are shown in Fig. 2 and Fig. 3, respectively. Because charge-dependent leptons mainly contribute at low densities, one can see that the onset of Σ^- is a little earlier than that of Λ, in spite of $M_{\Sigma^-} > M_\Lambda$.

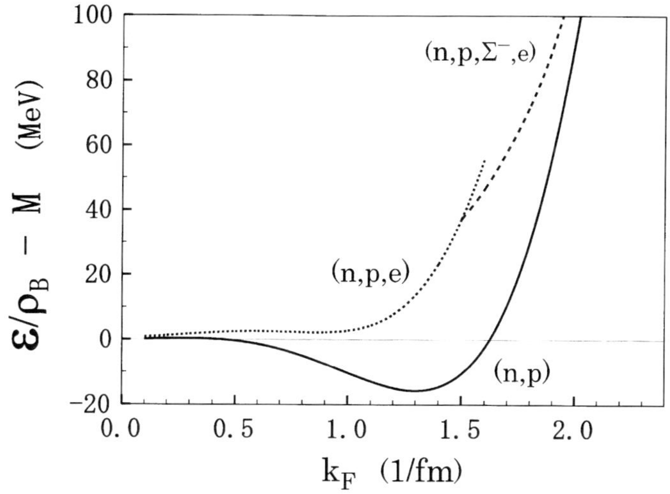

Figure 2. Binding energies of symmetric (n, p) matter (solid line) and isospin asymmetric (n, p, e)-(n, p, Σ^-, e) matter (dotted-dashed line). The onset density of Σ^- is about $k_F \sim 1.6$ fm^{-1} for the binding energy, $|\alpha_{\Sigma^-}| = 20.0$ MeV. The coupling ratios are: $r^{\sigma}_{\Sigma^- N} = 0.925$, $r^{\omega}_{\Sigma^- N} = 1.00$, and $r^{\rho}_{\Sigma^- N} = 1.00$.

The binding energies and phase transitions from (n, p, e) to (n, p, Σ^-, e) and to $(n, p, \Sigma^-, \Lambda)$ are shown in Fig. 4 and Fig. 5. The binding energies in Fig. 4 are calculated with the coupling ratios $r^{\omega}_{\Sigma N} = g_{\omega\Sigma}/g^*_{\omega N} = 1.0$ and $r^{\omega}_{\Lambda N} = g_{\omega\Lambda}/g^*_{\omega N} = 1.0$; whereas,

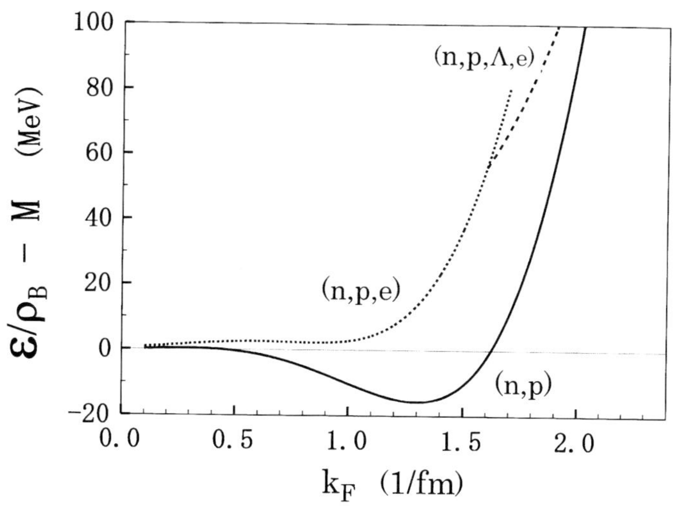

Figure 3. Binding energies of symmetric (n, p) matter (solid line) and isospin asymmetric (n, p, e)-(n, p, Λ, e) matter (dotted-dashed line). The onset density of Λ is about $k_F \sim 1.7$ fm^{-1} for the binding energy, $\alpha_\Lambda = -28.0$ MeV. The coupling ratios are: $r^{\sigma}_{\Lambda N} = 0.677$, and $r^{\omega}_{\Lambda N} = 2/3$.

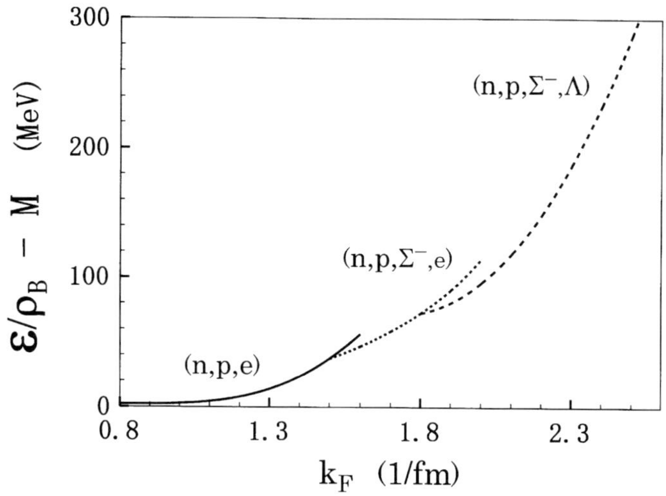

Figure 4. Binding energies of isospin asymmetric matter with phase transitions (n, p, e)-(n, p, Σ^-, e)-$(n, p, \Sigma^-, \Lambda)$ (solid-dotted-dashed line). The coupling ratios are: $(r^\sigma_{\Lambda N} = 0.964$, $r^\omega_{\Lambda N} = 1.00)$ and $(r^\sigma_{\Sigma^- N} = 0.925$, $r^\omega_{\Sigma^- N} = 1.00$, $r^\rho_{\Sigma^- N} = 1.00)$.

the binding energies in Fig. 5 are calculated with the coupling ratios $r^\omega_{\Sigma N} = 2/3$ and $r^\omega_{\Lambda N} = 2/3$, as required by the SU(6) quark model for vector coupling constants [20, 21]. By comparing binding energies in Fig. 4 and Fig. 5, it can be readily checked that binding energies with the coupling ratios, $r^\omega_{\Sigma N} = 2/3$ and $r^\omega_{\Lambda N} = 2/3$ in Fig. 5, increase slowly, which designates weak density-dependence of binding energies, resulting in soft equations

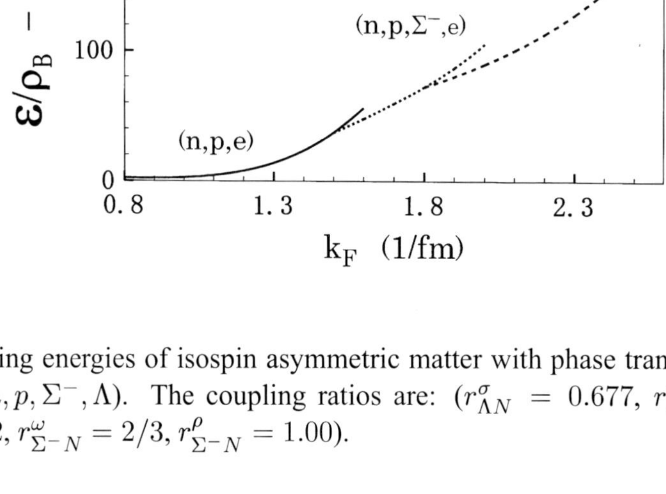

Figure 5. Binding energies of isospin asymmetric matter with phase transitions: (n, p, e)-(n, p, Σ^-, e)-$(n, p, \Sigma^-, \Lambda)$. The coupling ratios are: $(r^\sigma_{\Lambda N} = 0.677$, $r^\omega_{\Lambda N} = 2/3)$ and $(r^\sigma_{\Sigma^- N} = 0.632$, $r^\omega_{\Sigma^- N} = 2/3$, $r^\rho_{\Sigma^- N} = 1.00)$.

of state at high densities. The character of weak density-dependence with $r^{\omega}_{\Sigma N} = 2/3$ and $r^{\omega}_{\Lambda N} = 2/3$ is also shown by comparing effective masses, K and a_4 of hyperons.

The functional derivative of energy density, $\mathcal{E}_{NHA}(\phi_0, V_0, R_0, n_i)$, with respect to the baryon number distribution, n_i, is given by:

$$\frac{\delta \mathcal{E}_{NHA}}{\delta n_i} = E(k_i) + \sum_i \left(\frac{\delta \mathcal{E}_{NHA}}{\delta \phi_0} \frac{\delta \phi_0}{\delta n_i} + \frac{\delta \mathcal{E}_{NHA}}{\delta V_0} \frac{\delta V_0}{\delta n_i} + \frac{\delta \mathcal{E}_{NHA}}{\delta R_0} \frac{\delta R_0}{\delta n_i} \right). \tag{15}$$

Thermodynamic consistency is expressed as: $\frac{\delta \mathcal{E}_{NHA}}{\delta \phi_0} = 0$, $\frac{\delta \mathcal{E}_{NHA}}{\delta V_0} = 0$ and $\frac{\delta \mathcal{E}_{NHA}}{\delta R_0} = 0$ [16]. Now, by employing $\mathcal{E}_{NHA}$ in eq. (13), one can directly prove that self-energies calculated by both the method of propagators and the conditions of conserving approximations become equivalent, only if the effective masses and effective coupling constants of mesons are given by (4) and (12) [5–7]. One should note that the functional derivative, $\delta\mathcal{E}/\delta n_i$, becomes equivalent to the fundamental requirement of the density functional theory [36, 37] when conditions of thermodynamic consistency are applied to (15), and hence, one can also prove that self-consistent effective masses, effective coupling constants and self-energies should be derived as (4) $\sim$ (12) by employing the requirement of the density functional theory.

3. Effective Masses, Coupling Constants, and Hyperon Onset Densities

Quantitative calculations in papers [6, 7] show that the effective masses of hadrons and incompressibility should be $M^*_N/M_N \sim 0.70$, $m^*_\sigma/m_\sigma \sim 1.02$, $m^*_\omega/m_\omega \sim 1.01$, and $K \sim 320$ MeV at the saturation density of symmetric nuclear matter. The maximum mass of β-equilibrium (n, p, e) neutron stars should be $M_{\max}(n, p, e) = 2.50$ M$_\odot$, to appropriately reproduce the empirical values of nuclear matter and neutron stars. The effective mass of the ρ-meson is $m^*_\rho/m_\rho \sim 0.99$ at saturation, and monotonically decreases in isospin asymmetric nuclear matter.

The numerical procedure to determine nonlinear parameters is as follows. First, all nonlinear parameters must be adjusted to produce constraints at saturation (-15.75 MeV at $k_F = 1.30$ fm^{-1} and $a_4 = 30$ MeV), searching for the minimum value of incompressibility. Second, the equations of state for isospin asymmetric (n, p, e) matter at high densities must be constructed to calculate the maximum mass of neutron stars. If the EOS does not produce the required maximum mass of neutron stars, ($M_{\max}(n, p, e) = 2.50\ M_\odot$), parameters must be adjusted to produce the maximum mass. Subsequently, nuclear matter should be readdressed to adjust constraints at saturation, searching for the minimum value of incompressibility. Again, EOSs must be produced and the maximum mass of neutron stars must be calculated until both constraints are satisfied. As all physical quantities are correlated among others, a change in a nonlinear coupling constant will modify all physical quantities; the nonlinear coupling constants must be carefully adjusted to balance the variations. The upper bound of nonlinear coupling constants and the minimum value of incompressibility are determined in this iterative procedure. The current conserving nonlinear σ-ω-ρ mean-field approximation maintains conditions of thermodynamic consistency, and solutions for

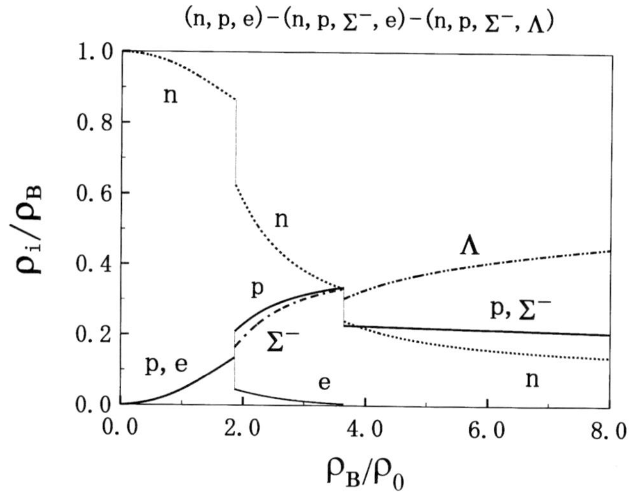

Figure 6. Ratios of particle densities: ρ_i/ρ_B ($i = n, p, \Lambda, \Sigma^-, e$) *v.s.* ρ_B/ρ_0 ($\rho_0 = 0.148$ fm^{-3}). The coupling ratios are: ($r^{\sigma}_{\Lambda N} = 0.964$, $r^{\omega}_{\Lambda N} = 1.00$) and ($r^{\sigma}_{\Sigma^- N} = 0.925$, $r^{\omega}_{\Sigma^- N} = 1.00$, $r^{\rho}_{\Sigma^- N} = 1.00$).

nonlinear coupling constants and Fermi-liquid properties of nuclear matter are listed in the Table 1.

The nonlinear coupling constants are employed to calculate EOSs and masses of neutron stars with phase transitions from (n, p, e) to (n, p, Σ^-, e) and to $(n, p, \Sigma^-, \Lambda)$, which is denoted as (n, p, e)-(n, p, Σ^-, e)-$(n, p, \Sigma^-, \Lambda)$. The ratios of particle densities, ρ_p/ρ_B, ρ_n/ρ_B, ρ_{Σ^-}/ρ_B, ρ_Λ/ρ_B, and ρ_e/ρ_B are shown in Fig. 6 and Fig. 7 for each fixed baryon

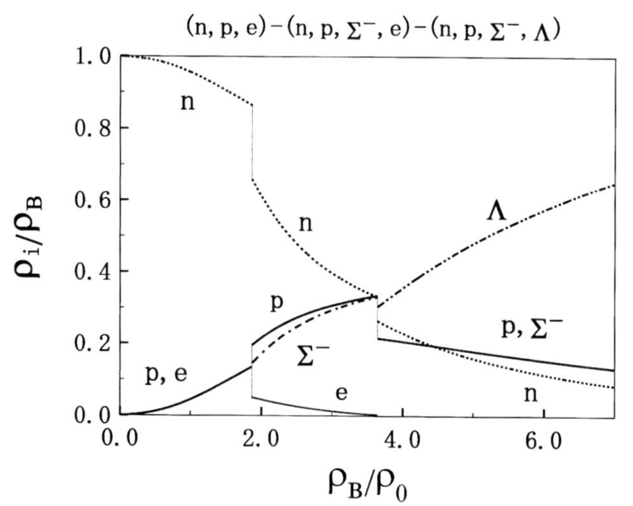

Figure 7. Ratios of particle densities. The coupling ratios are: ($r^{\sigma}_{\Lambda N} = 0.677$, $r^{\omega}_{\Lambda N} = 2/3$) and ($r^{\sigma}_{\Sigma^- N} = 0.632$, $r^{\omega}_{\Sigma^- N} = 2/3$, $r^{\rho}_{\Sigma^- N} = 1.00$).

density ρ_B. The particle densities become piecewise continuous lines in each phase, and the conservation of baryon density, charge neutrality can be checked in each phase. The densities of leptons, ρ_e and ρ_μ, contribute only at low densities and disappear rapidly at high densities.

Hyperon-onset densities at phase transitions are given by equilibrium conditions of the chemical potentials, $\mu_H = \mu_n - q_H\mu_e$, where μ_H, μ_n, and μ_e are hyperon, neutron, and electron chemical potentials; q_H is the hyperon charge with a unit of e. The conditions of (1) are given by minimizing the energy density $\mathcal{E}(n, p, H, e)$, with respect to baryons and leptons being restricted by baryon-number conservation and charge neutrality. The electron and muon densities slowly increase in a low density region. However, they decrease rapidly, disappearing at high densities, as the lepton energies are absorbed and used to produce higher energy hyperons in equilibrium infinite matter. The muon can be generated but restricted in a density region much narrower than that of the electron because of $\mu_{\mu^-} = \mu_{e^-}$ and $m_{\mu^-} \gg m_{e^-}$. Hence, the effect of muon chemical potential is smaller than that of the electron.

The hyperon coupling ratios, $r^{\sigma}_{HN} = g_{\sigma H}/g^*_{\sigma N}$ and $r^{\omega}_{HN} = g_{\omega H}/g^*_{\omega N}$, are related to each other and can be calculated in terms of effective masses, effective coupling constants, and binding energies of hyperons. For example, suppose that (n, p, H, e) hyperon matter is generated after (n, p, e) matter. Here, we denote (n, p, H, e) matter as hyperon generation on the support (n, p, e). The binding energy at the hyperon onset-density, α_H, should be the lowest energy level of the hyperon H, if the hyperon matter is self-bound. The Hugenholtz-Van Hove theorem of self-bound nuclear matter at the onset-density ($\rho_H = 0$ and $p_H = 0$)

Table 1. Coupling constants and properties of nuclear matter. The properties of symmetric nuclear matter connected to isospin asymmetric, β-equilibrium matter (n, p, e), whose EOS produces $M_{max}(n, p, e) = 2.50\ M_\odot$, are listed. The coupling constants are chosen from the data NHA$^{2.50}$ presented in the paper [16]. The effective mass of ρ-meson is $m^*_\rho/m_\rho \sim 0.99$ at $k_F = 1.30$ fm^{-1}, and monotonically decrease at high densities for (n, p, e) matter. Results with beta-equilibrium matter (n, p, e), whose EOS produces $M_{max}(n, p, e) = 2.00\ M_\odot$, are discussed in ref. [7]. Note that the values of coupling constants, ($g_{\sigma\sigma N}$, $g_{\sigma\omega N}$, and $g_{\sigma\rho N}$), correspond to $m_\sigma \times (g_{\sigma\sigma N}, g_{\sigma\omega N}$, and $g_{\sigma\rho N})$, respectively, in the paper [16].

g_σ	g_ω	g_ρ	$g_{\sigma 3}$ (MeV)	$g_{\sigma 4}$	$g_{\omega 4}$	$g_{\rho 4}$
9.326	10.421	4.765	10.0	20.0	20.0	4.00
$g_{\sigma\omega}$	$g_{\sigma\rho}$	$g_{\omega\rho}$	$g_{\sigma\sigma N}$	$g_{\sigma\omega N}$	$g_{\sigma\rho N}$	
18.0	−18.0	−18.0	−9.90	7.15	26.4	

g^*_σ	g^*_ω	g^*_ρ	M^*_N/M	m^*_σ/m_σ	m^*_ω/m_ω	K (MeV)	a_4 (MeV)
9.063	10.80	7.567	0.70	1.02	1.01	329	30.0

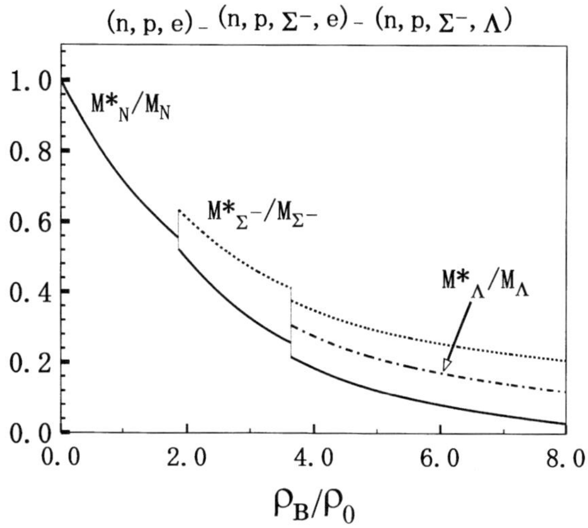

Figure 8. Effective masses of baryons. The coupling ratios are: ($r^{\sigma}_{\Lambda N} = 0.964$, $r^{\omega}_{\Lambda N} = 1.00$) and ($r^{\sigma}_{\Sigma^- N} = 0.925$, $r^{\omega}_{\Sigma^- N} = 1.00$, $r^{\rho}_{\Sigma^- N} = 1.00$).

leads to:

$$\alpha_H = ((E/\rho_B)_H - M_H)_{\rho_H=0} = E_{(k_{F_H}=0)} - \Sigma^0_{\omega H} - M_H = g_{\omega H}V_0 + M^*_H - M_H \,. \quad (16)$$

With the self-energy of ω-meson and $\Sigma^0_{\omega H} = -g^*_{\omega N}V_0$, carrying substitution of Σ^0_{ω} in (10) for $\Sigma^0_{\omega H}$ at the onset density, one can obtain:

$$r^{\omega}_{HN} = \frac{m^{*2}_{\omega}}{g_{\omega N}g^*_{\omega N}\rho_{\omega}}\left(r^{\sigma}_{HN}(M_N - M^*_N) + \alpha_H\right) = \frac{m^{*2}_{\omega}}{g_{\omega N}g^*_{\omega N}\rho_{\omega}}(M_H - M^*_H + \alpha_H)\,, \quad (17)$$

where $\rho_{\omega} = \rho_p + \rho_n$ because $\rho_H = 0$; the eq. (7) is used for the last expression of (17). Note that the particle densities behave piecewise continuously (see Fig. 6 and Fig. 7) and, hence, effective masses and coupling constants in hyperon coupling ratios, r^{ω}_{HN}, must be given at each onset density. The hyperon-onset densities are important to determine the maximum masses of neutron stars, as the generation of hyperons will soften the EOS of hyperonic nuclear matter. Thus, density-dependent interactions of hadrons are strongly correlated to properties of hadronic neutron stars.

The Λ-hyperon onset densities in (n,p,e)-(n,p,Λ,e) and (n,p,e)-(n,p,Σ^-,e)-(n,p,Σ^-,Λ) matter are different. They are, respectively, $k_{F_\Lambda} \sim 1.7$ fm^{-1} and $k_{F_\Lambda} \sim 2.0$ fm^{-1}, as shown in Fig. 4 and Fig. 5. In general, the onset densities of new hyperons will be pushed up to higher densities. This can be understood from Fermi energies and phase equilibrium conditions of baryons and leptons. This is denoted as the *push-up phenomena* of many-fold hyperon onset densities [6]. As EOSs are softened with generations of hyperons, the push-up phenomena shows that the relevant hyperonic matter to determine the maximum mass of neutron stars will be confined mainly to $(n,p,\Sigma^-,\Lambda,e^-,\mu^-)$. Quantitative results will be shown in Sec. 5. The muon can be included; however, muon effects are negligible compared to those of electrons. The nonlinear σ-ω-ρ mean-field approximation agrees with the result of BHF calculations [17, 18] and suppressions of hyperon generations in high density matter.

The effective masses of hyperons demonstrate a distinct density-dependent property with respect to the values of hyperon coupling ratios, r^{σ}_{HN} and r^{ω}_{HN} [6]. We have compared the coupling ratios required by a self-consistent model of hadrons with those required by the SU(6) quark model for the vector coupling constants [20, 21]. The SU(6) quark model requires $r^{\omega}_{\Sigma N} = 2/3$ and $r^{\omega}_{\Lambda N} = 2/3$; however, the self-consistent σ-ω-ρ mean-field model of hadrons requires $r^{\omega}_{\Sigma N} \sim 1.0$ and $r^{\omega}_{\Lambda N} \sim 1.0$ to reproduce properties of nuclear matter and neutron stars. The differences in coupling ratios have been shown to exhibit essentially different density-dependent results [6, 7]. The density-dependent effective masses of hyperons are shown in Fig. 8 ($r^{\omega}_{\Sigma N} = 1.0$ and $r^{\omega}_{\Lambda N} = 1.0$) and Fig. 9 ($r^{\omega}_{\Sigma N} = 2/3$ and $r^{\omega}_{\Lambda N} = 2/3$). The strong density-dependent behavior of baryon effective masses is model-independently known as a rapid decrease, $M^*_H/M_H \to 0$, as the baryon density ρ_B increases. By comparing Figs. 5a and 5b, the hyperon coupling constants, $r^{\omega}_{\Sigma N} = 2/3$ and $r^{\omega}_{\Lambda N} = 2/3$, clearly generate weak density-dependent relations at high densities. This is not consistent with the expected strong density-dependent relations among hadron properties at high densities [38, 39]. The density-dependent characters of effective meson masses, m^*_σ, m^*_ω, and m^*_ρ, become piecewise continuous with the onset of hyperons, but they are quantitatively similar to the results NHA$^{2.50}(n, p, e)$ in the paper [5].

The density-dependent properties of incompressibility, K, are shown in Fig. 10 and Fig. 11. The ratios in Fig. 10, $r^{\omega}_{\Sigma N} = 1.0$ and $r^{\omega}_{\Lambda N} = 1.0$, produce strong density-dependent behavior, resulting in large values of incompressibility at high densities. On the other hand, the ratios in Fig. 11, $r^{\omega}_{\Sigma N} = 2/3$ and $r^{\omega}_{\Lambda N} = 2/3$ exhibit weak density-dependence and lower values at high densities, compared with Fig. 10. The density-dependent properties of symmetry energy a_4 are shown in Fig. 12 and Fig. 13. The symmetry energy shows a monotonic increase above nuclear matter saturation density, but a sudden decrease at the onset

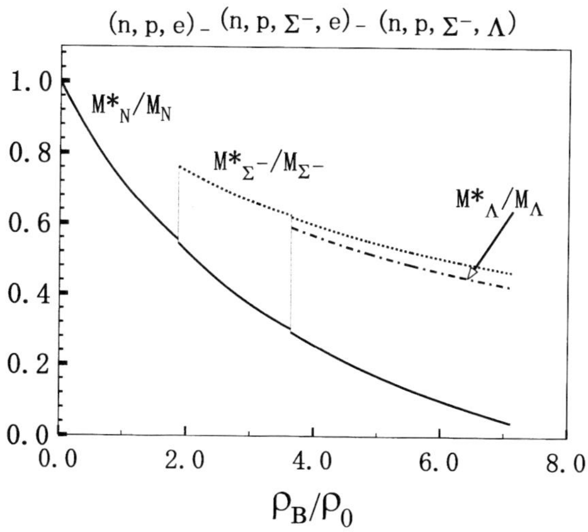

Figure 9. Effective masses of baryons. The coupling ratios are: ($r^{\sigma}_{\Lambda N} = 0.677$, $r^{\omega}_{\Lambda N} = 2/3$) and ($r^{\sigma}_{\Sigma^- N} = 0.632$, $r^{\omega}_{\Sigma^- N} = 2/3$, $r^{\rho}_{\Sigma^- N} = 1.00$).

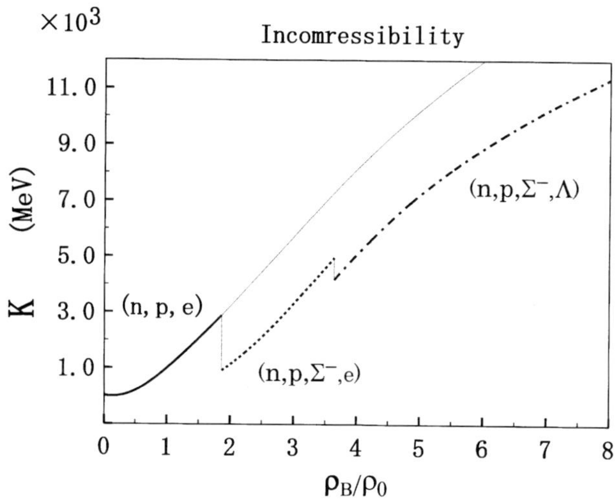

Figure 10. Incompressibility with phase transitions: (n, p, e)-(n, p, Σ^-, e)-$(n, p, \Sigma^-, \Lambda)$. The coupling ratios are: ($r^{\sigma}_{\Lambda N} = 0.964$, $r^{\omega}_{\Lambda N} = 1.00$) and ($r^{\sigma}_{\Sigma^- N} = 0.925$, $r^{\omega}_{\Sigma^- N} = 1.00$, $r^{\rho}_{\Sigma^- N} = 1.00$).

density of (Σ, Λ)-generation. It saturates at high densities [5–7]. The density-dependence of symmetry energy in other nuclear models [39, 40] which still depend on undetermined parameters is qualitatively similar to the current calculation. As incompressibility and symmetry energy depend on derivatives of energy density with respect to ρ_B or ρ_3, they produce finite discontinuous changes at onset densities of hyperons. However, we are not emphasizing the discontinuous behavior but suggesting the fact that K and a_4 become softer when

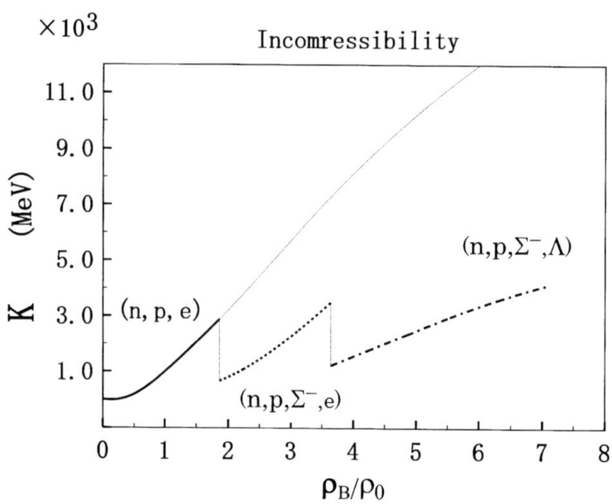

Figure 11. Incompressibility. The coupling ratios are: ($r^{\sigma}_{\Lambda N} = 0.677$, $r^{\omega}_{\Lambda N} = 2/3$) and ($r^{\sigma}_{\Sigma^- N} = 0.632$, $r^{\omega}_{\Sigma^- N} = 2/3$, $r^{\rho}_{\Sigma^- N} = 1.00$).

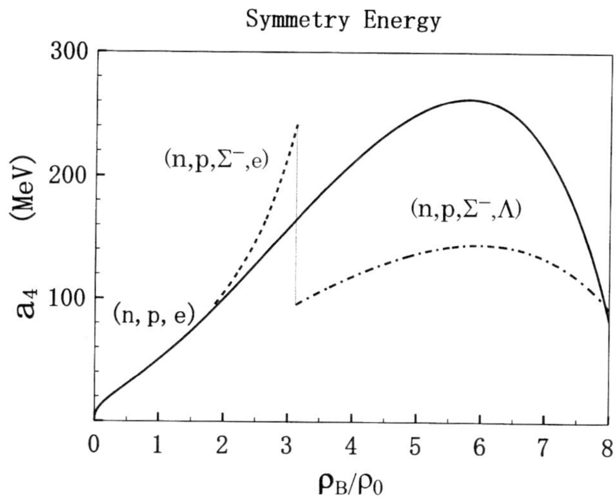

Figure 12. Symmetry energies with phase transitions: (n, p, e)-(n, p, Σ^-, e)-$(n, p, \Sigma^-, \Lambda)$. The coupling ratios are: ($r^\sigma_{\Lambda N} = 0.964$, $r^\omega_{\Lambda N} = 1.00$) and ($r^\sigma_{\Sigma^- N} = 0.925$, $r^\omega_{\Sigma^- N} = 1.00$, $r^\rho_{\Sigma^- N} = 1.00$).

phase transitions occur. The softening of K and a_4 should be explored, for example, in heavy-ion collision experiments [41].

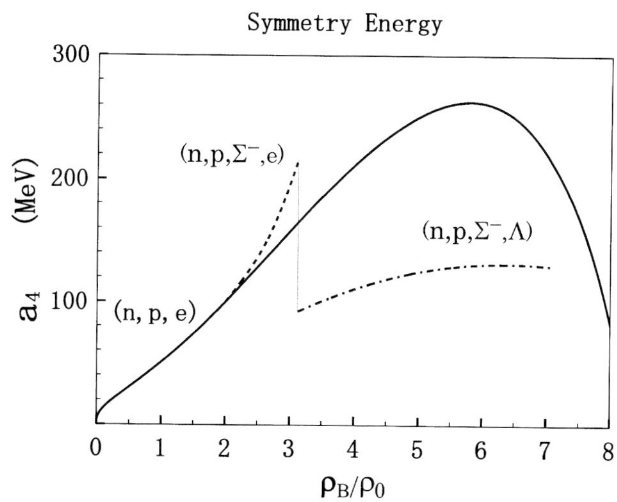

Figure 13. Symmetry energies. The coupling ratios are: ($r^\sigma_{\Lambda N} = 0.677$, $r^\omega_{\Lambda N} = 2/3$) and ($r^\sigma_{\Sigma^- N} = 0.632$, $r^\omega_{\Sigma^- N} = 2/3$, $r^\rho_{\Sigma^- N} = 1.00$).

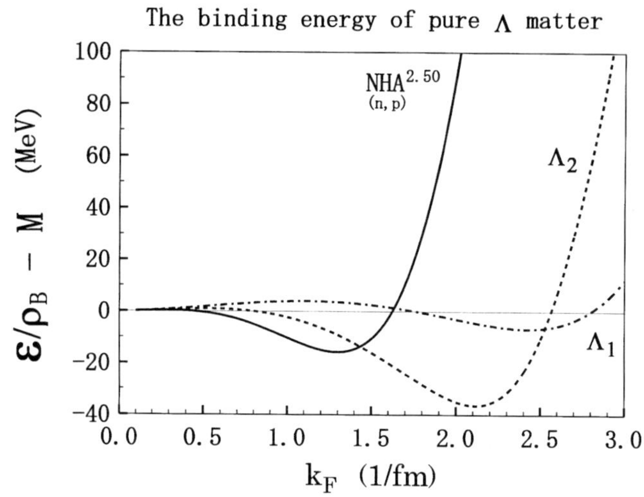

Figure 14. The binding energies of symmetric nuclear matter and pure-Λ matter. $\mathrm{NHA}^{2.50}_{(n,p)}$: -15.75 MeV, at $k_F = 1.30\ \mathrm{fm}^{-1}$. The binding energy of Λ_2 ($r^{\sigma}_{\Lambda N} = 0.964$, $r^{\omega}_{\Lambda N} = 1.00$) produces -36.37 MeV at $k_F = 2.10\ \mathrm{fm}^{-1}$. The binding energy of Λ_1 ($r^{\sigma}_{\Lambda N} = 0.677$, $r^{\omega}_{\Lambda N} = 2/3$) produces -6.69 MeV at $k_F = 2.40\ \mathrm{fm}^{-1}$.

4. The Binding Energies of Hyperonic Matter

The σ and ω hyperon coupling constants, r^{σ}_{HN} and r^{ω}_{HN}, are related to each other by eq. (17), which connects hyperon onset density and binding energy with density-dependent effective masses and coupling constants. The condition of thermodynamic consistency with constraints of nuclear and neutron matter will confine nonlinear coupling constants and induce density-dependent relations among physical quantities. Therefore, the determination of binding energy, effective masses, and coupling constants of symmetric nuclear matter will simultaneously determine binding energies, saturation properties, and effective masses of hyperons [6, 7]. Although hyperon-hyperon (YY) interactions are not readily obtained from experiments in free space, properties at saturation ($p_H = 0$) and the Hugenholtz-Van Hove (HV) theorem of hyperonic matter are essential to check self-consistency for interactions of symmetric nuclear matter, hyperonic matter, and hadronic stars.

We examine self-consistency of hyperon binding energy as follows. (1) Suppose a binding energy, α_I, as an input. For example, we tentatively choose $\alpha_I \approx \pm 20$ or ± 30 MeV (repulsive or attractive) for Λ or Σ^-. (2) With coupling ratios for hyperons which are theoretically restricted by the eq. (17) and effective masses and effective coupling constants, an approximation must generate a theoretical binding energy α_H. Therefore, if the input and output binding energies become equal: $\alpha_I = \alpha_H$, the approximation is considered to be self-consistent.

Historically, self-consistency was investigated in symmetric nuclear matter [42, 43]; K. A. Brueckner found necessity of *rearrangement diagrams* and J. D. Walecka introduced σ-ω mean-field approximation [1] in order to resolve self-consistency at saturation. Similarly, self-consistency must be examined in hyperonic matter as discussed in symmetric nuclear matter. We employed conditions of thermodynamic consistency at saturation of hyperon binding energy (16) and derived the relation for coupling constants: (17). The hyperon coupling constant r^{ω}_{HN} is a function of r^{σ}_{HN} given explicitly in (17), which is different from (g_σ, g_ω) free-parameter approach in the σ-ω mean-field approximation. We applied thermodynamic consistency for hyperon binding energy in order to examine consistent hyperon coupling constants in case of $r^{\omega}_{\Lambda N} = 1.0$ and $r^{\omega}_{\Lambda N} = 2/3$.

The binding energies of pure Λ matter are shown in Fig. 14 (the assumed input binding energy: $\alpha_{I,\Lambda} = -30$ MeV), and are calculated with coupling constants, $r^{\omega}_{\Lambda N} = 1.0$ and $r^{\omega}_{\Lambda N} = 2/3$. The pure Λ-matter with coupling ratios ($r^{\sigma}_{\Lambda N} = 0.677, r^{\omega}_{\Lambda N} = 2/3$) is shown with a dash-dotted line and is denoted as Λ_1. The pure Λ-matter with coupling ratios ($r^{\sigma}_{\Lambda N} = 0.987, r^{\omega}_{\Lambda N} = 1.00$) is shown with a dotted line, Λ_2. The coupling constant $r^{\omega}_{\Lambda N} = 2/3$, which is required by the SU(6) quark model [20, 21], generates $\alpha_\Lambda \simeq -6.70$ MeV at $k_F \simeq 2.4$ fm^{-1}. It is not possible to produce the assumed input binding energy $\alpha_{I,\Lambda} = -30$ MeV. On the contrary, the coupling ratio, $r^{\omega}_{\Lambda N} = 1.0$, can be compatible with the self-bound binding energy of the Λ-hyperon. The binding energy saturates at $k_F \simeq 2.1$ fm^{-1} and $\alpha_\Lambda \simeq -36.3$ MeV, and the binding energy at onset density, $k_F \simeq 1.7$ fm^{-1}, is $\alpha_\Lambda \simeq -26.1$ MeV. This is close to the assumed input value of $\alpha_{I,\Lambda} = -30$ MeV. Hence, the hyperon binding energy can be self-consistent when coupling ratios are $r^{\sigma}_{\Lambda N} \simeq 1.0$ and $r^{\omega}_{\Lambda N} \simeq 1.0$.

The binding energies of pure Σ^- matter are shown in Fig. 15. The binding energy of Σ^- is tentatively assumed to be $\alpha_{\Sigma^-} = -20$ MeV. We have checked the numerical calculation with $|\alpha_{\Sigma^-}| \leq 20$, and the results change little for the current numerical analysis of neutron stars. The Σ^- binding energy produced with $r^{\sigma}_{\Sigma^- N} = 0.632$ and $r^{\omega}_{\Sigma^- N} = 2/3$ is denoted as Σ_1 (dash-dotted line), which is repulsive at all densities. This is theoretically inconsistent

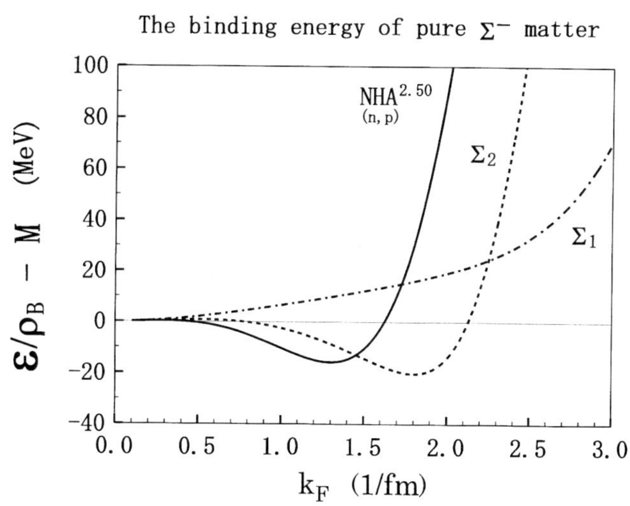

Figure 15. The binding energies of pure-Σ^- matter. Note that the binding energy of Σ_2 ($r^{\sigma}_{\Sigma^- N} = 1.210$, $r^{\omega}_{\Sigma^- N} = 1.323$) produces -20.25 MeV at $k_F = 1.80$ fm^{-1}, which is consistent with the assumed onset density and binding energy. The binding energy of Σ_1 ($r^{\sigma}_{\Sigma^- N} = 0.632$, $r^{\omega}_{\Sigma^- N} = 2/3$) produces no saturation.

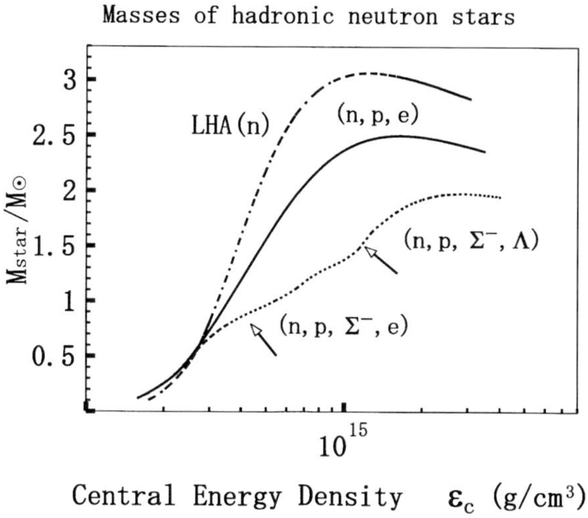

Figure 16. Masses of hadronic neutron stars. Pure-neutron matter in linear σ-ω: $M(n)_{max} = 3.06\ M_{\odot}$. Isospin asymmetric β-equilibrium matter: $M(n,p,e)_{max} = 2.50\ M_{\odot}$. (n,p,e)-(n,p,Σ^-,e)-(n,p,Σ^-,Λ) matter with the coupling ratios: ($r^{\sigma}_{\Lambda N} = 0.964$, $r^{\omega}_{\Lambda N} = 1.00$) and ($r^{\sigma}_{\Sigma^- N} = 0.925$, $r^{\omega}_{\Sigma^- N} = 1.00$, $r^{\rho}_{\Sigma^- N} = 1.00$) give $M_{max} = 1.98\ M_{\odot}$.

as the current theoretical input of binding energy is attractive: $\alpha_{\Sigma^-} = -20$ MeV. The binding energy of Σ^- consistent with the assumed value, $\alpha_{\Sigma^-} = -20$ MeV, is denoted as Σ_2 ($r^{\sigma}_{\Sigma^- N} = 1.210$ and $r^{\omega}_{\Sigma^- N} = 1.323$, dash-dotted line). Although the nonlinear σ-ω-ρ mean-field (Hartree) approximation should be improved by including higher-order corrections, the numerical results suggest that hyperon coupling ratios, $r^{\omega}_{\Lambda N} \simeq 1.0$, should be consistent with saturation and thermodynamic consistency for hyperons.

Hyperon coupling constants required by the hadronic mean-field model and SU(6) quark model for the vector coupling constants exhibit essentially different results on the problem of density-dependent interactions (*i.e.*, effective masses of hyperons, binding energies, and saturation of hyperonic matter). This may indicate another important discrepancy between the hadronic model and effective quark model for hadrons. This should be investigated further in this model and other many-body approximations for nuclear physics. The different density-dependent properties will also be shown in the numerical analysis for hadron and H-Q stars.

5. The Hyperon Coupling Ratios and Properties of Pure-Hadronic and Pure-Quark Stars

The vacuum fluctuation correction (VFC) to the conserving nonlinear mean-field approximation is discussed in chap. 4, and in this section, the nonlinear σ-ω-ρ mean-field approximation with a VFC is applied to properties of hadronic neutron stars and pure-quark stars.

The properties of pure hadronic stars are produced by employing hadronic equations of state, phase transition conditions, coupling ratios (17), and the TOV equation. The masses v.s. central densities of hadronic stars produced by EOSs for (n,p,e)-(n,p,Σ^-,e)-(n,p,Σ^-,Λ) matter are shown in Figs. 16 ($r^{\omega}_{\Lambda N} = 1.0, r^{\omega}_{\Sigma N} = 1.0$) and 17 ($r^{\omega}_{\Lambda N} = 2/3, r^{\omega}_{\Sigma N} = 2/3$). The solid line shows (n,p,e) matter which produces the maximum mass of hadronic neutron stars, $M_{max} = 2.50$ M$_\odot$ [6]. LHA(n) (dot-dashed line) is the result of the linear σ-ω mean-field approximation [1] ($M_{max} = 3.06$ M$_\odot$). The arrows respectively indicate phase transitions from (n,p,e)-(n,p,Σ^-,e) and (n,p,Σ^-,e)-(n,p,Σ^-,Λ) matter. The EOS of hadronic matter with $r^{\omega}_{\Lambda N} = 1.0, r^{\omega}_{\Sigma N} = 1.0$ generates the maximum mass $M_{max}(n,p,\Sigma^-,\Lambda) = 1.98$ M$_\odot$ (Fig. 16). However, the EOS with $r^{\omega}_{\Lambda N} = 2/3, r^{\omega}_{\Sigma N} = 2/3$ as suggested by the SU(6) quark model, produces the small maximum mass of neutron stars, $M_{max}(n,p,\Sigma^-,\Lambda) = 1.33$ M$_\odot$ (Fig. 17). This is unable to support the observed masses of neutron stars. The discrepancy between $r^{\omega}_{HN} = 1.0$ and $r^{\omega}_{HN} = 2/3$ calculations is clearly recognized by comparing EOSs as well as masses of neutron stars.

All curves of $\mathcal{E}_c - M_{\rm star}$ in Fig. 16 and 17 have simple positive slopes, $dM/d\mathcal{E}_c > 0$, and one maximum point that is typical for lines of one critical (inflection) point, $d^2M/d\mathcal{E}_c^2 < 0$. In this case, stability can be examined clearly for these simple saturating curves by the condition $dM_{star}/d\mathcal{E}_c > 0$ [14, 44], respectively. Hence, positive slopes of $\mathcal{E}_c - M_{\rm star}$ curves are solutions of stable neutron stars, but the decreasing curve, $dM_{star}/d\mathcal{E}_c < 0$, after the maximum point indicates unstable neutron stars. These characters of $\mathcal{E}_c - M_{\rm star}$ are important to examine stability of hadron-quark stars. The properties of pure hadronic stars including a VFC are shown in Figs. 18 and 19. The VFC effects are not very significant both at low and high densities, compared with those of nonlinear many-body interactions. The pressure difference at the phase transition region becomes small, making the EOS smooth (Figs.. 4 and 5). Note that the stable hadronic stars with $M_{star} \geq 1.0$ M$_\odot$ are limited in the central energy density range: $5.0 \times 10^{14} \lesssim \mathcal{E}_c \lesssim 10^{15}$ g/cm^3, as shown

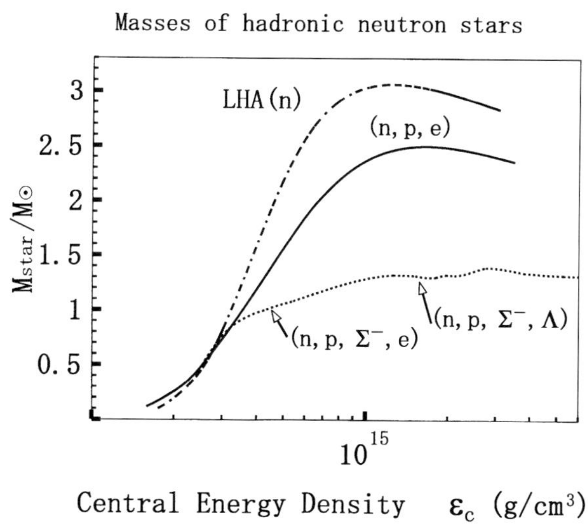

Figure 17. Masses of hadronic neutron stars. (n,p,e)-(n,p,Σ^-,e)-(n,p,Σ^-,Λ) matter with the coupling ratios: ($r^{\sigma}_{\Lambda N} = 0.677$, $r^{\omega}_{\Lambda N} = 2/3$) and ($r^{\sigma}_{\Sigma^- N} = 0.632$, $r^{\omega}_{\Sigma^- N} = 2/3$, $r^{\rho}_{\Sigma^- N} = 1.00$) give $M_{max} = 1.40\ M_\odot$.

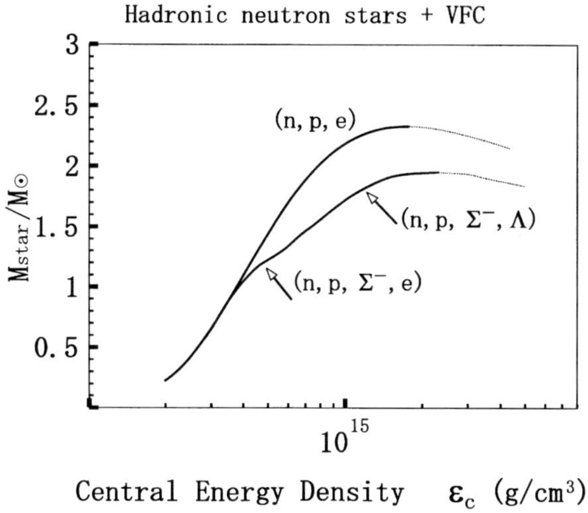

Figure 18. Hadronic neutron stars with the VFC. (n, p, e)-(n, p, Σ^-, e)-$(n, p, \Sigma^-, \Lambda)$ matter with the coupling ratios: ($r^{\sigma}_{\Lambda N} = 0.964$, $r^{\omega}_{\Lambda N} = 1.00$) and ($r^{\sigma}_{\Sigma^- N} = 0.925$, $r^{\omega}_{\Sigma^- N} = 1.00$, $r^{\rho}_{\Sigma^- N} = 1.00$) give $M_{max} = 1.95\ M_{\odot}$.

in bold, solid lines. The maximum masses are, respectively, $M_{max}(n, p, e) = 2.33$ M$_{\odot}$, and $M_{max}(n, p, \Sigma^-, \Lambda) = 1.95$ M$_{\odot}$ for the coupling ratios, $r^{\omega}_{\Lambda N} = 1.0$ and $r^{\omega}_{\Sigma N} = 1.0$. The maximum mass produced by SU(6) coupling ratios, $r^{\omega}_{\Lambda N} = 2/3$ and $r^{\omega}_{\Sigma N} = 2/3$, is $M_{max}(n, p, \Sigma^-, \Lambda) = 1.50$ M$_{\odot}$.

The energy density and pressure of the MIT-bag model (at $T = 0$) are derived as [29–

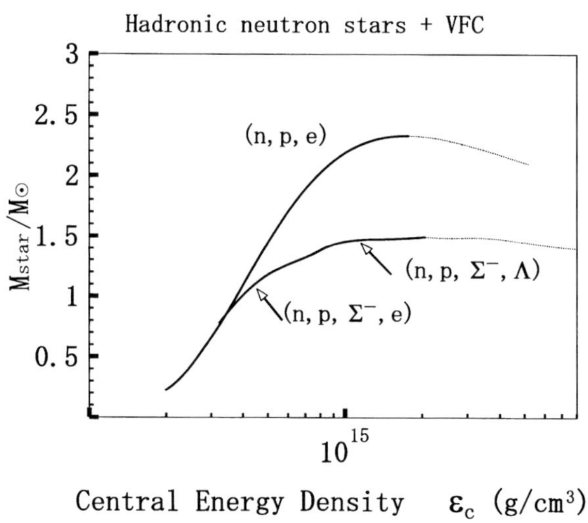

Figure 19. Hadronic neutron stars with the VFC. (n, p, e)-(n, p, Σ^-, e)-$(n, p, \Sigma^-, \Lambda)$ matter with the coupling ratios: ($r^{\sigma}_{\Lambda N} = 0.677$, $r^{\omega}_{\Lambda N} = 2/3$) and ($r^{\sigma}_{\Sigma^- N} = 0.632$, $r^{\omega}_{\Sigma^- N} = 2/3$, $r^{\rho}_{\Sigma^- N} = 1.00$) give $M_{max} = 1.50\ M_{\odot}$.

31]:

$$\mathcal{E}_q = \frac{3}{8\pi^2}\sum_f \Big[2k_f E_f^3(k_f) - m_f^2 k_f E_f(k_f) - m_f^4 \log|\frac{k_f + E_f(k_f)}{m_f}|\Big] + B\,, \quad (18)$$

$$P_q = \frac{1}{8\pi^2}\sum_f \Big[(2k_f^3 - 3k_f m_f^2)E_f(k_f) + 3m_f^4 \log|\frac{k_f + E_f(k_f)}{m_f}|\Big] - B\,, \quad (19)$$

where $E_f(k_f) = (k_f^2 + m_f^2)^{1/2}$, $(f = u, d, s)$, and B is the bag-constant (MeV/fm^3); k_f is the Fermi-momentum of flavor f. The bag model is a very simplified model as recognized from eq. (18) and (19). In terms of energy density, the bag constant means energy density to create the bag in which quarks are confined, whereas at low densities, pressure becomes negative, $P_q \approx -B < 0$ (for densities less than $\rho_B/\rho_0 \sim 4.0$, when $B = 150$ MeV/fm^3), which designates confinement of quarks. The bound state of quarks inside the bag (pressure $P_q < 0$) is important when hadron-quark phase transition is assumed, because if one assumes continuous phase transition from hadron matter to quark matter, it is imperative to produce the bound state of hadrons (negative pressure, $P < 0$) at a high energy density so that hadrons must be compressed to form the quark bag.

As the pressure of the ground state of hadronic matter increases monotonically ($P > 0$), the discontinuous phase transition from hadron to quark matter can be consistently obtained from Maxwell construction method. However, if hadron-quark phase transition can occur in neutron stars, continuous phase transition may be expected, and hence, it is often discussed that Maxwell construction method is inconsistent and not applicable in phase transition in neutron stars [14]. We will show that Maxwell construction method is consistently applicable to continuous phase transition by employing the idea of Fermi-liquid theory in nonequilibrium system. In this approach, all conservation laws and thermodynamic relations are naturally preserved; it is not necessary to impose an artificial volume ratio on hadrons and quarks to define the sum of averaged energy densities and particle densities of hadrons and quarks [14].

The energy density and pressure with the following given baryon density, charge neutrality, and phase equilibrium conditions:

$$\rho_B = \frac{2k_F^3}{3\pi^2} = \frac{1}{3}(\rho_u + \rho_d + \rho_s)\,, \quad (20)$$

$$0 = 2\rho_u - \rho_d - \rho_s\,, \quad (21)$$

$$E_d(k_d) = E_s(k_s)\,, \quad (22)$$

determine (u, d, s)-quark matter uniquely (note: $\rho_f = k_f^3/\pi^2$). The thermodynamic potential for the flavor f, to the first-order in the strong coupling constant, is given by [14,30,31]:

$$\begin{aligned}\Omega_f = &-\frac{\gamma_f}{24\pi^2}\Big\{k_f E_f(k_f)(k_f^2 - \frac{3}{2}m_f^2) + \frac{3}{2}m_f^4 \log|\frac{k_f + E_f(k_f)}{m_f}| \\ &- \frac{2\alpha_s}{\pi}\Big[3\Big(E_f(k_f)k_f - m_f^2 \log|\frac{k_f + E_f(k_f)}{E_f(k_f)}|\Big)^2 - 2k_f^4 - 3m_f^4 \log^2\Big(\frac{m_f}{E_f(k_f)}\Big) \quad (23) \\ &+ 6\log\Big(\frac{\sigma}{E_f(k_f)}\Big)\Big(k_f E_f(k_f) m_f^2 - m_f^4 \log|\frac{k_f + E_f(k_f)}{m_f}|\Big)\Big]\Big\}\,,\end{aligned}$$

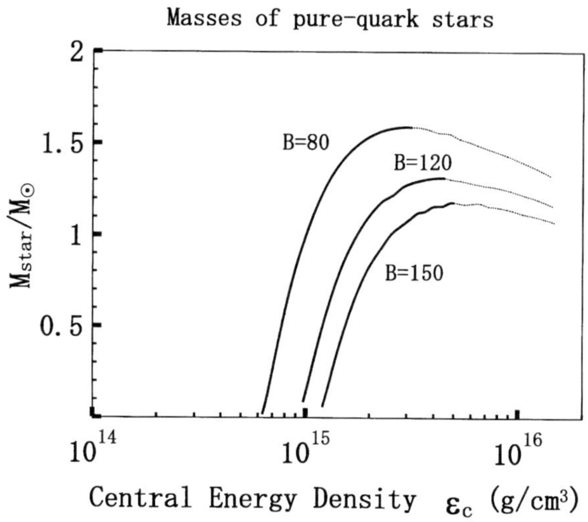

Figure 20. Masses of pure-quark stars with $B = 80, 120, 150$ MeV/fm^3, ($\alpha_s = 0$). The stable pure-quark stars with $M_{max} \geq 1.0$ are in $10^{15} \sim \mathcal{E}_c \sim 5.0 \times 10^{15}$ g/cm^3.

where $\gamma_f = spin \times color$ degeneracy. α_s is the strong interaction coupling constant and σ is the renormalization scale constant chosen as ~ 300 MeV [14], which is considered as a typical chemical potential of the problem. However, the EOS of hadron-quark compact stars is not sensitive to the parameter σ, which is numerically checked and theoretically expected from (23) in the high density region. It is suggested that the EOS of quark matter interconnected with hadronic matter is mainly sensitive to parameters (B, α_s) in the current analysis. The expected value and restriction to (B, α_s) can be numerically extracted from properties of pure-hadron and hadron-quark stars in Sec. 6.

The energy density and pressure are now expressed as:

$$\mathcal{E}_q = B + \sum_f (\Omega_f + \mu_f n_f) \, ,$$
$$P_q = -B - \sum_f \Omega_f \, , \tag{24}$$

where $\mu_f = E_f(k_f)$ and n_f are the chemical potential and particle distribution, respectively, for the flavor f. The EOSs for hadronic and quark matter, phase transition conditions, and TOV equation [45] are employed to calculate pure-hadron, pure-quark, and hadron-quark compact stars.

The EOS of pure-quark matter is used to calculate masses of pure-quark stars, as shown in Fig. 20. The bag constant should be $B \gtrsim 80$ MeV/fm^3, as normal nuclear matter becomes quark matter in the case of $B \simeq 80$ MeV/fm^3. The maximum masses of quark stars are $M_{max} = 1.59$ M$_\odot$ ($B = 80$ MeV/fm^3), $M_{max} = 1.31$ M$_\odot$ ($B = 120$ MeV/fm^3) and $M_{max} = 1.18$ M$_\odot$ ($B = 150$ MeV/fm^3). The $\mathcal{E}_c - M_{\rm star}$ curves in Fig. 20 and 21 for the current pure hadron and quark stars are smoothly increasing and saturating. These simple curves are classified as those of one critical (inflection) point. Therefore, stability of neutron stars in pure-hadron and pure-quark matter can be checked respectively by the condition:

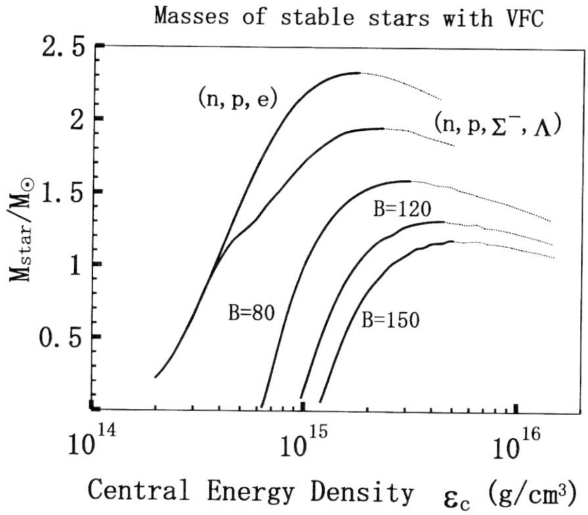

Figure 21. Hadron stars with the VFC and pure-quark stars. The stable hadronic stars are in $5.0 \times 10^{14} \sim \mathcal{E}_c \sim 10^{15}$ g/cm^3.

$dM_{\rm star}/d\mathcal{E}_c > 0$, though one needs to check other conditions if $\mathcal{E}_c - M_{\rm star}$ curves exhibit complicated phase transition and instability, which results in curves with critical points larger than 2 [44].

Stable quark stars with $M_{star} \geq 1.0$ M$_\odot$ are limited in the range of central energy densities: $10^{15} \lesssim \mathcal{E}_c \lesssim 5.0 \times 10^{15}$ g/cm^3, as shown in bold, solid lines in Fig. 20. Stable hadronic neutron stars with the VFC and pure quark stars are shown in Fig. 21, to remark on the energy-density regions of stable stars. The results of hadronic and quark stars in Fig. 21 suggest that H-Q stars are possible about the central energy density, $\mathcal{E}_c \sim 10^{15}$ g/cm^3, as the EOS of quark matter could be energetically preferable at high densities, compared to that of hadrons. However, the overlap in central energy density between stable hadronic stars ($M^H_{max} \geq 1.0$) and pure-quark stars ($M^Q_{max} \geq 1.0$) is very narrow. If stable central energy densities are the same during the H-Q phase transition, this indicates that H-Q stars would become immediately unstable [29].

6. The Properties of Hadron-Quark (H-Q) Stars

Many-component hyperonic matter $(n, p, H_1, H_2, \cdots, e^-, \mu^-)$ is equivalent to one-component nuclear matter, which is proved by employing baryon conservation, charge neutrality and phase equilibrium conditions. The fact is important to understand that thermodynamic treatment of phase transition in terms of Maxwell construction can be consistent with piecewise continuous phase transitions in hadron and hadron-quark matter.

Suppose that many-component hyperonic matter $(n, p, \Lambda, \Sigma^-, e^-, \mu^-)$ is in equilibrium, so that pressure P and energy density $\mathcal{E}$ satisfy in general:

$$P + \mathcal{E} = \mu_p \rho_p + \mu_n \rho_n + \mu_\Lambda \rho_\Lambda + \mu_{\Sigma^-} \rho_{\Sigma^-} + \mu_{e^-} \rho_{e^-} + \mu_{\mu^-} \rho_{\mu^-}\,. \quad (25)$$

The minimum condition of energy density, $\mathcal{E}$, with respect to particles yields phase equilibrium conditions: $\mu_n = \mu_p + \mu_{e^-}$, $\mu_n = \mu_\Lambda$, $\mu_{\Sigma^-} = \mu_n + \mu_{e^-}$, and $\mu_{e^-} = \mu_{\mu^-}$. The charge-neutrality is given by $\rho_p = \rho_{\Sigma^-} + \rho_{e^-} + \rho_{\mu^-}$. Substitution of these relations into (25) results in:

$$P + \mathcal{E} = \mu_n \rho_B \; . \tag{26}$$

where ρ_B is the baryon density: $\rho_B = \rho_p + \rho_n + \rho_\Lambda + \rho_{\Sigma^-}$. All baryons have common chemical potential to constitute the total pressure and energy density. This proves that many-component hyperon matter produced by way of phase transition can be considered equivalent to one-component matter. Hence, there is no inconsistency or difficulty to use the method of Maxwell construction to determine critical density, pressure, and chemical potential for phase transition in hyperonic matter.

Maxwell construction gives a constant pressure and discontinuous phase transition, and hence, it is often said that the method is not appropriate for realistic picture inside neutron stars because pressure is expected to change continuously in neutron stars. It is expected that phase transitions from (n, p, e) to $(n, p, H_1, H_2, \cdots, e, \mu)$ in hadronic matter should be continuous, but phase transition from hadron to quark matter may not be continuous, because hadrons must generate strong bindings ($p_H < 0$) for themselves, confinement and subsequently deconfinement ($p_H > 0$) as quark matter. The process requires certain amount of energy at a high density, before quark matter equation of state produces the mass of quark stars. In the current calculations, the ground state equations of state for hadron and quark matter at $T = 0$ are linked in order to examine high density matter, and hence, the binding and change of hadronic phase must be generated phenomenologically by introducing a finite temperature Fermi distribution function. By creating excited Fermi energy state, the appearance of the strong binding of hadrons ($p_H < 0$) is displayed, which suggests certain phase transition of hadronic matter.

The mean-field theory of quantum hadrodynamics is primarily for studying hadronic interactions in nuclear physics so that approximations are constructed at $T = 0$ formalism, and neutron stars are regarded as cold matter ($T \sim 0$) if the temperature is compared with Fermi-energies of hadrons ($\sim 10^3$ MeV $= 10^{14}\,^\circ K$). In addition, it should be expected that nuclear matter would be unstable for high density and energy because of $N\bar{N}$-generations, vacuum fluctuations, and many-body effects,..., resulting in excited, nonequilibrium state. The nonequilibrium state can be considered characterized, for instance, by the *smearing effect of Fermi-surface* , and thermodynamic quantities are determined with the smeared density distribution. The discontinuous and continuous changes of equations of state are well known for finite temperature isotherms [1]. The constant pressure and discontinuous phase transition are produced by the sharp Fermi-surface which is given by a θ-function distribution function at $T = 0$. However, the sharp Fermi-surface becomes smooth and continuous as temperature increases, and nonequilibrium excited state arises at a high density. Therefore, we replace the θ-function distribution function with the one analogous to finite-temperature, excited Fermi-energy state:

$$\theta(k_F - k) \longrightarrow 1/(1 + e^{(E(k)-\mu)/a}) \; . \tag{27}$$

where a is a constant and μ is assumed to be $E(k_F)$ at $T = 0$.

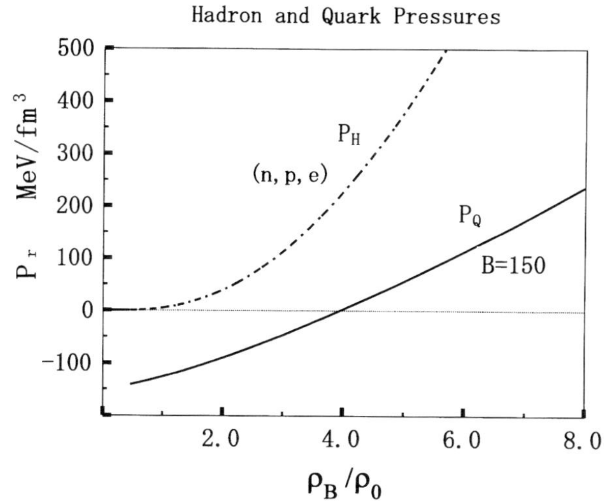

Figure 22. The pressures of isospin asymmetric (n, p, e) and quark matter (MIT-bag model), P_H and P_Q which do not intersect with each other.

Employing the distribution function (27), the appropriate value of a is determined by checking phase transition conditions, $p_H = p_Q$, $\mu_H = \mu_Q$, and if it is possible to connect hadron-quark phase transition density continuously. The appropriate value of a is about $a \lesssim 50.0$ MeV, when it is linked with the quark EOS ($B = 150$ MeV/fm^3), but it was not possible to find the hadron-quark *continuous phase-transition* density. The hadron-quark phase transition is still discontinuous. If a is chosen arbitrary large, it produces softer pressure than that of quark matter, and so, hadron-quark phase transition becomes quark-hadron phase transition, which is in contradiction. It suggests that it may be inappropriate to assume simple continuous phase transition from hadron to quark phase.

It is not necessary to require additional assumptions to define average particle density and energy density by assuming hadron and quark volume ratios in mixing phase [14]. The Fig. 22 shows pressures of (n, p, e) hadronic matter and quark matter ($B = 150$ MeV) do not intersect and that the hadron pressure, p_H, increases rapidly, while the quark pressure, P_Q, becomes negative (attractive) and bounded. The artificial excitation (27) at a high density is introduced to examine how densities of hadron-quark phase transition change. The pressure is shown in the Fig. 23. The numerical analyses by employing the artificial excitation (27) indicate that the phase transition from hadron to quark matter cannot be smoothly connected and consequently, a simple picture of hadron-quark mixed phase may not be possible. The dynamical mechanism of hadron-quark phase transition is so complex and unknown that EOS at phase transition should be assumed and discussed as a piecewise continuous line in thermodynamical point of view as shown by Maxwell construction method.

The excited, nonequilibrium state of hadrons defined by (27) suggests that hadron and quark phase cannot be smoothly connected because certain amount of energy is required for hadrons to generate a strong bound state of hadrons and subsequently deconfinement to produce quark matter. Therefore, the simple picture of hadron-quark hybrid matter could not be produced by employing hadron and quark matter approximations. The energy density and pressure at phase transition should be piecewise continuous in terms of thermodynamics.

We assumed that the H-Q phase transition is the first-order, and solved phase transition conditions for chemical potential and pressure ($\mu_H = \mu_Q$ and $P_H = P_Q$) by employing the double-tangent method numerically. The TOV equation, as well as hadronic and quark EOSs with phase transition conditions, is applied to calculate the properties of neutron stars. The discontinuous character of thermodynamic quantities, which depend on derivatives of energy density naturally appears at a critical density of phase transition.

First, we examined H-Q stars in case of (n, p, e) + VFC and quark matter with phase transitions connected with Maxwell construction. The analysis is extended to hyperonic matter + VFC and quark matter. In Fig. 24, stable pure-hadron stars exist in the central energy range, $2.0 \times 10^{14} \sim 6.0 \times 10^{14}$ g/cm^3 with $dM_{star}/d\mathcal{E}_c > 0$. The total mass of H-Q stars decreases (solid line with $dM_{\text{H-Q}}/d\mathcal{E}_c < 0$), but a stable quark-core is developing (dotted line with $dM_{core}/d\mathcal{E}_c > 0$). The hadron-quark stars composed of a quark-core and a hadron mantle exist in central energy range, $10^{15} \sim 10^{16}$ g/cm^3.

The stability criterion, $dM_{star}/d\mathcal{E}_c > 0$, is for single phase compact stars, such as pure hadronic stars or quark stars. As H-Q stars are 2-phase compact stars (*i.e.*, the quark phase for a star's core and hadron phase for a mantle), the stability of H-Q stars should be considered carefully. One can see that the quark-core (monotonically increasing dotted line in Fig. 24) has a simple positive slope, $dM_{core}/d\mathcal{E}_c > 0$, with one critical point, $d^2M_{core}/d\mathcal{E}_c^2 < 0$. Hence, the stability criterion, $dM_{core}/d\mathcal{E}_c > 0$, can be applicable to the quark-core, but not to the total mass of the H-Q star. It indicates that a stable quark-core is developing; however, the total mass of H-Q star (a quark-core and a hadron mantle) shows a decreasing curve (solid line in Fig. 24), which suggests that only the hadron mantle would be unstable because of hadron-quark phase transition.

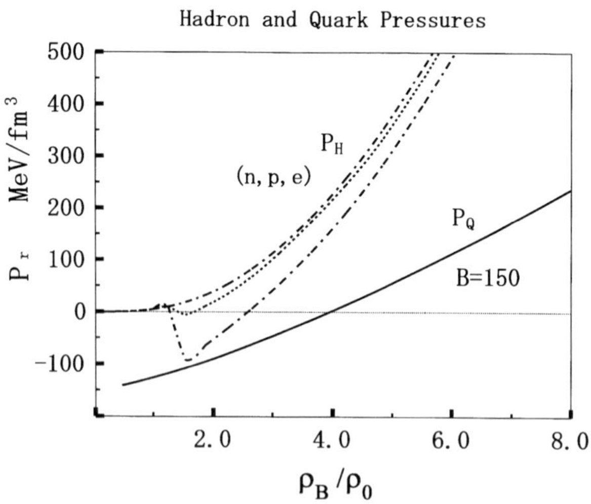

Figure 23. The pressure of isospin asymmetric (n, p, e) with the smearing Fermi surface effect. The parameter a is changed from 10.0 to 50.0 MeV. The negative (attractive) pressure is gradually produced, which may indicate a change of phase or a bound state of hadrons.

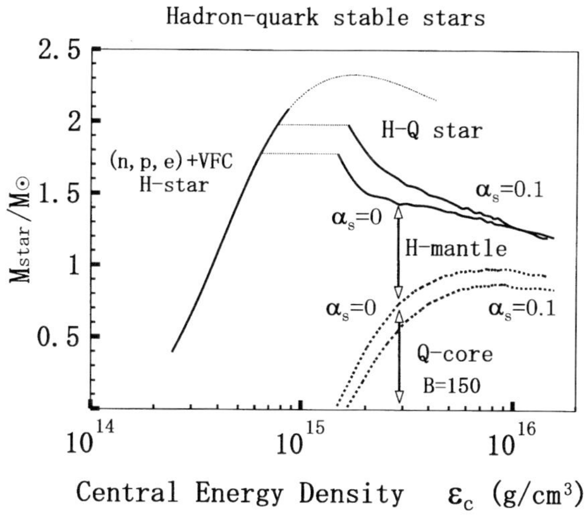

Figure 24. Hadron-quark stars based on isospin asymmetric (n, p, e)+VFC matter. Note that the quark-core of H-Q stars is stable.

The Fig. 24 indicates that stable pure hadronic stars ($dM_H/d\mathcal{E}_c > 0$) in the central energy range, $\mathcal{E}_c = 2.0 \times 10^{14} \sim 6.0 \times 10^{14}$ g/cm^3, will undergo a phase transition (dotted horizontal line), reaching H-Q stars ($B = 150$ MeV/fm^3). The total mass of H-Q stars decreases (solid line), but a stable quark-core develops (monotonically increasing dotted line, $dM_{core}/d\mathcal{E}_c > 0$). The hadron mantle provides the star-core with quark particles by way of hadron-quark phase transition, resulting in the development of a stable quark-core. Therefore, the H-Q star is stable in the sense that a stable quark-core is developing, although the total mass of the star decreases. The decreasing mass of hadron-mantle or equivalently the decreasing energy of hadron-mantle could be considered used to convert hadrons to produce free quark particles in the deconfined vacuum relative to the confined vacuum, which is devised as the bag-constant, B, in MIT-bag model (see, eq. (18)).

By comparing stable energy densities of the quark phase in Fig. 21 with those of H-Q stars, the central energy density of stable H-Q stars is found to be more extended for higher densities than that of single phase stars. This suggests that compact stars consisting of a mantle and a high density core are more stable than stars in a homogeneous single phase structure [31,46]. When the QCD coupling constant, α_s, increases, the H-Q phase transition density and quark-core will shift to higher densities, but the quark-core is stable and extends to higher densities. If the bag constant is small, such as $B \sim 100$ MeV/fm^3, one will obtain a saturation curve (an inflection point for stability) within $10^{15} \sim 10^{16}$ g/cm^3, as in Fig. 21.

The H-Q stars calculated with the EOS for (n, p, e)-(n, p, Σ^-, e)-$(n, p, \Sigma^-, \Lambda)$ + VFC to quark matter ($B = 150$ MeV/fm^3) are shown in Figs. 25 ($r^{\omega}_{HN} = 1.0$) and 26 ($r^{\omega}_{HN} = 2/3$), respectively. The maximum masses in Fig. 25 are $M_{max} = 1.61$ M$_\odot$ ($\alpha_s = 0.0$) and $M_{max} = 1.77$ M$_\odot$ ($\alpha_s = 0.1$); whereas, in Fig. 26, they are $M_{max} = 1.37$ M$_\odot$ ($\alpha_s = 0.0$) and $M_{max} = 1.48$ M$_\odot$ ($\alpha_s = 0.1$). A remarkable feature of H-Q stars in Fig. 25 ($r^{\omega}_{HN} = 1.0$) is that they can reasonably explain central energy densities and maximum mass configurations [47] for stable compact stars, ranging from $10^{14} \sim 10^{16}$ g/cm^3.

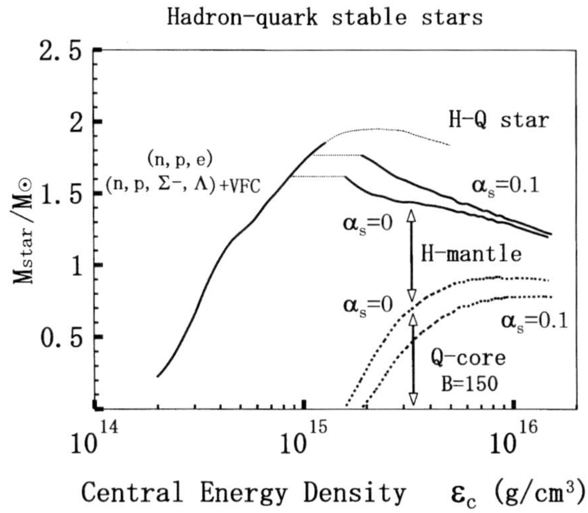

Figure 25. Hadron-quark stars based on (n, p, e)-(n, p, Σ^-, e)-$(n, p, \Sigma^-, \Lambda)$ + VFC matter. The coupling ratios are: ($r^{\sigma}_{\Lambda N} = 0.964$, $r^{\omega}_{\Lambda N} = 1.00$) and ($r^{\sigma}_{\Sigma^- N} = 0.925$, $r^{\omega}_{\Sigma^- N} = 1.00$, $r^{\rho}_{\Sigma^- N} = 1.00$).

The results are not sensitive to the parameter σ of (23), which is checked by changing $\sigma = 200 \sim 800$ MeV. However, the H-Q stars are sensitive to the values of B and α_s, $B = 100 \sim 150$ MeV/fm^3, $\alpha_s = 0.2 \sim 0.1$ to produce the observed masses of neutron stars ($M_{star} \gtrsim 1.30$ M$_\odot$). This is because the quark-EOS shifts to high energy densities if B is increased, meaning that the EOS becomes softer and results in compact stars with

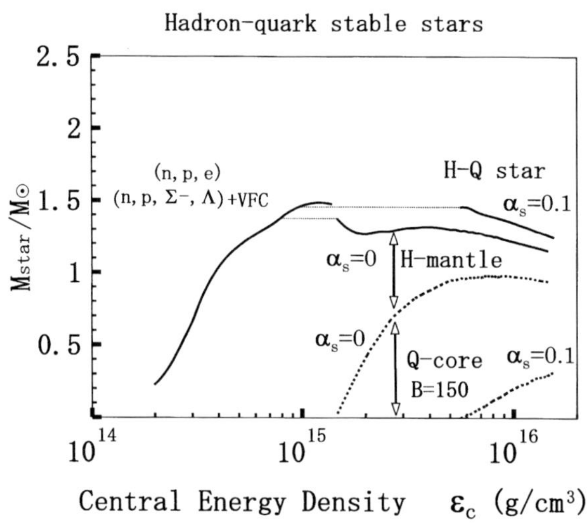

Figure 26. Hadron-quark stars based on (n, p, e)-(n, p, Σ^-, e)-$(n, p, \Sigma^-, \Lambda)$ + VFC matter. The coupling ratios are: ($r^{\sigma}_{\Lambda N} = 0.677$, $r^{\omega}_{\Lambda N} = 2/3$) and ($r^{\sigma}_{\Sigma^- N} = 0.632$, $r^{\omega}_{\Sigma^- N} = 2/3$, $r^{\rho}_{\Sigma^- N} = 1.00$).

a smaller mass and radius at high densities. In addition, if the QCD coupling constant α_s is increased as $\alpha_s = 0.1 \rightarrow 0.2$, the masses of stable H-Q stars become small and shifted to high densities. Hence, appropriate values of B and α_s to produce the observed data of neutron stars in the current EOS are correlated to each other so that if B is increased, α_s should be decreased, such as ($B \sim 100$ MeV/fm^3, $\alpha_s \sim 0.2$) and ($B \sim 150$ MeV/fm^3, $\alpha_s \sim 0.1$). The property of (B, α_s) in the analysis of H-Q infinite matter is similar to the results of bag-model fits to light-hadron spectra and renormalization group analyses in the paper by Farhi and Jaffe [31], which should be investigated in other approximations for infinite matter.

In the case of the softer hadronic EOS in Fig. 26 ($r^{\omega}_{HN} = 2/3$) with $B = 100 \sim 150$ and α_s-correction, phase transitions occur in relatively low densities. This results in smaller maximum masses of H-Q stars, $M_{max} \lesssim 1.4\,\mathrm{M}_{\odot}$. Thus, the EOS with the coupling constants ($r^{\omega}_{HN} = 2/3$) may not be appropriate to explain the observed masses of neutron stars. Hence, the coupling ratios, $r_{HN} = 2/3$ or $r_{HN} = 1/3$, suggested by SU(6) quark effective model may not be appropriate when the values are directly employed for mean-field approximations of hadrons. A physical reason from the hadronic sector is clear in terms of the EOS; however, it is very interesting to investigate how hadronic and quark models reconcile the problems of coupling constants and discrepancies pointed out in this chapter.

If the QCD coupling constant is set large ($\alpha_s \gtrsim 0.1$) in the soft EOS of Fig. 26 ($r^{\omega}_{HN} = 2/3$), generation of the quark-phase will move up to a very high density, separating the energy densities of hadron and quark phases completely. In this case, no stable hadron-phase can exist or coexist even in a crust of the surface of quark stars, resulting in literally pure-quark stars. Although more precise and detailed analyses are needed for many-body interactions of hadronic and quark matter, the coupling constants for hadrons ($r^{\omega}_{HN} \sim 1.0$) would be appropriate to examine properties of nuclear and hyperonic matter, as well as maximum masses of high density, hadronic compact stars.

7. Conclusions

The linear and nonlinear σ-ω-ρ mean-field approximations are interesting to use for describing hadronic many-body systems at large distances. However, they will ultimately break down at short distances where QCD is valid. Thus, it is important to investigate when hadron and quark predictions will become different and how they influence and interconnect to each other dynamically [1, 2]. We have applied the conserving nonlinear σ-ω-ρ mean-field approximation to study self-consistent, density-dependent interactions of hadrons and have examined hadronic dynamics by employing coupling constants required from the SU(6) quark effective model. Discrepancies between hadronic and QCD-model predictions are shown in terms of effective masses, effective coupling constants, incompressibility and symmetry energy, saturation properties of hyperon binding energies, and maximum masses of high density compact stars. One purpose of this work was to compare predictions indicated by effective hadronic and quark models in order to clarify interrelations and distinctions between them.

As saturation properties of symmetric nuclear matter are self-consistently related to those of hyperons, reproducing saturation properties of hyperons will be one of the fundamental problems of nuclear many-body approximations. Although more quantitative and model-independent analyses are needed, hyperon coupling ratios required by the SU(6) quark model for vector coupling constants [20, 21] may not be appropriate to generate the saturation properties of hyperons. The coupling ratios of hyperons ($r^{\omega}_{HN} \sim 1.0$) are appropriate to explain Fermi-liquid properties, saturation of hyperons [6, 7], and maximum masses of neutron stars. The difference of coupling ratios seems small, but their effects to properties of nuclear matter and neutron stars are large. Discrepancies of hyperon coupling ratios required by the hadronic and SU(6) quark models indicate that both effective approaches should be improved each other as consistent theories for hadronic physics.

Heavy-ion collision experiments, as well as neutron stars, are useful for examining the applications of quark models and constraints for hadronic calculations [48, 49], as conditions of the H-Q phase transition depend on both EOSs for hadronic and quark matter. The appropriate values of (B, α_c) can be independently suggested from properties of infinite matter, which will qualitatively support properties discussed in the paper [31]. The stable pure-hadronic stars and pure-quark stars based on the MIT-bag model exist at different energy densities; however, if the first-order phase transition is assumed, the energy density of stable H-Q stars expands from 10^{14} to 10^{16} g/cm^3. In the range of energy density, pure-hadron stars are less than $M(n, p, e)_{max} \lesssim 2.5$, but H-Q stars are $M^{H-Q}_{max} \gtrsim 1.0\ M_{\odot}$. This reasonably explains the expected masses and energy density relations, $M - \mathcal{E}_c$ [47]. Therefore, existing high density compact stars are more likely to be regarded as H-Q stars rather than pure-hadron and pure-quark stars. These results should be further examined quantitatively from empirical data and effective theories of hadrons.

It is noteworthy that density-dependent many-body effects simulated by nonlinear σ-ω-ρ interactions contribute significantly compared to those of the renormalized VFC effect in the nonlinear σ-ω-ρ Hartree approximation, whereas the VFC effects contributed to observables significantly in the linear σ-ω-ρ Hartree approximation [1]. Hence, it is interesting to investigate contributions of the renormalized VFC effects further in more complicated nonlinear σ-ω-ρ HF, BHF approximations so as to quantitatively study vacuum-fluctuation and density-dependent corrections.

The EOSs constructed with (n,p,e)-(n, p, H, e)-(n, p, H_1, H_2) and VFC to the current nonlinear approximation were linked to MIT-bag model. It is important to build such EOSs by employing sophisticated approximations to analyze properties of nuclear matter and compact stars. The role of chiral symmetry in hadronic models should be examined by extending the current nonlinear σ-ω-ρ mean-field (Hartree) approximation via the inclusion of the π-meson. This would reveal the physical meaning and significance of chirality in hadronic and quark models for hadrons.

References

[1] B. D. Serot and J. D. Walecka, *Advances in Nuclear Physics*, edited by J. W. Negele and E. Vogt (Plenum, New York, 1986), Vol. 16.

[2] B. D. Serot, *Rep. Prog. Phys.* 55 (1992) 1855.

[3] H. Uechi, *Nucl. Phys.* A501 (1989) 813; H. Uechi, *Nucl. Phys.* A541 (1992) 397.

[4] R. J. Furnstahl and B. D. Serot, *Phys. Rev.* C44 (1991) 2141; *Phys. Rev.* C41 (1990) 262.

[5] H. Uechi, *Nucl. Phys.* A780 (2006) 247; H. Uechi, *Nucl. Phys.* A 799 (2008) 181.

[6] S. T. Uechi and H. Uechi, *Adv. in High Ene. Phys.*, (2009) 640919.

[7] H. Uechi and S. T. Uechi, *The Open Nucl. and Part. Phys. J.*, V2, (2009) 47.

[8] J. Boguta, *Phys. Lett.* V106B (1981) 250.

[9] H. Müller and B. D. Serot, *Nucl. Phys.* A606 (1996) 508.

[10] R. J. Furnstahl, B. D. Serot and Hua-Bin Tang, *Nucl. Phys.* A598 (1996) 539; Nucl. Phys. A615, (1997) 441.

[11] R. J. Furnstahl, *Nucl. Phys.* A706, (2002) 85.

[12] Y. Takada, *Phys. Rev.* B52 (1995) 12708.

[13] A. Sedrakian, Prog. in Part. and *Nucl. Phys.,* 58 (2007) 168-246.

[14] N. K. Glendenning, *Compact Stars*, Springer-Verlag New York, 2000.

[15] G. Baym and L. P. Kadanoff, *Phys. Rev.* 124 (1961) 287; G. Baym, *Phys. Rev.* 127 (1962) 1391.

[16] H. Uechi, *Prog. Theor. Phys.* 111 (2004) 525.

[17] I. Vidaña, A. Polls, A. Ramos, L. Engvik, M. Hjorth-Jensen, *Phys. Rev.* C62 (2000) 035801.

[18] M. Baldo, G. F. Burgio, H. -J. Schulze, *Phys. Rev.* C61 (2000) 055801.

[19] G. Gelmini and B. Ritzi, *Phys. Lett.* B 357 (1995) 431.

[20] J. Schaffner and I. N. Mishustin, *Phys. Rev.* C53 (1996) 1416.

[21] G.-yun Shao, and Y.-xin Liu, *Phys. Rev.* C79 (2009) 025804.

[22] R. J. Furnstahl, B. D. Serot and Hua-Bin Tang, *Nucl. Phys.* A615 (1997) 441.

[23] J. McIntire, Y. Hu and B. D. Serot, *Nucl. Phys.* A794 (2007) 187.

[24] J. D. Walecka, *Theoretical Nuclear and Subnuclear Physics*, Oxford University Press, 1995.

[25] T. K. Jha and H. Mishra, *Phys. Rev.* C 78 (2008) 065802.

[26] G. E. Brown and M. Rho, *Phys. Rev. Lett.*, 66 (1991) 2720.

[27] M. Chiapparini, et.al., *Nucl. Phys.* A826 (2009) 178.

[28] H. Uechi and S. T. Uechi, *Meeting Abst. Phys. Soc. of Japan,* V63, Issue 1, Part1, (2008) 49.

[29] B. D. Serot and H. Uechi, *Ann. Phys.* 179 (1989) 272.

[30] A. Chodos, R. L. Jaffe, K Johnson, C. B. Thorn and V. F. Weisskopf, *Phys. Rev.* D9 (1974) 3471.

[31] E. Farhi and R. L. Jaffe, *Phys. Rev.* D30 (1984) 2379.

[32] N. M. Hugenholtz and L. Van Hove, *Physica,* 24 (1958) 363.

[33] L. D. Landau, *Sov. Phys. JETP* 3 (1956) 920; 5 (1957) 101.

[34] D. Pines and P. Nozières, *The Theory of Quantum Liquids*, Addison-Wesley, 1966.

[35] P. Nozières, *Theory of Interacting Fermi Systems*, Perseus Publishing, 1964.

[36] W. Kohn and L. J. Sham, *Phys. Rev.* 140 (1965) A1133.

[37] W. Kohn, *Rev. Mod. Phys.*, 71 (1999) 1253.

[38] D. N. Basu and T. Mukhopadhyay, *Acta. Phys. Polon.* B38 (2007) 169.

[39] D. V. Shetty, S. J. Yennello, G. A. Souliotis, *Phys. Rev.* C75 (2007) 034602.

[40] P. Danielewicz and J. Lee, *Nucl. Phys.* A 818 (2009) 36.

[41] H. H. Wolter, et. al, Prog. in Part. and *Nucl. Phys.*, 62 (2009) 402.

[42] B. D. Day, *Reviews of Modern Physics,* 50 (1978) 495.

[43] K. A. Brueckner and C. A. Levinson, *Phys. Rev.* 97 (1955) 1344; K. A. Brueckner, *Phys. Rev.* 110 (1958) 597; K. A. Brueckner and D. T. Goldman, *Phys. Rev.* 117 (1960) 207; K. A. Brueckner, J. L. Gammel and J. T. Kubis, *Phys. Rev.* 118 (1960) 1438.

[44] S. L. Shapiro and S. A. Teukolsky, *Black Holes, White Dwarfs, and Neutron Stars*, John Wiley & Sons Inc, 1983.

[45] C. W. Misner, K. S. Thorne and J. W. Wheeler, *Gravitation*, W. H. Freeman and Company, New York, 1973.

[46] E. Witten, *Phys. Rev.* D30 (1984) 272.

[47] J. M. Lattimer and M. Prakash, *Phys. Rep.,* 442 (2007) 109.

[48] V. A. Dexheimer and S. Schramm, *Nucl. Phys.* A 827 (2009) 579c.

[49] I. Bombaci, et. al, *Phys. Lett* B 680 (2009) 448.

In: Neutron Stars
Editors: Hiroshi Uechi et al, pp. 171-198

ISBN: 978-1-62081-123-8

Chapter 7

SIGNALS OF DYNAMICAL PHENOMENA (MESON CONDENSATIONS) AND EOS OF NEUTRON STARS

Takumi Muto*
Department of Physics, Chiba Institute of Technology
2-1-1 Shibazono, Narashino, Chiba 275-0023, Japan

Abstract

Mesons (pions and kaons) are ordinarily regarded as mediators of baryon-baryon interaction. In the inner core region of neutron stars, where baryon number density is several times larger than the standard nuclear matter density ρ_0 (~ 0.17 fm^{-3}), they are expected to be condensed. In this chapter, we recapitulate recent theoretical development on phase transitions of normal neutron-star matter to a new phase including condensation of pions and kaons, the latter of which have been stimulating recent studies of strangeness nuclear physics. Onset mechanisms and ground-state properties of pion and kaon condensations are outlined. Coexistence problem of pion and kaon condensations is also discussed. Implications for neutron star phenomena such as static and dynamic properties of neutron stars and rapid cooling through neutrino emission are recapitulated. We consider coexistence of kaon condensation and hyperons in highly dense matter and discuss the effects of the kaon-condensed phase in hyperonic matter on the EOS and internal structure of neutron stars. In particular, a possible existence of self-bound objects with kaon condensates and hyperons on any scale is suggested. Finally, relevance of meson condensations in hadronic matter to those in color superconductivity are briefly mentioned.

1. Meson Condensation and Chiral Symmetry

Meson condensation in high-density hadronic matter is characterized as macroscopic occupation of Nambu-Goldstone (NG) bosons in the lowest energy state. Its possible existence has been extensively investigated mainly with relevance to astrophysical phenomena associated with neutron stars[1, 2, 3, 4, 5, 6, 7, 8, 9].

In case of atomic systems such as ideal Bose gases and dilute alkali Bose gases, atoms accompany wave packets with an extent of the thermal de-Broglie wave length, λ_T =

*E-mail address: takumi.muto@it-chiba.ac.jp

$h/(2\pi mk_BT)^{1/2}$. At extremely low temperatures, e.g., T=($10^{-7}-10^{-6}$) K for alkali Bose gases, the λ_T becomes larger than the inter-atomic distance d, and the wave packet of the atom overlaps each other, where the quantum statistics becomes responsible for coherent behaviors of the atoms as Bose-Einstein condensation (BEC). Thus formation of the BEC in atomic Bose gas system occurs as a result of competition between thermal fluctuation and quantum statistics.

In contrast to the dilute Bose gas system, hadronic matter in neutron stars is the strongly interacting system of mesons and baryons. The temperature of hadronic matter, which may be realized in the inner core region of neutron stars, is $O(10^8)$ K$\sim$ 0.01MeV. Such an energy scale is much smaller than the strong interaction energy scale of nuclear Fermi energy (several tens of MeV), so that the hadronic matter in neutron stars is considered to be at extremely low temperature. However, even if T=0 K, the NG bosons have a rest mass, so that spontaneous creation of the NG bosons is kinematically inhibited in normal matter. Nevertheless the NG bosons receive attractive meson-baryon interaction, and the energy of the NG bosons decreases in nuclear medium as baryon density increases. At certain high baryon density, the NG bosons are kinematically allowed to be condensed. Therefore high-density hadronic matter is a decisive environment for realization of meson condensation.

The order parameter of meson condensation can be chosen as the ground-state expectation value of the pseudoscalar meson field, $\langle\hat{M}\rangle$ ($\equiv\langle g|\hat{M}|g\rangle$ with $|g\rangle$ being the ground state). In normal neutron-star matter, the expectation value $\langle\hat{M}\rangle$ vanishes since the normal state, denoted as $|g\rangle$=$|x\rangle$ is an eigenstate of parity. The meson-condensed state, $|g\rangle$=$|M\rangle$, is a kind of coherent state, so that it has a nonvanishing expectation value. Condensations of the NG bosons can be naturally described on the basis of chiral symmetry, which is inherent in low energy hadron dynamics. In the context of chiral symmetry, for instance, a charged pion-condensed state $|\pi\rangle$ is represented as a chirally- rotated one from the normal state $|x\rangle$ such that $|\pi\rangle=\hat{U}(\theta,\chi)|x\rangle$ with a unitary operator $\hat{U}(\theta,\chi)$[3, 11] :

$$\hat{U}(\theta,\chi)=\exp(i\int d^3r\chi V_3^0)\exp(iQ_1^5\theta)\ . \tag{1}$$

In Eq. (1), χ and θ (the chiral angle) are quantities characterizing a type of condensation, and V_3^μ and Q_1^5 the vector current and the axial charge, respectively. With the help of current algebra and PCAC (partial conservation of axial-vector current), the classical pion field is given by

$$\langle\pi\rangle=\langle\pi|\hat{\pi}|\pi\rangle=\langle 0|\hat{U}^\dagger(\theta,\chi)\hat{\pi}\hat{U}(\theta,\chi)|0\rangle=\frac{f}{\sqrt{2}}\sin\theta e^{i\chi}\ , \tag{2}$$

where f is the meson decay constant. The classical field [Eq. (2)] specifies an order parameter of condensation.

Based on chiral symmetry, a charged pion condensation was considered by Campbell, Dashen, and Manassah by the use of the σ model which embodies chiral dynamics including p-wave π-N interaction and π-π interaction[10]. It was also demonstrated by Tatsumi that a neutral pion condensation can be described self-consistently with the Alternating-Layer-Spin (ALS) structure for baryonic system within the σ model[11]. After much work has been done concerning pion condensation from viewpoints of nuclear and astrophysics, a possible existence of kaon condensation was suggested by Kaplan and Nelson by the use of the effective chiral Lagrangian[5], which opened up subsequent theoretical and experimental development of strangeness nuclear physics. Kaon condensation was formulated model-independently within current algebra and PCAC by Tatsumi in the case of s-wave K-N interaction[12].

Here we survey chiral dynamics relevant to meson condensation in baryonic matter in terms of the effective chiral SU(3)$_L\times$ SU(3)$_R$ Lagrangian density : [1]

$$\begin{aligned}\mathcal{L}_{M,B} &= \frac{1}{4}f^2\,\mathrm{Tr}\partial^\mu\Sigma^\dagger\partial_\mu\Sigma + \frac{1}{2}f^2\Lambda_{\chi\mathrm{SB}}(\mathrm{Tr}M(\Sigma-1)+\mathrm{h.c.})\\ &+\mathrm{Tr}\overline{\Psi}(i\partial\!\!\!/ - M_\mathrm{B})\Psi + \mathrm{Tr}\overline{\Psi}i\gamma^\mu[V_\mu,\Psi] + D\mathrm{Tr}\overline{\Psi}\gamma^\mu\gamma^5\{A_\mu,\Psi\} + F\mathrm{Tr}\overline{\Psi}\gamma^\mu\gamma^5[A_\mu,\Psi]\\ &+a_1\mathrm{Tr}\overline{\Psi}(\xi M^\dagger\xi+\mathrm{h.c.})\Psi + a_2\mathrm{Tr}\overline{\Psi}\Psi(\xi M^\dagger\xi+\mathrm{h.c.}) + a_3(\mathrm{Tr}M\Sigma+\mathrm{h.c.})\mathrm{Tr}\overline{\Psi}\Psi\ ,\end{aligned} \quad (3)$$

where f is the meson decay constant. Throughout this chapter, f is set to be the lowest-order value in chiral perturbation theory, $f = f_\pi$=93 MeV. In (3), the nonlinear meson field Σ is given by $\Sigma \equiv e^{2i\Pi/f}$ in terms of the octet meson field Π :

$$\Pi = \pi^a T_a = \frac{1}{\sqrt{2}}\begin{pmatrix} \pi^0/\sqrt{2}+\eta/\sqrt{6} & \pi^+ & K^+ \\ \pi^- & -\pi^0/\sqrt{2}+\eta/\sqrt{6} & K^0 \\ K^- & \overline{K}^0 & -\sqrt{\frac{2}{3}}\eta \end{pmatrix} \quad (4)$$

with the Nambu-Goldstone bosons π^a and the SU(3) generators T_a ($a = 1 \sim 8$). $\Lambda_{\chi\mathrm{SB}}$ is the chiral symmetry breaking scale, $\sim$ 1 GeV, M the mass matrix which is defined as $M \equiv \mathrm{diag}(m_u, m_d, m_s)$ with the quark masses m_i, and M_B is the baryon mass generated as a consequence of spontaneous chiral symmetry breaking. The baryon fields are given as the spin 1/2 octet $\Psi = \sqrt{2}B^a T_a$:

$$\Psi = \begin{pmatrix} \Sigma^0/\sqrt{2}+\Lambda/\sqrt{6} & \Sigma^+ & p \\ \Sigma^- & -\Sigma^0/\sqrt{2}+\Lambda/\sqrt{6} & n \\ \Xi^- & \Xi^0 & -\sqrt{\frac{2}{3}}\Lambda \end{pmatrix}. \quad (5)$$

Baryons couple with mesons through the mesonic vector current V_μ, which is defined by $V_\mu \equiv 1/2(\xi^\dagger\partial_\mu\xi + \xi\partial_\mu\xi^\dagger)$ with $\xi \equiv \Sigma^{1/2}$, and the axial vector current A_μ, defined by $A_\mu \equiv i/2(\xi^\dagger\partial_\mu\xi - \xi\partial_\mu\xi^\dagger)$. In the nonrelativistic limit, the fourth term including V_μ in Eq. (3) reduces to the s-wave vector interaction, and the fifth and sixth terms to the p-wave interaction. The last three terms in Eq. (3) represent the s-wave meson-baryon scalar interaction, which explicitly breaks chiral symmetry.

We take the same set of parameters as those in Ref. [5]: The axial vector coupling constants D and F are determined from the weak decays of the baryon octet states and are chosen as D=0.81 and F=0.44. The quark masses are taken to be m_u=6 MeV, m_d=12 MeV and m_s=240 MeV. With these values for m_i, the parameters a_1 and a_2 are fixed so as to give the empirical octet baryon mass splittings and given as a_1=−0.28, a_2=0.56.

The parameter a_3 is related to the "πN and KN sigma terms" which simulate the s-wave meson-baryon scalar interactions. For instance, the πN sigma term $\Sigma_{\pi N}$ is given in the quark model as $\Sigma_{\pi N} = \frac{1}{2}(m_u + m_d)\langle N|\bar{u}u + \bar{d}d|N\rangle$, obtained from the next to leading

[1] We use the units in which $\hbar$=c=k_B=1 throughout this chapter.

order of the chiral expansion $[O(m_\pi^2)]$. By the use of the relation, $\partial M_i/\partial m_q = \langle i|\bar{q}q|i\rangle$ ($i = p, n$ and $q = u, d, s$) with M_i being the nucleon mass, one obtains $\Sigma_{\pi N} = -(a_1 + 2a_3)(m_u + m_d)$[13]. The value of $\Sigma_{\pi N}$ is empirically estimated to be ~ 45 MeV[14]. Likewise, the KN sigma terms are given by $\Sigma_{Kp} = -(a_1+a_2+2a_3)(m_u+m_s)$, and $\Sigma_{Kn} = -(a_2 + 2a_3)(m_u + m_s)$. Considering that there is some ambiguity about these values, the value of a_3 is chosen to be $a_3 = -0.9$ and -1.1, which yields $\Sigma_{\pi N}$=37 MeV, Σ_{Kp}=374 MeV, Σ_{Kn}=305 MeV for $a_3 = -0.9$, and $\Sigma_{\pi N}$=45 MeV, Σ_{Kp}=472 MeV, Σ_{Kn}=403 MeV for $a_3 = -1.1$.

The energy of meson-condensed phase is obtained from the ground-state expectation value of the Hamiltonian density, $\mathcal{H} = \mathcal{H}_{M,B} + \mathcal{H}_l$, where $\mathcal{H}_{M,B}$ is the mesonic and baryonic part defined by $\mathcal{H}_{M,B} = (\partial\mathcal{L}_{M,B}/\partial\dot{B}^a)\dot{B}^a + (\partial\mathcal{L}_{M,B}/\partial\dot{\pi}^a)\dot{\pi}^a - \mathcal{L}_{M,B}$, and $\mathcal{H}_l$ is the leptonic part. The chemically equilibrated matter in neutron stars follows both the electric charge neutrality and baryon number conservation. These conditions are assured by considering the effective Hamiltonian density, which is expressed as

$$\mathcal{H}_{\text{eff}} = \mathcal{H}_{M,B} + \mathcal{H}_l + \mu(\hat{\rho}_B^c + \hat{\rho}_M^c + \hat{\rho}_l^c) + \nu\hat{\rho}_B \ , \tag{6}$$

where the charge chemical potential μ and the baryon number chemical potential ν are introduced as the Lagrange multipliers associated with these conditions. In (6), $\hat{\rho}_B^c$, $\hat{\rho}_M^c$, and $\hat{\rho}_e^c$ are charge density operators for baryons, mesons, and leptons, and $\hat{\rho}_B$ is the total baryon number density operator. It is to be noted that $\hat{\rho}_M^c = i\phi\partial\mathcal{L}_{M,B}/\partial\dot{\phi}$ for the negatively charged meson (π^-, K^-) field ϕ. The charge density of the mesons includes a contribution from the meson-baryon interaction. Together with the time dependence of ϕ, $\dot{\phi} = -i\mu\phi$[3, 5], one obtains $(\partial\mathcal{L}_{M,B}/\partial\dot{\pi}^a)\dot{\pi}^a \rightarrow (\partial\mathcal{L}_{M,B}/\partial\dot{\phi})\dot{\phi} = (\partial\mathcal{L}_{M,B}/\partial\dot{\phi})(-i\mu\phi)$ $= -\mu\hat{\rho}_M^c$. Therefore the effective Hamiltonian is expressed in a reduced form as

$$\mathcal{H}_{\text{eff}} = (\partial\mathcal{L}_{M,B}/\partial\dot{B}^a)\dot{B}^a - \mathcal{L}_{M,B} + \mu\hat{\rho}_B^c + \nu\hat{\rho}_B + (\mathcal{H}_l + \mu\hat{\rho}_l^c) \ . \tag{7}$$

2. P-wave Pion Condensation

2.1. Charged Pion Condensation

We begin with obtaining the energy expression of the charged pion (π^c)-condensed phase in neutron-star matter. The classical π^c field in the nonlinear representation is set to be a plane wave type as

$$\pi^{\pm}(\mathbf{r}, t) = \frac{f}{\sqrt{2}}\theta e^{\pm i(\mu_\pi t - \mathbf{p}_\pi\cdot\mathbf{r})} \ , \tag{8}$$

where θ is identical to the chiral angle in the linear representation [(2)], and μ_π and $\mathbf{p}_\pi$ are the chemical potential and spatial momentum of the π^c, respectively[2] . After other mesons and baryon fields are discarded, the π^c and nucleon parts of the Lagrangian read

$$\begin{aligned}\mathcal{L}_{\pi,N} &= \bar{\psi}(i\,\not{\partial} - m_N)\psi + \bar{\psi}\gamma^\mu\frac{\tau_3}{2}\psi(p_\pi)_\mu(1-\cos\theta) + \Sigma_{\pi N}\bar{\psi}\psi(1-\cos\theta) \\ &+ i\frac{g_A}{2}\left(\bar{\psi}\gamma^\mu\gamma_5\tau_-\psi - \bar{\psi}\gamma^\mu\gamma_5\tau_+\psi\right)(p_\pi)_\mu\sin\theta\end{aligned}$$

[2] In a chemically equilibrated matter, the chemical potentials for the π^- , K^- mesons, and the electron are equal to the charge chemical potential μ, i.e., $\mu \equiv \mu_\pi = \mu_K = \mu_e$.

$$+ \frac{1}{2} f^2 (p_\pi)^2 \sin^2\theta - \frac{1}{2} f^2 m_\pi^2 (1 - \cos\theta) , \qquad (9)$$

where ψ is the nucleon isodoublet redefined by multiplying a phase factor $\exp(i\tau_3 p_\pi \cdot x/2)$ to the original one with $(p_\pi)^\mu = (\mu_\pi, \mathbf{p}_\pi)$ being the four momentum of the π^c and τ_3 the third component of the isospin operator. The first term on the r.h.s. of (9) is the free nucleon part with m_N being the nucleon mass. The second and third terms represent the s-wave π-N vector interaction and scalar interaction, respectively. The fourth term in (9), proportional to $\boldsymbol{\sigma} \cdot \boldsymbol{\nabla}$ couplings in the nonrelativistic limit, gives the p-wave π-N attractive interaction, with g_A (=$D+F$=1.25) being the axial-vector coupling strength, leading to the main driving force of pion condensation. The remaining two terms in (9) are kinetic and mass terms of the π^c condensate, respectively, with m_π being the pion mass.

In the case of π^c condensation, the s-wave vector interaction, which corresponds to the Tomozawa-Weinberg term related to the πN forward scattering amplitude, is proportional to the nucleon isospin density $\psi^\dagger \tau_3 \psi$ in Eq. (9). The s-wave $\pi^- p$ vector interaction has an attractive contribution to the pion self-energy, while the $\pi^- n$ vector interaction is repulsive, so that the net vector interaction works repulsively in neutron-star matter. At certain density, the p-wave πN attractive interactions overcome the pion mass and repulsive s-wave vector interactions. It is to be noted that the s-wave πN interactions given by $\Sigma_{\pi N}$ is relatively weak, which is in contrast to the case of kaon condensation.

The ground-state energy density $\mathcal{E}$ is obtained from the expectation value of the effective Hamiltonian density (7) with respect to the π^c-condensed state $|\pi^c\rangle$:

$$\mathcal{E} = \mathcal{E}_{\rm eff} - \nu\rho_{\rm B} = \langle \pi^c | \mathcal{H}_{\rm eff} | \pi^c \rangle - \nu\rho_{\rm B} . \qquad (10)$$

2.2. Onset Mechanisms of Charged Pion Condensation and Ground-State

We obtain dispersion relations of charged pions (π^c) in neutron matter to illustrate how in-medium properties of mesons are responsible for onset mechanisms of meson condensation. The π^c inverse propagator $D_\pi^{-1}(\omega, \mathbf{p}_\pi; \rho_{\rm B})$ is derived from expansion of the effective energy density with respect to the classical π^c field on the assumption that a transition from normal phase to π^c condensed phase is continuous[3]:

$$\mathcal{E}_{\rm eff}(\theta) = \mathcal{E}_{\rm eff}(\theta = 0) - \frac{f^2}{2} D_\pi^{-1}(\omega = \mu, \mathbf{p}_\pi; \rho_{\rm B})\theta^2 + O(\theta^4) . \qquad (11)$$

The lowest energy ω for π^c as a function of $|\mathbf{p}_\pi|$, $\omega = \omega(|\mathbf{p}_\pi|)$, is obtained from zero points of $D_\pi^{-1}(\omega, \mathbf{p}_\pi; \rho_{\rm B})$. In Fig. 1-(a), the dispersion relations for π^c in neutron matter is shown at $\rho_{\rm B}$=0.25 fm^{-3}. In Fig. 1-(b), the pion inverse propagator $D_\pi^{-1}(\omega, \mathbf{p}_\pi; \rho_{\rm B})$ is shown as a function of ω at $\rho_{\rm B}$=0.25 fm^{-3} and $|\mathbf{p}_\pi| = 1.8 m_\pi$. There appears a spin-isospin sound mode with a π^+ quantum number (π_s^+) in addition to the usual $\pi^\pm$ branches. At a critical density $\rho_{\rm B} = \rho_{\rm B}^c$, the π^- and π_s^+ branches merge with a pion momentum $\mathbf{p}_\pi = \mathbf{p}_\pi^c$. The critical density $\rho_{\rm B}^c$ and the momentum $\mathbf{p}_\pi^c$ are determined from a double-pole condition for the pion inverse propagator,

$$D_\pi^{-1}(\omega, \mathbf{p}_\pi; \rho_{\rm B}) = 0 \, , \partial D_\pi^{-1}(\omega, \mathbf{p}_\pi; \rho_{\rm B})/\partial\omega = 0 \, , \qquad (12)$$

together with $\partial D_\pi^{-1}(\omega, \mathbf{p}_\pi; \rho_{\rm B})/\partial|\mathbf{p}_\pi| = 0$. At this critical point, the π^--π_s^+ pairs are created with no cost of energy, which implies that the system becomes unstable with respect to condensation of the π^--π_s^+ pairs[2, 3]. The critical density $\rho_B^c(\pi^c)$ is estimated to be around $1.5\rho_0$ with ρ_0 ($\simeq$0.17 fm^{-3}) being the standard nuclear saturation density.

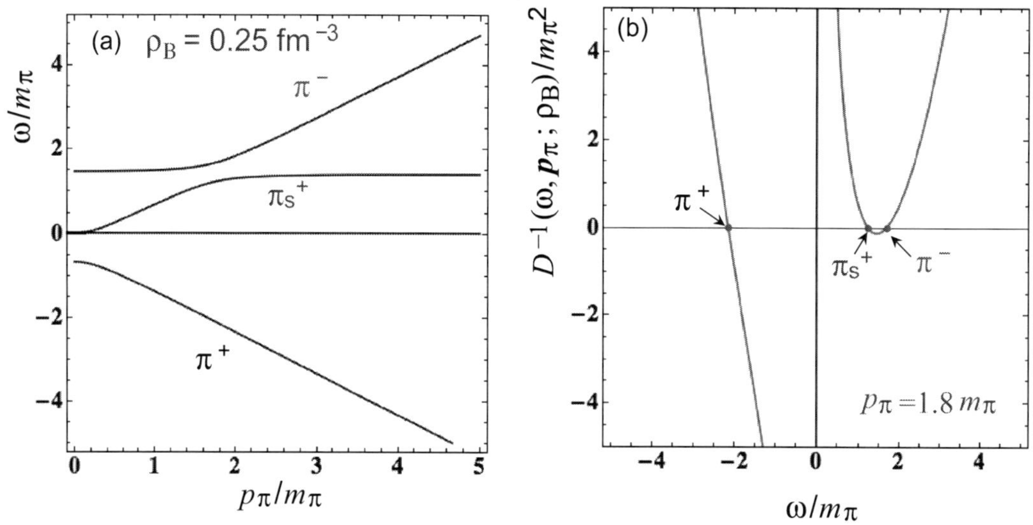

Figure 1. (a) The dispersion relations of the π^c at ρ_B=0.25 fm^{-3} in a simple model. The lowest energy ω and spatial momentum $|\mathbf{p}_\pi|$ for the π^c are scaled by the free pion mass m_π. (b) The pion inverse propagator $D^{-1}(\omega, \mathbf{p}_\pi; \rho_B)$ as a function of ω at ρ_B=0.25 fm^{-3} and $|\mathbf{p}_\pi| = 1.8m_\pi$.

In the π^c-condensed phase, protons and neutrons are strongly coupled due to the isospin-flip through the p-wave πN interaction. As a result, the baryons exist as quasiparticles η, the state of which is superposition of the neutron and proton states, i.e., $|\eta_s\rangle = \cos\phi|n_s\rangle \pm i\sin\phi|p_s\rangle$ for s=$\pm$1/2, where s denotes the spin state. The neutrino emissions through the β decay processes of the quasiparticles become kinematically possible even in highly dense degenerate matter owing to the presence of the classical π^c field, and the resulting rapid cooling mechanisms by neutrino emission in the π^c-condensed phase bring about important consequences on thermal evolution of neutron stars[15] (see also section 6).

2.3. Neutral Pion Condensation and Alternating Layer Spin (ALS) Structure

In the case of neutral pion (π^0) condensation, the spin-isospin sound mode appears around $\omega \sim 0$, $|\mathbf{p}_\pi| \sim 0$ in addition to the usual π^0 branch. But it enters into a kinematical region of particle-hole continuum and it disappears inside the continuum region. The π^0 condensation occurs as a consequence of softening of the spin-isospin sound branch, and the critical density is given by the condition that the excitation energy ω for the spin-isospin sound becomes zero : $D_{\pi^0}^{-1}(\omega = 0, \mathbf{p}_\pi; \rho_B) = 0$, together with $\partial D_{\pi^0}^{-1}(\omega, \mathbf{p}_\pi; \rho_B)/\partial|\mathbf{p}_\pi| = 0$ at $\omega = 0$. Since the chemical potential of the π^0 is zero, the classical π^0 field can be taken as a standing wave type such that $\langle\varphi_0\rangle = A\sin|\mathbf{p}_\pi|z$, where A is the amplitude of π^0 condensation, and the direction of the π^0 momentum is taken to be that of the z axis.

For the fully developed phase of π^0 condensation beyond the critical density, Takatsuka, Tamiya, Tatsumi, and Tamagaki presented a model for structure change of the baryonic system in the existence of π^0 condensation as shown in Fig. 2[16, 17]. Under the potential U_π generated by π^0 condensation ($U_\pi = -\tilde{f}\tau_3\boldsymbol{\sigma}\cdot\nabla\langle\varphi_0\rangle$), baryons are localized along one dimensional (z) direction in a specific spin-isospin order, and behave as a Fermi gas in the remaining two-dimensional layer. The spin-isospin of baryons changes alternately layer by layer.

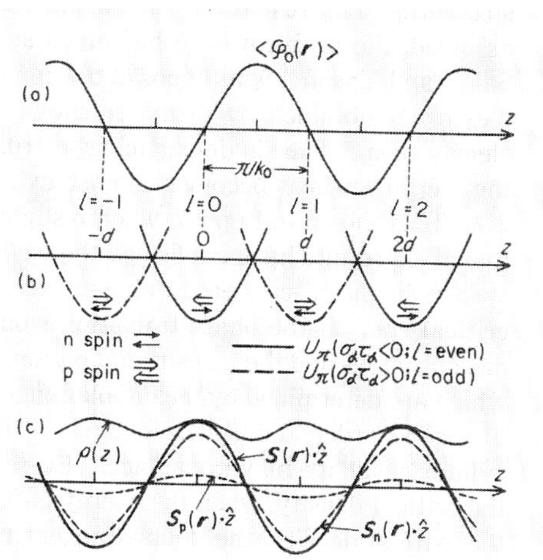

Figure 2. The ALS structure under π^0 condensation, taken from Ref. 15).

Such an ordered structure of the baryonic system, called the Alternating Layer Spin (ALS) structure, is determined selfconsistently with the classical π^0 field. It has been shown that the ALS structure is the most favorable baryonic configuration, stemming from attractive first-order effect of the one-pion-exchange tensor force within a potential description. The critical density $\rho_{\rm B}^{\rm c}$ (ALS) is estimated to be $\sim 3\rho_0$.

The ALS structure provides us with a new type of quantum solid for high-density hadronic matter just like liquid crystals (smectic A phase) in solid state physics. The quantum solid brought about by the ALS structure gives important aspects in high-density hadronic matter in relation to neutron star phenomena in particular, pulsar glitches (see 4.3).

As the most favorable type of pion condensation, possible coexistence of π^0 and π^c condensations has been considered[18, 19]. In the combined condensation of π^0 and π^c, the ALS structure develops in the z direction accompanying π^0 condensation with a standing wave type, and in the remaining two-dimensional layer, π^c condensation exists with a plane wave type. The combined π^0-π^c condensation is energetically most favorable, and takes over the characteristic features of both π^0 and π^c condensations.

2.4. Medium Effects on Pion Condensation

Pions are easily excited in nuclear matter, and they couple to particle-hole states. The critical density and EOS of pion condensation depend sensitively on the medium effects in the pion channel. Phenomenologically, the following competing effects have been considered in a realistic approach of pion condensation. One is the excitation of the isobar Δ (1232), which gives strong attraction through p-wave $\pi N\Delta$ coupling strength

$f_{\pi N\Delta} = \sqrt{72/25} \cdot f_{\pi NN}$ in the quark model and is favorable for pion condensation. Another is short-range correlation between baryons, which is simulated by the Landau-Migdal (LM) parameter in the spin-isospin channel and is unfavorable for pion condensation. The LM parameters are denoted as g'_{NN} $(=\widetilde{g}'_{NN}\widetilde{f}^2)$ with $\widetilde{f} = f_{\pi NN}/m_\pi$ for correlations between the NN^{-1}-NN^{-1} particle-hole pairs ('-1' denotes a hole state), $g'_{N\Delta}$ $(=\widetilde{g}'_{N\Delta}\widetilde{f}\widetilde{f}_\Delta)$ with $\widetilde{f}_\Delta = f_{\pi N\Delta}/m_\pi$ for NN^{-1}-ΔN^{-1} particle-hole pairs, and $g'_{\Delta\Delta}$ $(=\widetilde{g}'_{\Delta\Delta}\widetilde{f}^2_\Delta)$ for ΔN^{-1}-ΔN^{-1} particle-hole pairs. The value of the LM parameter was evaluated phenomenologically or based on the reaction matrix calculations. However, in most of the calculations in nuclear matter, the universality was assumed, i. e., $\widetilde{g}' \equiv \widetilde{g}'_{NN} = \widetilde{g}'_{N\Delta} = \widetilde{g}'_{\Delta\Delta}$, and a rather large value, $\widetilde{g}'$=0.7$-$0.9, was obtained[20]. The empirical values of the LM

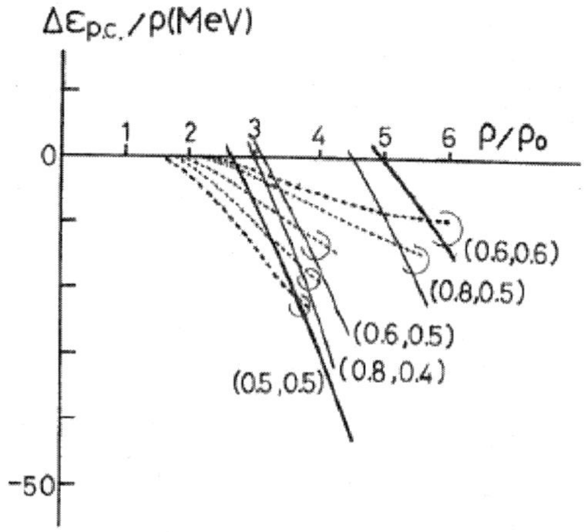

Figure 3. The energy differences per particle between the combined π^0-π^c phase and normal phase (solid lines), and those between the pure π^c-condensed phase and normal phase (dashed lines), taken from Ref. [11].

parameters have been improved in the recent experiment of the giant Gamow-Teller (GT) resonances, which are relevant to the spin-isospin sound mode of the kinematical region $\omega \sim 0$, $|\mathbf{p}_\pi| \sim 0$[21]. From the analysis of the GT sum rule for the reaction, ^{90}Zr (p,n) ^{90}Nb at 295 MeV, the quenching factor has been estimated to be $Q = 90 \pm 5\%$, which leads to constraint on the LM parameters as[22]

$$\widetilde{g}'_{N\Delta} = 0.18 \quad +0.05\widetilde{g}'_{\Delta\Delta} \text{ and } \widetilde{g}'_{NN} \simeq 0.59 \quad \text{for } \widetilde{g}'_{\Delta\Delta} < 1. \tag{13}$$

This result shows that the universality is largely broken. By the use of the LM parameters given by Eq. (13), the critical density of π^0 condensation in symmetric nuclear matter and that of π^c condensation in neutron matter have been estimated to be less than $2\rho_0$ for both condensations[23].

By accounting for the breaking of universality on the LM parameters, the EOS of pion-condensed phase in neutron matter was obtained in a chiral-symmetry approach with relaxing the universality by the use of the parameter sets $(\widetilde{g}'_{NN}, \widetilde{g}'_{N\Delta})$=(0.6$-$0.8, 0.4$-$0.5) together with the assumption $\widetilde{g}'_{N\Delta}$=$\widetilde{g}'_{\Delta\Delta}$[11] (Fig. 3). It has been shown that pure π^c condensation appears at lower densities $\rho_{\rm B} \sim 2\rho_0$ from the normal phase as a continuous phase transition (dashed lines in Fig. 3). At higher densities, depending much upon the effects of short-range correlations, the combined π^c-π^0 phase is realized as the first-order phase transition (solid lines in Fig. 3). The resulting EOS lie in the intermediate region bounded by the previous two curves based on the universality and smaller values of $\widetilde{g}' = 0.5 - 0.6$[19]. One can conclude that the previous results on the EOS and discussion about neutron star phenomena, based on the universality and with use of the smaller values of $\widetilde{g}'$, keep their validity.

3. S-wave Kaon Condensation

3.1. Role of Weak Interaction

The driving force for kaon condensation is the s-wave kaon-nucleon (KN) interactions, which consist of the scalar and vector interactions[5]. The former is simulated by the KN sigma term Σ_{KN} [$=\frac{1}{2}(m_u+m_s)\langle N|\bar{u}u+\bar{s}s|N\rangle$], which is larger than the πN sigma term by an order of magnitude due to the large strangeness quark mass m_s, reflecting the violation of chiral symmetry in SU(3) case (see Sec. 1). The vector interaction is proportional to the nucleon V-spin density, which is written as $\psi^\dagger\frac{1}{2}(V_3 + \sqrt{3}V_8)\psi = \frac{1}{4}\psi^\dagger(\tau_3 + 3)\psi$ with V_a being the vector charge. The V-spin of the proton (the neutron) is 1 (1/2), so that the antikaon (K^-) feels attractive from both proton and neutron in the vector channel, while K^+ feels repulsive. Empirically, the s-wave KN scattering lengths are obtained as $a(K^+p)$=-0.33 fm, $a(K^+n)$=-0.16 fm, $a(K^-p)$=(-0.67 +i 0.64) fm, $a(K^-n)$=(0.37 +i 0.60) fm[24]. The K^-n interaction is attractive. The repulsive K^-p scattering length is due to the pole contribution of the $\Lambda(1405)$ which lies below the KN threshold by 30 MeV.

In kaon-condensed phase in neutron stars, the net negative strangeness is kept abundant through chemical equilibrium for weak interaction processes, $n \rightleftharpoons p + K^-$, $n \rightleftharpoons p + e^- + \bar{\nu}_e$, where the chemical potentials of participating particles are related by $\mu_n = \mu_p + \mu_K$, $\mu_n = \mu_p + \mu_e$. At threshold, the onset condition of kaon condensation is given as

$$\omega(\rho_{\rm B}) = \mu_e = \mu_K = \mu \ , \tag{14}$$

where $\omega(\rho_{\rm B})$ is the lowest energy of K^- obtained at $\rho_{\rm B}$ as the zero point of the K^- inverse propagator, $D_K^{-1}(\omega; \rho_{\rm B}) = 0$.

In neutron-star matter, the lowest excitation energy ω of the K^- decreases with increase in baryon number density $\rho_{\rm B}$ due to the s-wave KN attractions. In Fig. 4, we show the kaon excitation energy ω with zero momentum in neutron-star matter, taken from Ref. [25]. We base our model of the EOS on current algebra and PCAC for description of the s-wave K-N dynamics and on the meson-exchange for the N-N interaction within the relativistic mean-field (RMF) theory. Solid curves denote the results with inclusion of the range terms and $\Lambda(1405)$-pole contribution to the s-wave $\bar{K}$-N and K-N scattering lengths. We call this energy correction the second-order effect (SOE) (see 3.2). The dashed curves denote the results without the SOE. The dash-dotted line denotes the charge chemical potential μ ($=\mu_K = \mu_e$). At a critical density, ω becomes equal to the kaon chemical potential μ_K.

As a result, the distribution function $f_K(\mathbf{p}_K)$ for the K^- diverges at $\omega = \mu$, and the K^- appears macroscopically as a Bose-Einstein condensate (BEC)[26, 27]. The critical density for K^- condensation, $\rho_{\rm B}^{\rm C}(K^-)$, is estimated as $\rho_{\rm B}^{\rm C}(K^-) = (3-4)\rho_0$ depending on the value of Σ_{KN}. Formation of kaon condensates is mainly controlled by *weak* interactions,

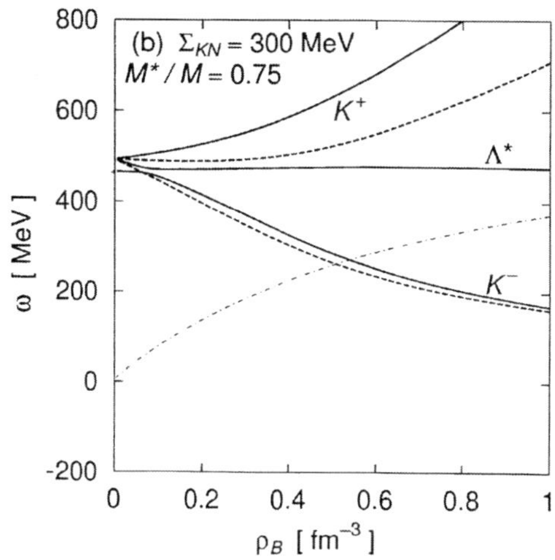

Figure 4. The kaon excitation energy ω with zero momentum in neutron-star matter in the RMF calculation, taken from Ref. [25].

in particular, by the thermal kaon process, $n + n \rightarrow n + p + K^-$, which gives the most efficient reaction rate among the nonequilibrium weak reaction processes at high temperature and density relevant to protoneutron stars. Kinetics of kaon condensation in newly born neutron stars has been investigated through obtaining time scales of saturation of thermal kaons, growth and subsequent saturation of kaon condensates and relaxation to chemically equilibrated state[28]. It has been found that the time scale of the development of kaon condensates is much smaller than that of the cooling (several tens of seconds) for the relevant temperatures of protoneutron stars. The time delay for the formation of kaon condensates may not have much influence on the delayed collapse of a neutron star.

3.2. Phenomenological $\bar{K}$-N Interaction

Motivated by studies of kaon condensation, the in-medium kaon properties have been investigated from both theoretical and experimental viewpoints.

The kaon inverse propagator D_K^{-1} includes the self energy which is related to the forward KN scattering amplitudes. It is to be noted that there are different ways of off-shell extrapolation from the KN scattering amplitudes in obtaining the kaon self energy: One is derived from chiral perturbation theory and the other is based on low energy theorems of current algebra and PCAC. These approaches lead to different off-shell properties of the kaon excitation energy[29, 30, 7, 25]. The relations between these approaches have also been discussed[29, 31]. Based on the former way with the chiral effective Lagrangian (3), the s-wave KN scattering amplitudes near threshold cannot be reproduced empirically without higher order terms in chiral expansion and a pole contribution from the $\Lambda(1405)$ (abbreviated to Λ^*)[30, 7, 32]. These corrections enter into the kaon self energy. Nevertheless, it has been shown that part of the higher order terms for K^-, the range term [$\propto \omega(K^-)^2$] , becomes small in high-density matter, since the K^- excitation energy $\omega(K^-)$ decreases with density owing to the s-wave KN attractive interaction. Furthermore, the pole contribution from the Λ^* becomes negligible since the K^-N threshold lies far below the pole of the Λ^* at high density due to the decrease in $\omega(K^-)$. Hence these corrections become irrelevant to the s-wave K^- excitation energy in high-density matter[7, 25, 32, 30], while the K^+ excitation energy is largely changed due to the repulsive effects from the higher-order corrections.

The kaon optical potential in nuclear matter has been deduced by the use of chiral models[33], a G-matrix method with consistently taking into account the modification of kaons in medium[34], dispersion relation approach[35] and so on. Several authors obtained the K^- optical potential phenomenologically from kaonic atom data[36, 37, 38]. There is still controversy on the strength of the kaon optical potential, although recent experimental results on the subthreshold K^+K^- production in relativistic heavy-ion collisions and proton-nucleus collisions suggest a substantial decrease in the antikaon effective mass[39].

It should be noted that deeply bound kaonic nuclei have been recently proposed based on the strongly attractive kaonic potential[40, 41, 42]. In Ref. [40], a very high-density state comparable to that of neutron stars is predicted to form inside the nuclei. If experimental achievement of such nuclei is established, it will provide us with information on kaon dynamics in high-density matter at zero temperature, especially, the possible existence of kaon condensation.

4. Static and Dynamic Properties of Meson-Condensed Neutron Stars

The existence of meson condensations has important implications for static and dynamic properties of neutron stars.

4.1. Internal Structure of Neutron Stars

A phase transition to a meson-condensed phase leads to softening of the EOS, making a neutron star compact with higher central density and smaller radius than normal neutron stars. As a result, a maximum gravitational mass of a neutron star becomes small[43]. Here we show a result of the EOS of the kaon-condensed phase and the effect of the EOS on the gross structure of neutron stars[25]. The energy per baryon (upper half curves) and the energy difference from the normal phase (lower half curves) in the kaon-condensed phase (solid lines) are shown in Fig. 5 (a). The parameter sets PM1$\sim$ PM3 are distinguished by the choice of the effective nucleon mass at $\rho_{\rm B} = \rho_0$. In Fig. 5 (b), the gravitational mass (M in the unit of the solar mass $M_\odot$) to radius (R) relations of neutron stars with kaon condensates are shown.

From the observation of PSR 1913+16, its mass was obtained as $1.442 \pm 0.003 M_\odot$[44]. The statistical analysis of mass determination of binary pulsar systems gave the upper limit of the mass $\sim 1.6 M_\odot$[45], and measurements of masses of neutron stars in X-ray binaries

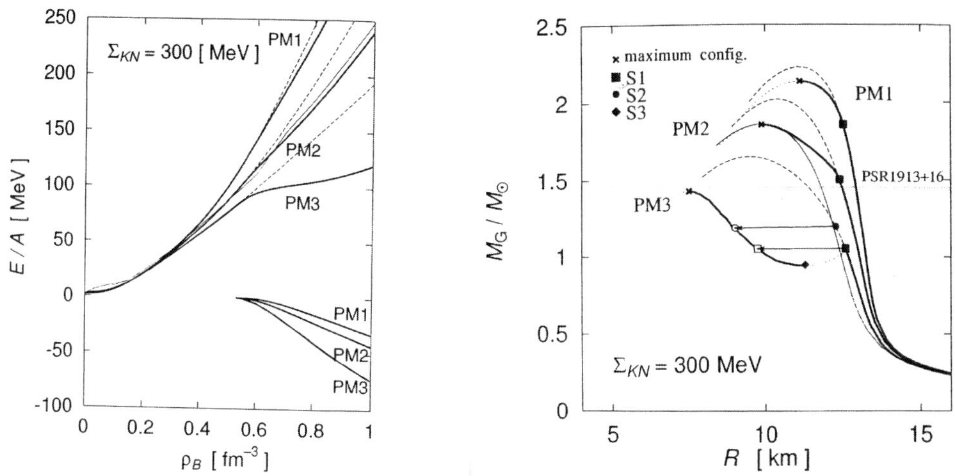

Figure 5. (a) Left panel: The energy per baryon in the kaon-condensed phase in the RMF theory (solid lines). The corresponding ones for the normal phase are shown in dashed lines. The lower half curves show the energy difference from the normal phase. (b) Right panel: The gravitational mass to radius relation of neutron stars with kaon condensates. Branches with normal EOS are shown with dashed curves. The maximum mass configuration is denoted by crosses. Both figures (a) and (b) are taken from Ref. [25].

led to a narrow mass range with $1.35\pm0.27M_\odot$[46]. The mass should be even larger ($\sim 2.0M_\odot$), if the recent analyses from the observations of the quasiperiodic oscillations are confirmed[47]. The mass-radius relation was constrained from the analyses of the gravitationally redshifted absorption lines of the X-ray bursts[48]. Recently the observation of the mass M=$1.97\pm0.04M_\odot$ for the binary millisecond pulsar PSR J1614-2230 has been reported by Shapiro delay measurement[49]. If this result is confirmed, it will give a stringent constraint on the allowable EOS. The mass inferred from the EOS should be consistent with these observations.

The significant softening will make phase transition of a first order. Here we mention consequences from a first-order phase transition induced by kaon condensation, following recent development of studies of kaon condensation. We can also see common properties in case of pion condensation[15].

Effects of first-order phase transition on the internal structure of neutron stars have been discussed by the use of the Maxwell's construction by which the phase equilibrium was obtained. However, it has been pointed out by Glendenning that the Gibb's condition determines the appropriate phase equilibrium in case more than one chemical potentials are present corresponding to conserved charges such as the electromagnetic charge, baryon number and so on[50]. According to the Gibb's condition, the first-order phase transition may imply a mixed phase where droplets of a kaon condensate are immersed in the normal phase. Recently the effects of Coulomb energy and surface tension on the mixed phase have been discussed[51, 52, 53, 54, 55].

There are several works on a possible formation of the crystalline mixed phase on a Coulomb lattice and on the stability of the lattice[52, 58]. The nucleation of a kaon droplet has also been considered[54]. In Ref. [56], the nonuniform structure of kaonic pasta in the mixed phase has been studied by the use of a density functional theory within the RMF. The part of the numerical results is shown in Fig. 6. It has been shown through a series of their works that the Maxwell's construction is effectively justified as a result of the charge screening effects[55, 56, 57].

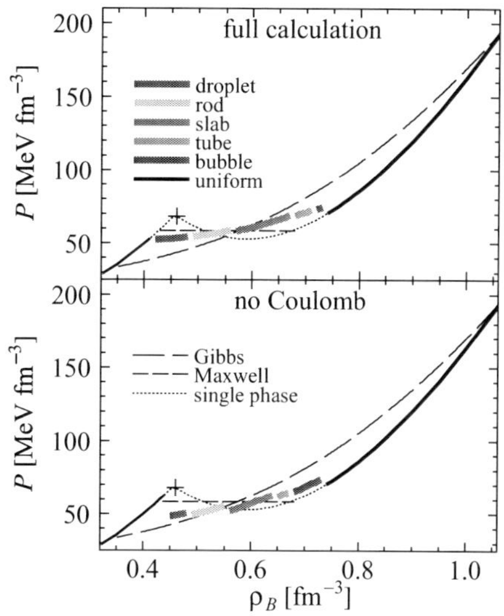

Figure 6. The pressure P versus the baryon number density ρ_B including the kaonic pasta in the mixed phase after the Gibbs condition is applied, taken from Ref. [56].

4.2. Dynamic Properties Associated with the First-Order Phase Transition

The first-order phase transition has also important effects on the dynamical properties of neutron stars. A minicollapse from a supercompressed metastable neutron star to a stable kaon-condensed star has been considered within a relativistic mean field model[25] (see also Fig. 5). The energy release after the minicollapse has been estimated, and it has been found that the released energy, $10^{49} - 10^{52}$ erg depending on a scale of the first-order phase transition, is sufficient for explaining the observed anomalous gamma-ray burst energies.

In relation to an early stage of neutron stars, a formation scenario of a low-mass black hole by a delayed collapse of a hot neutron star after a phase transition to a kaon-condensed star has been presented by Brown and Bethe[59], and a numerical simulation has been elaborated[60].

Delayed collapse of protoneutron stars has been discussed in detail by several authors[61, 62]. Tatsumi and Yasuhira obtained the EOS with kaon condensates at finite temperature including thermal effects and neutrino degeneracy. Based on this EOS, they discussed the gravitational stability of the protoneutron stars against a phase transition to a kaon-condensed neutron star during the deleptonization era[62]. The EOS of kaon condensation at finite-temperature has also been elucidated in the Brueckner-Hartree-Fock theory[63].

4.3. Pulsar Glitches

Pulsar glitches give information on the inner structure of neutron stars. They are characterized by sudden discontinuous spin up of rotation and its subsequent relaxation to a steady state of radio pulsars. From the temporal behavior of the angular velocity after the spin up, one can see two kinds of relaxation time scales, one is long with $\tau_1 \sim 80$ days and the other is short with $\tau_2 \sim$ several days. According to the usual glitch scenario, sudden spin up is attributed to a catastrophic unpinning of vortices of the nucleon $^1S_0$ superfluids from nuclei in the inner crust region of a neutron star and the gradual decrease of rotation results from creeping of vortex toward the outside region of the star[64, 65]. The vortex creep theory has been considered as one of the standard postglitch models[64].

On the other hand, Takatsuka and Tamagaki presented a corequake model with combined π^0-π^c condensation for giant glitches like the Vela pulsar[66]. This model is based on the π^0-π^c condensed phase in a central core region of a neutron star. The ALS structure accompanying π^0 condensation presents a solid core. As the pulsar spins down, stress of the core is accumulated, and finally the core cracks. At that time, the pulsar suddenly spins up as a result of angular momentum conservation. Large energy release due to the corequake is carried away through a rapid cooling by π^c condensation before the next glitch. The two time scales τ_1 and τ_2 which are observed in the postglitch timing behavior can be explained by the two relaxation times of the $^1S_0$ superfluids in the crust and the $^3P_2$ superfluids in the core. They also succeeded in descriminating giant glitches like the Vela pulsar and small ones (smaller ratio of change of angular velocity to stationary angular velocity than that of giant glitches by two-orders of magnitude) like the Crab pulsar: The Vela pulsar is assumed to have a large mass $\sim 1.6 M_\odot$. Then the central density in the inner core exceeds the critical density of π^0-π^c condensation, so that the core has the π^0-π^c-condensed phase, which induces the giant glitch by a corequake. On the other hand, the Crab pulsar is assumed to have a smaller mass $\sim 1.2 M_\odot$, where the central density does not reach the critical density of the π^0-π^c condensation, so that there is no ALS solid core. In this case, only a crustquake, which is smaller in magnitude than the giant glitches, occurs. Thus the new corequake model based on pion condensation has a possibility of consistently explaining the observed glitch phenomena.

5. Coexistence of Pion and Kaon Condensations

According to the preceding results, pion condensation appears at a lower baryon number density ($\sim 2\rho_0$) as compared with that for kaon condensation ($\sim 3-4\rho_0$). Possible coexistence of pion and kaon condensations (π-K condensation) in neutron-star matter has been studied in a simple model based on the $SU(3)_L \times SU(3)_R$ chiral effective Lagrangian[5, 67]. The classical meson fields in the nonlinear representation are taken to be plane-wave forms:

$$\pi^-(\mathbf{r},t) = \frac{f}{\sqrt{2}}\theta \cdot \cos\beta e^{-i(\mu t - \mathbf{p}_\pi \cdot \mathbf{r})} \ , \ K^-(\mathbf{r},t) = \frac{f}{\sqrt{2}}\theta \cdot \sin\beta e^{-i(\mu t - \mathbf{p}_K \cdot \mathbf{r})} \tag{15}$$

with a parameter β representing the relative strengh of pion and kaon condensations.

It has been shown that charged pion condensation is suppressed as kaon condensation grows with increase in baryon number density[5]. However, there was a problem in the original works that the pion spatial momentum diverges beyond a scope of chiral perturbation theory at a density where pion condensation disappears in a kaon-condensed phase. In Ref. [68], the coexistence problem of π-K condensation has been reconsidered by taking into account the recoil terms between mesons and baryons in the nonrelativistic expansion of the baryonic Hamiltonian with respect to $1/m_N$ in order to avoid the divergence of the pion momentum. Only the charged pion and kaon have been considered.

The result is that kaon condensates begin to appear in the π^c condensed phase at the baryon number density $\sim$0.5 fm^{-3} and that the energy gain due to kaon condensates is added to that due to pion condensates as density increases. Pion and kaon condensations can coexist over the relevant densities in neutron stars, increasing their amplitudes monotonically with increase in density. In the ground state of the π-K condensed phase, only lower energy eigenstate of quasiparticles, $|\eta\rangle = \alpha|n\rangle + \beta|p\rangle$ which tends to $|n\rangle$ in the limit $\theta \to 0$, fills the Fermi sea ("one-Fermi sea" approximation is valid), which is a different picture from a pure kaon-condensed case: In the K^- condensed phase, both protons and neutrons fill independent Fermi seas. Toward realistic studies of π-K condensation, one has to take into account the effecs of baryonic potentials in matter, short-range correlations between baryons, vertex renormalizations at the meson-baryon vertices, hyperon-mixing and other subthreshold resonances such as $\Lambda(1405)$, $\Sigma(1385)$, $\Delta(1232)$ etc.

6. Rapid Cooling Processes in Meson Condensations

6.1. Neutrino Emission in Meson Condensates

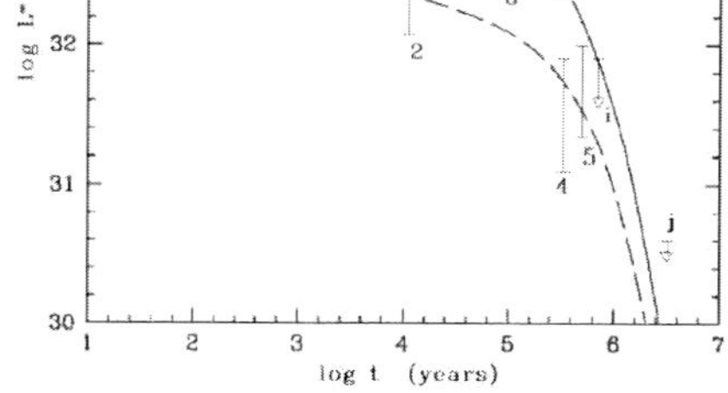

Figure 7. The photon luminosities as functions of the age of a neutron star, taken from Tsuruta et al., Ref. [72]. The solid line denotes the standard cooling without any exotic agents and the dashed line including the extra rapid cooling under pion condensation.

In ordinary neutron-star matter without exotic states, the nucleons and electrons are strongly degenerate, and the energy-momentum conservation is not satisfied for the following neutrino and antineutrino emission processes as one-nucleon processes, $n \to$

$p + e^- + \bar{\nu}_e$, $p + e^- \rightarrow n + \nu_e$, which is called the direct Urca process[3]. Therefore, a spectator nucleon participates in the reaction so as to meet the kinematical condition, and the two-nucleon process, $n + n \rightarrow n + p + e^- + \bar{\nu}_e$, $n + p + e^- \rightarrow n + n + \nu_e$, becomes possible. This reaction is the standard cooling process in the normal phase and is called the modified Urca process[71]. In Fig. 7, we show a standard cooling curve of neutron stars without any exotic phase with the gravitational mass M=1.2$M_\odot$ ($M_\odot$ being the solar mass) by a solid line, which is taken from Tsuruta et al.[72]. The photon luminosity observed at infinity, which is connected with the surface temperature of a neutron star, is given as a function of the age of a neutron star. It should be noted that there are several point sources observed, (c): PSR J0205+6449 in 3C58, (2): the Vela pulsar, (4) Geminga of which the surface temperatures are too low to be explained by any standard cooling theories. Extra cooling mechanisms leading to rapid cooling of neutron stars are needed for some observed point sources.

In the presence of meson condensates, the classical meson field supplies the system with energy and momentum, and the one-nucleon process becomes possible. The following reactions have been mainly considered. One is the quasiparticle (η) Urca process ,

$$\eta + \langle\pi^-\rangle \rightarrow \eta + e^- + \bar{\nu}_e \ , \ \ \eta + e^- \rightarrow \eta + \langle\pi^-\rangle + \nu_e \ , \tag{16}$$

for pion condensation[73], and another is the Kaon-induced Urca (KU) process,

$$n\ (p) + \langle K^-\rangle \rightarrow n\ (p) + e^- + \bar{\nu}_e \ , \ \ n\ (p) + e^- \rightarrow n\ (p) + \langle K^-\rangle + \nu_e \ , \tag{17}$$

for the s-wave kaon condensation[74, 70].

These weak processes can be treated in the framework of chiral symmetry.[4]. In the left panel of Fig. 8, we show the relation between the baryon and lepton momenta on the Fermi surface contributing to momentum conservation, and the lowest order diagrams for the quasiparticle Urca process in charged pion condensates. The right panel of Fig. 8 is the same but for the s-wave K^- condensation.

Here we take up only the KU process associated with the s-wave K^- condensation. The relevant weak Hamiltonian for KU is given by the current×current type, $H_{\rm W} = \frac{G_F}{\sqrt{2}} J_{\rm h}^\mu \cdot l_\mu + {\rm h.c.}$ with l_μ [$=\bar{\psi}_e\gamma_\mu(1-\gamma_5)\psi_\nu$] being the charged leptonic current and the charged hadronic current, $J_{\rm h}^\mu = \cos\theta_{\rm C}(V_{1+i2}^\mu - A_{1+i2}^\mu) + \sin\theta_{\rm C}(V_{4+i5}^\mu - A_{4+i5}^\mu)$, where $\theta_{\rm C}(\simeq 0.24)$ is the Cabibbo angle , and V_a^μ and A_a^μ are the vector and axial-vector currents, respectively. The matrix elements are given from the transformed Hamiltonian:

$$\widetilde{H}_{\rm W} = \hat{U}_K^{-1} H_{\rm W} \hat{U}_K = \frac{G_F}{\sqrt{2}} \widetilde{J}_{\rm h}^\mu l_\mu + {\rm h.c.}\ , \tag{18}$$

where $\hat{U}_K = \exp(i\mu_K Q)\exp(i\theta Q_4^5)$, which is a unitary operator generating a kaon-condensed state, $|K\rangle = \hat{U}_K|0\rangle$. By the use of current algebra , the tranformed hadronic

[3]The possibility of the direct Urca process depends on the density-dependence of the nuclear symmetry energy[69, 70].

[4]Neutrino absorption processes in meson condensates, such as $\nu_e + \eta + \langle\pi^-\rangle \rightarrow e^- + \eta$ and $\nu_e + n(p) + \langle K^-\rangle \rightarrow e^- + n(p)$, which are relevant to neutrino opacities, can also be considered within this framework[75, 76]

current is shown to have a form,

$$\begin{aligned} \widetilde{J}_{\rm h}^{\mu} &= \hat{U}_K^{-1} J_{\rm h}^{\mu} \hat{U}_K \\ &= e^{-i\mu_K t}\Bigg[\cos\theta_c\Big\{(V_{1+i2}^{\mu} - A_{1+i2}^{\mu})\cos(\theta/2) + i(V_{6-i7}^{\mu} - A_{6-i7}^{\mu})\sin(\theta/2)\Big\} \\ &+ \sin\theta_c\Big\{(V_4^{\mu} - A_4^{\mu}) + i\cos\theta(V_5^{\mu} - A_5^{\mu}) - i\sin\theta\Big(V_V^{\mu} - A_V^{\mu}\Big)\Big\}\Bigg], \end{aligned} \quad (19)$$

where $V_V^{\mu} \equiv (V_3^{\mu} + \sqrt{3}V_8^{\mu})/2$ and $A_V^{\mu} \equiv (A_3^{\mu} + \sqrt{3}A_8^{\mu})/2$. The matrix elements for KU come from the term proportional to $\sin\theta$ in the last term of Eq. (19). One can see that the kaon chemical potential μ_K is supplied to energy conservation of the reaction, stemming from the time dependent phase factor $\exp(-i\mu_K t)$.

The weak reactions (16) and (17) give large neutrino emissivities since the available phase space is large as compared with that of two-nucleon reactions like the modified Urca process. As an example, the ratio of the emissivity for KU to that for the modified Urca process is roughly $O(\mu_{\rm B}/T)^2 \sim 10^{6-8}$ with $\mu_{\rm B}$ being the nucleon chemical potential depending on the temperature. Once such an exotic phase appears, it has a significant effect on thermal evolution of neutron stars[77, 78].

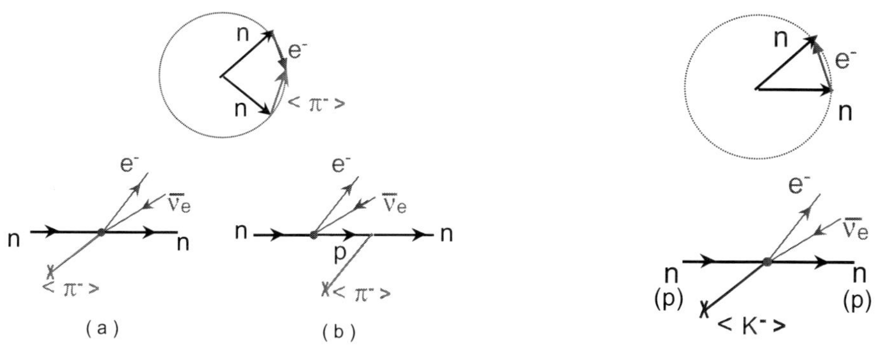

Figure 8. Left panel: The upper part representing the momentum conservation between the participating degenerate nucleons, leptons, and the classical π^c field. The lower part showing the lowest order diagrams for the quasiparticle Urca process in charged pion condensates. Right panel: The same as the left panel but for the Kaon-induced Urca process in the s-wave kaon condensates.

6.2. Effects of Superfluidity

The neutrino emissivities themselves in meson condensates lead to too rapid cooling of the neutron stars that need an enhanced cooling scenario, as other cooling agents such as quarks, hyperons, and the direct Urca process do[78]. Thus suppression mechanisms are needed in order to counterbalance the too rapid cooling. Nucleon superfluidity suppresses both specific heat of nuclear matter and neutrino emissivities owing to the existence of an energy gap Δ. For the neutrino emissivities, the available phase space is largely restricted due to the energy gap Δ.

In an inner core of a neutron star, the $^3\mathrm{P}_2$ pairing is most favorable for the neutrons, while $^1\mathrm{S}_0$ pairing is for the protons, which is deduced from the nucleon-nucleon scattering phase shifts for the isospin $T = 1$ pair[79]. For the normal phase, Takatsuka and Tamagaki studied the neutron $^3P_2$ ($+^3F_2$) pairing by generalizing the Bogoliubov transformation to the pairings with nonzero angular momentum[80].

They also studied pairing of quasineutrons (η) under pion condensation[81]. They found that the energy gap is sensitive to the effective nucleon mass m^*_η at a high density. They obtained $m^*_\eta(\pi^c) \simeq m^*_n$(Normal)=(0.8−0.7)$m_N$, and $m^*_\eta(\pi^0-\pi^c) \simeq m^*_n(\pi^0)$ ~0.9 m_N for $\rho_B = 1 - 3\rho_0$ with m_N being the free nucleon mass. Note that the effective masses in the π^0-condensed and combined π^0-π^c condensed phases do not decrease much even at high densities $\sim 3\rho_0$, reflecting the localization picture of the ALS structure. The large effective mass results in large pairing energy gap. Thus the nucleon superfluid in the $^3P_2$ ($+^3F_2$) channel can exist at rather high densities with the existence of π^0 condensation. On the other hand, the quasineutron pairing in the π^c-condensed phase has been shown to be less favorable owing to the attenuation effect coming from a factor $1 - \cos\phi \cdot \sin\phi$ with the mixing angle ϕ.

With the effects of quasineutron superfluidity in π^0-π^c condensed phase in the inner core of a neutron star, the cooling behavior of the pion-condensed neutron star has been obtained as shown by a dashed line in Fig. 7. It has been shown that the anomalously low temperatures for several neutron stars including the three candidates in this figure can be consistently explained with the rapid cooling under π^c condensation and its moderate suppression due to the coexistent quasineutron superfluidity under π^0 condensation.

7. Coexistence of Kaon Condensation and Hyperons

7.1. Outline of the Model

As shown in Chapter 6, hyperons (Λ, Σ^-, Ξ^-, $\cdots$) are expected to be mixed in the ground state of neutron-star matter [82, 83, 84, 85, 86, 87, 88, 89, 90, 91, 92, 93, 94]. We call the hyperon-mixed neutron-star matter *hyperonic matter* throughout this chapter. With regard to coexistence or competition of kaon condensation with hyperons in neutron stars, it has been pointed out that onset density of the s-wave kaon condensation subject to the condition (14) in hyperonic matter is shifted to a higher density[9, 83, 84] : The negative charge carried by the electron is replaced by that of the negatively charged hyperon, so that the charge chemical potential μ (=$\mu_e = (3\pi^2\rho_e)^{1/3}$) is diminished as compared with the case of neutron-star matter without hyperons. As a result, the lowest K^- energy $\omega(\rho_B)$ meets the charge chemical potential μ at a higher density. Subsequently, several works on the onset density and the EOS for the kaon-condensed phase in hyperonic matter have been elaborated with the relativistic mean-field theory[95, 96] and the quark-meson coupling models[97, 98]. Recently the in-medium kaon dynamics and mechanisms of kaon condensation stemming from the s and p-wave kaon-baryon interactions in hyperonic matter have been investigated[99, 100, 101].

We reexamine coexistence of kaon condensation and hyperonic matter. We consider the s-wave kaon condensation and incorporate the kaon-baryon interaction within the effective chiral Lagrangian (3). The classical kaon field is chosen to be spatially uniform:

$$\langle K^- \rangle = \frac{f}{\sqrt{2}}\theta e^{-i\mu t} . \tag{20}$$

We take into account the proton, neutron, Λ, Σ^-, and Ξ^- for the octet baryons, and ultrarelativistic electrons e^- for leptons. By imposing the charge neutrality condition and baryon number conservation, we introduce the effective Hamiltonian density following Eq. (6) as

$$\mathcal{E}_{\text{eff}} = \mathcal{E} + \mu(\rho_p - \rho_{\Xi^-} - \rho_{\Sigma^-} - \rho_K - \rho_e) + \nu(\rho_p + \rho_\Lambda + \rho_{\Xi^-} + \rho_n + \rho_{\Sigma^-}) \,, \quad (21)$$

where $\mathcal{E}$ is the total energy density of the kaon-condensed phase, and ρ_a (a= p, Λ, Ξ^-, n, Σ^-, K^-, e^-) are the number densities of the particles a. From the extremum conditions for $\mathcal{E}_{\text{eff}}$ with respect to variation of ρ_a, one obtains the following relations,

$$\mu_K = \mu_e = \mu_n - \mu_p = \mu_{\Sigma^-} - \mu_n = \mu_{\Xi^-} - \mu_\Lambda = \mu \,, \quad (22a)$$

$$\mu_\Lambda = \mu_n = -\nu \,, \quad (22b)$$

where μ_a (a= p, Λ, Ξ^-, n, Σ^-, K^-, e^-) are the chemical potentials, $\mu_a = \partial\mathcal{E}/\partial\rho_a$. Equations (22a) and (22b) imply that the system is in chemical equilibrium for the weak interaction processes, $n \rightleftharpoons p + K^-$, $n \rightleftharpoons p + e^-(+\bar{\nu}_e)$, $n + e^- \rightleftharpoons \Sigma^-(+\nu_e)$, $\Lambda + e^- \rightleftharpoons \Xi^-(+\nu_e)$, and $n \rightleftharpoons \Lambda(+\nu_e + \bar{\nu}_e)$.

The baryon-baryon interaction is taken into account in a potential picture. The baryon potential in hyperonic matter is defined by $V_i = \partial\mathcal{E}_{\text{pot}}/\partial\rho_i$ ($i = p, \Lambda, \Xi^-, n, \Sigma^-$), where $\mathcal{E}_{\text{pot}}(\rho_p, \rho_\Lambda, \rho_{\Xi^-}, \rho_n, \rho_{\Sigma^-})$ is the potential energy density parameterized in terms of ρ_i[91]. and the parameters are determined so as to reproduce the nuclear saturation properties and baryon potential depths deduced from the recent hypernuclear experiments[42]. The energy density of the kaon-condensed phase in hyperonic matter is written as

$$\begin{aligned}\mathcal{E} &= \sum_{i=p,\Lambda,\Xi^-,n,\Sigma^-} \frac{2}{(2\pi)^3}\int_{|\vec{k}|\le p_F(i)} d^3|\vec{k}|\sqrt{|\vec{k}|^2 + M_i^{*2}} + \mathcal{E}_{\text{pot}} \\ &+ \frac{1}{2}f^2\mu^2\sin^2\theta + f^2 m_K^2(1-\cos\theta) + \frac{\mu^4}{4\pi^2} \,, \end{aligned} \quad (23)$$

where M_i^* ($i = p, \Lambda, \Xi^-, n, \Sigma^-$) is the effective baryon mass: $M_i^* = M_i - \Sigma_{Ki}(1-\cos\theta)$ with Σ_{Ki} being the "K-baryon sigma term" [99, 101]. The $\bar{K}$-baryon scalar interaction is absorbed into M_i^*. The first and second terms on the r.h.s. of Eq. (23) are the baryonic parts, the third and fourth terms are kinetic and free mass parts of the condensed kaons, and the last term is the electron kinetic energy.

The classical K^- field equation is derived from $\partial\mathcal{E}_{\text{eff}}/\partial\theta = 0$. One obtains

$$\mu^2\cos\theta - m_K^2 + \frac{1}{f^2}\sum_{i=p,\Lambda,\Xi^-,n,\Sigma^-}\rho_i^s\cdot\Sigma_{Ki} + \frac{\mu}{f^2}\left(\rho_p + \frac{1}{2}\rho_n - \rho_{\Xi^-} - \frac{1}{2}\rho_{\Sigma^-}\right) = 0 \,, \quad (24)$$

where ρ_i^s is the scalar density of baryon i given by $\rho_i^s = \frac{2}{(2\pi)^3}\int_{|\vec{k}|\le p_F(i)} d^3|\vec{k}|\frac{M_i^*}{\sqrt{|\vec{k}|^2 + M_i^{*2}}}$. The third and fourth terms on the l. h. s. of Eq. (24) come from the s-wave $\bar{K}$-baryon scalar and vector interactions, respectively.

The ground state energy density $\mathcal{E}$ is obtained under the constraints of the chemical equilibrium conditions (22), together with the charge neutrality condition, $\partial\mathcal{E}_{\text{eff}}/\partial\mu = 0$, and the baryon number conservation, $\partial\mathcal{E}_{\text{eff}}/\partial\nu = \rho_{\text{B}}$.

7.2. The Ground State Properties and Possibility of Self-Bound Objects

The interplay between K^- condensates and hyperons reveals in the EOS and characteristic features of the fully-developed kaon-condensed phase such as density-dependence of particle fractions. In Fig. 9, we show the total energy per baryon $\mathcal{E}/\rho_{\rm B}$ [(a)] and the particle fractions $\rho_a/\rho_{\rm B}$ [(b)] for the kaon-condensed phase in hyperonic matter as functions of the baryon number density $\rho_{\rm B}$ for Σ_{Kn} = 305 MeV. The bold solid line (the thin dash-dotted line) stands for the case of the kaon-condensed phase with Λ, Σ^-, and Ξ^--mixing (Λ, Σ^--mixing). For comparison, the EOS for noncondensed hyperonic matter with and without mixing of Ξ^- hyperons are depicted in the bold and thin dotted lines, respectively. One

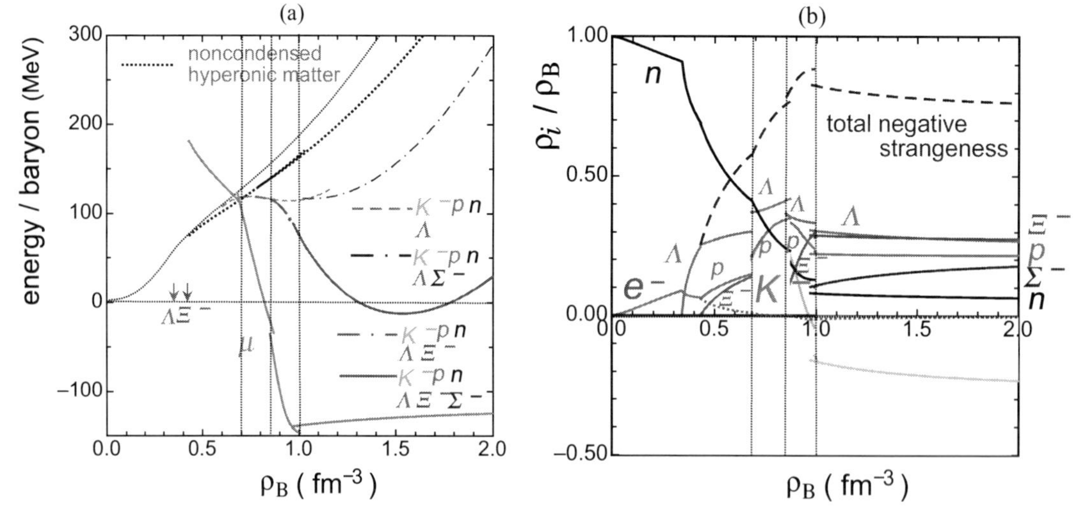

Figure 9. (a) The total energy per baryon, $\mathcal{E}/\rho_{\rm B}$, and (b) the particle fractions, $\rho_a/\rho_{\rm B}$ (a= p, Λ, Ξ^-, n, Σ^-, K^-, e^-), for the kaon-condensed phase in hyperonic matter as functions of the baryon number density $\rho_{\rm B}$ for Σ_{Kn} = 305 MeV. In (a), the curve for the charge chemical potential μ is also attached in the figure.

can see that in the kaon-condensed phase realized in hyperonic matter, the EOS becomes considerably soft. In particular, there appears a local energy minimum (a density isomer) at a certain density $\rho_{\rm B,min}$ (=1.54 fm^{-3}) in the case of Ξ^--mixing, as seen in Fig. 9 (a) (solid line), and the pressure [$P = \rho_{\rm B}^2 \partial(\mathcal{E}/\rho_{\rm B})/\partial\rho_{\rm B}$] becomes negative at some density intervals below $\rho_{\rm B,min}$.

From Fig. 9 (b), one finds that appearance of both kaon condensates and the Ξ^- and Σ^- hyperons leads to considerable suppression of the electron fraction, since the negative charge of the electron is replaced by that of kaon condensates, Ξ^- and Σ^- hyperons. Accordingly, the charge chemical potential μ decreases with increase in the baryon number density through the relation $\rho_e = \mu^3/(3\pi^2)$ [see Fig. 9 (a)]. It becomes even negative above the density $\rho_{\rm B}$ = 0.80 fm^{-3}. In case of the negative μ, the $\bar{K}$-Ξ^- vector interaction works attractively, as seen in Eq. (24), and its attractive contribution to the energy is large owing to the coefficient in front of ρ_{Ξ^-} (proportional to the V-spin charge). Hence at high densities where $\mu < 0$, mixing of the Ξ^- hyperons leads to a significant energy decrease in the kaon-condensed phase.

For $\mu < 0$, the positrons (e^+) appear in place of the electrons. It is to be noted that the K^- number density ρ_{K^-}, which is written as

$$\rho_{K^-} = \mu f^2 \sin^2\theta + \left(\rho_p + \frac{1}{2}\rho_n - \rho_{\Xi^-} - \frac{1}{2}\rho_{\Sigma^-}\right)(1 - \cos\theta) , \tag{25}$$

becomes negative at high densities, $\rho_B \gtrsim 1.0$ fm^{-3}. In this density region, kaon condensates carry the positive charge and positive strangeness. Also the chiral angle θ has a maximal value of π (rad), so that the free part of ρ_{K^-} [the first term in (25)] vanishes, and the negative contribution to ρ_{K^-} stems from the $\bar{K}$-baryon vector interaction [the second term in (25)].

The energy minimum appears as a consequence of considerable softening of the EOS due to both kaon condensation and hyperon-mixing and recovering of the stiffness of the EOS at very high densities due to the increase in the repulsive interaction between baryons as well as kaons.

In the preceding studies[101], it has been shown that the dependence of the total energy on the baryon number density is mainly determined by the two contributions: (I) the contribution from the classical kaons as the sum of the s-wave scalar kaon-baryon interaction and the free parts of the condensed kaon energy, and (II) the baryon potential energy $\mathcal{E}_{\rm pot}/\rho_{\rm B}$. The contribution (I) decreases with increase in density, while the contribution (II) increases as density increases. The attractive effect from the contribution (I) is pronounced due to mixing of hyperons as compared with the case without hyperons, lowering the total energy at a given density. In addition, the repulsive effect from the contribution (II) is much weakened due to mixing of hyperons at a given density, since the repulsive interaction between nucleons is avoided by lowering the relative nucleon density through mixing of hyperons.[5] As a result, the total energy is much reduced, leading to significant softening of the EOS for the kaon-condensed phase in hyperonic matter.

At higher densities above $\rho_{\rm B,min}$, the decrease of the contribution (I) gets slightly moderate, while the repulsive interaction between baryons becomes strong and so the increase of the contribution (II) with density gets more remarkable. As a result, the total energy per baryon increases rapidly with density, and the EOS recovers the stiffness at high densities. Thus the stiffness of the EOS depends on the quantitative behavior of the repulsive interaction between baryons at high densities, which has ambiguity depending upon the many-body hyperon-nucleon and hyperon-hyperon repulsive interaction terms in the potential energy density $\mathcal{E}_{\rm pot}$.

At high densities, the protons, Λ, Ξ^-, and Σ^- hyperons are equally populated in the kaon-condensed phase, i. e., $\rho_p/\rho_{\rm B}$, $\rho_\Lambda/\rho_{\rm B}$, $\rho_{\Sigma^-}/\rho_{\rm B}$ = 20–30 %, whereas the neutrons almost disappear. The total negative strangeness ratio, $\rho_{\rm strange}/\rho_{\rm B}$, gets larger with increase in density, and it reaches 80% at high densities. Such a high strangeness fraction implies a close connection between the kaon-condensed phase in hyperonic matter and strange matter where u, d and s quarks are almost equally populated in quark matter.

[5] This suppression mechanism of the repulsive interaction between nucleons is essentially the same as that for softening of the EOS in the noncondensed hyperonic matter as pointed out in Ref. [92, 93].

It has been suggested in Ref. [101] that a density isomer state with kaon-condensates in hyperonic matter implies the existence of self-bound objects, which can be bound essentially without gravitation on a scale of an atomic nucleus to a neutron star just like a strangelet and a strange star or other exotic matter. The density isomer state is located at a local energy minimum with respect to the baryon number density as a metastable state, but it decays only through multiple weak processes, so that it is regarded as substantially stable.

7.3. P-wave Kaon Condensation

Most of the studies on kaon condensation has been based on the s-wave $\bar{K}$-baryon interaction. In the presence of hyperons in matter, there are necessarily p-wave $\bar{K}NY$ interaction, as seen in the effective chiral Lagrangian (3). In ordinary neutron-star matter, the p-wave contribution may have a minor role as compared with s-wave $\bar{K}$-N interaction . However, it may become important in highly dense matter where hyperons are largely mixed. Here we consider the interplay between kaon condensation and hyperons by the use of the $\mathrm{SU(3)_L \times SU(3)_R}$ chiral effective Lagrangian where both p-wave $\bar{K}NY$ and s-wave kaon-baryon interactions are incorporated[13, 99]. The baryonic potentials $\mathcal{E}_{\rm pot}$ in hyperonic matter are taken into account in the same form as that in Sec. 7.1.

In Fig. 10, the kaon minimal excitation energies ω are shown as functions of baryon number density $\rho_{\rm B}$ at a typical kaon momentum $|\mathbf{k}|$ and for the Kn sigma term $\Sigma_{Kn} =$ 305 MeV. In this calculation, the p-wave KNY interactions relevant to the Λ, Σ^-, and Ξ^-

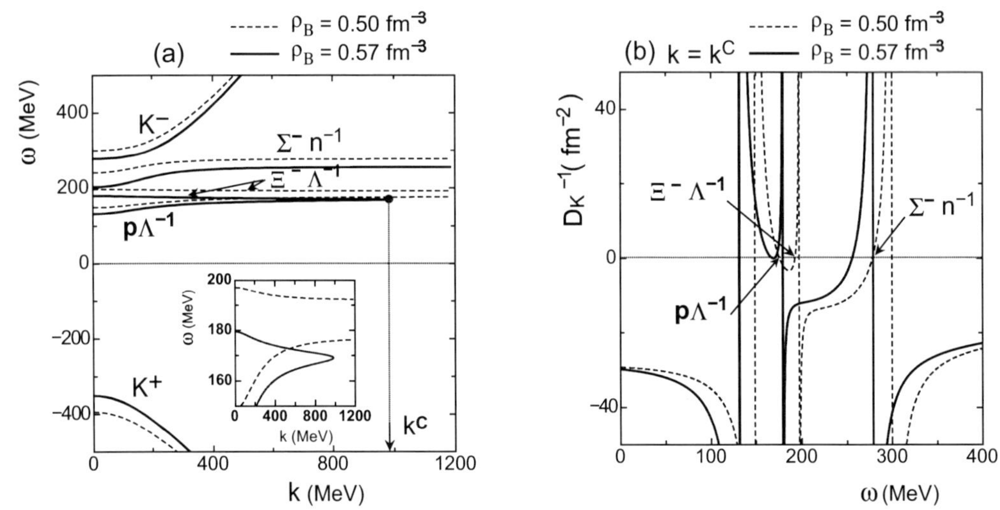

Figure 10. (a) The excitation energies ω for kaonic modes in hyperonic matter as functions of the kaon momentum $|\mathbf{k}|$ for $a_3 = -0.9$ and $\rho_{\rm B}$=0.50 fm^{-3} (dashed lines) and $\rho_{\rm B}$=0.57 fm^{-3} (solid lines). (b) The inverse kaon propergator $D_K^{-1}(\omega, \mathbf{k}; \rho_{\rm B})$ as a function of ω at $|\mathbf{k}| = k^c$ for th same set of a_3 and $\rho_{\rm B}$ as in (a). These figures are taken from Ref. [99].

hyperons have been taken into account. For these high densities, the $p\Lambda^{-1}$ branch which has a quantum number of the K^+ appears instead of the Λp^{-1} branch.

As the baryon number density increases, the locations of the particle-hole branches become lower owing to the p-wave interactions, as shown in Fig. 10 (a). In particular, the $\Xi^-\Lambda^{-1}$ and $p\Lambda^{-1}$ branches get close to each other, and they merge at certain density

($\rho_B \simeq 0.57$ fm^{-3}) with a critical momentum k^C (=984 MeV). At $|\mathbf{k}| = k^C$, these two excitation modes merge at the ω axis in the $D_K^{-1} - \omega$ plane At this density and momentum, the double-pole condition for the kaon inverse propagator D_K^{-1}, $D_K^{-1}(\omega, \mathbf{k}; \rho_B) = 0$, $\partial D_K^{-1}(\omega, \mathbf{k}; \rho_B)/\partial\omega = 0$ is satisfied. It means that a pair of the two modes, $\Xi^-\Lambda^{-1}$ and $p\Lambda^{-1}$, are created spontaneously with no cost of energy because the energy of $p\Lambda^{-1}$ mode with the quantum number K^+ is to be reversed in sign. Hence the system is unstable with respect to a pair creation of $[\Xi^-\Lambda^{-1}]$ and $[p\Lambda^{-1}]$ modes. This instability originates from the p-wave kaon-baryon interaction, and we call this instability p-wave kaon condensation. The onset mechanism of the p-wave kaon condensation is similar to that of pion condensation, where a driving force is given by the p-wave πN interaction[2, 3, 20, 4] (see 2.2).

In the p-wave kaon-condensed phase, baryons exist as quasiparticles, $|\tilde{p}_s\rangle = \alpha|p_s\rangle + \beta|\Lambda_s\rangle + \gamma|\Sigma_s^0\rangle$, $|\tilde{n}_s\rangle = \delta|n_s\rangle + \epsilon|\Sigma_s^-\rangle$, etc. with the spin state $s(=\pm 1/2)$. In case of p-wave kaon-condensation, enhanced cooling will occur through the weak processes, $\tilde{Y} + \langle K^-\rangle \to \tilde{Y} + e^- + \bar{\nu}_e$, $\tilde{Y} + e^- \to \tilde{Y} + \langle K^-\rangle + \nu_e$ (for $\tilde{Y} = \tilde{p}$, $\tilde{\Lambda}$, $\tilde{n}$, $\tilde{\Sigma}^-$), where $\tilde{Y}$ is a quasiparticle whose state is given by superposition of the nucleon and hyperon states. These weak processes may also lead to the increase of the absolute value of total negative strangeness of the system[11].

8. Summary and Concluding Remarks

We have surveyed the present studies of meson condensations as a typical hadronic phase in high-density matter and discussed implications for neutron-star phenomena. Meson-condensed phase leads to softening of the EOS of hadronic matter, affecting static and dynamic properties of neutron stars. The ALS structure accompanying π^0 condensation represents solid-like aspects. The existence of the quantum solid is essential to explain the giant glitches of pulsars in a corequake model. The crystalline mixed phase of kaon condensation on a Coulomb lattice may have an influence on pinning of nucleon superfluid vortices formed in neutron star core region, which may have a relevance to a vortex creep model for pulsar glitches. Neutrino emission processes are enhanced in meson condensates, which leads to rapid cooling of neutron stars. The existence of meson condensation in high-density hadronic matter provides us with important consequences on the EOS and phenomenology related with neutron stars.

Towards understanding of the QCD phase diagram, close connection of the meson-condensed phase with other hadronic and/or quark-gluon phases should be clarified through possible phase transitions and phase equilibrium between these phases. In this respect, the kaon-condensed phase in hyperonic matter leads to large (negative) strangeness fraction. This result suggests that kaon-condensed phase in the hadronic picture may be considered as a pathway to strange quark matter, where quark degrees of freedom may appear through chiral phase transitions. In a quark picture, a variety of quark phases including color superconductivity have been elaborated[102]. In particular, kaonic modes may be condensed in the color-flavor locked phase. It is interesting to clarify the relationship between kaon condensation in the hadronic phase and that in the quark phase and the possible transition between the two phases.

References

[1] R. F. Sawyer, *Phys. Rev. Lett.* 29 (1972), 382. D. J. Scalapino, ibid. 29 (1972), 392.

[2] A. B. Migdal, *Rev. Mod. Phys.* 50 (1978), 107.
A. B. Migdal, E. E. Saperstein, M. A. Troitsky and D. N. Voskresensky, *Phys. Rep.* 192 (1990), 179.

[3] G. Baym and D. K. Campbell, in Meson and Nuclei, ed. M. Rho and D. H. Wilkinson, (North Holland, Amsterdam, 1979), Vol. III, p. 1031.

[4] T. Kunihiro, T. Muto, T. Takatsuka, R. Tamagaki and T. Tatsumi, *Prog. Theor. Phys. Suppl.* 112 (1993).

[5] D. B. Kaplan and A. E. Nelson, *Phys. Lett.* B175 (1986), 57; 179 (1986), 409(E).

[6] T. Tatsumi, *Prog. Theor. Phys. Suppl.* 120 (1995), 111.

[7] C. -H. Lee, G. E. Brown, D. -P. Min and M. Rho, *Nucl. Phys.* A 585 (1995), 401.

[8] C. -H. Lee, *Phys. Rep.* 275 (1996), 197 .

[9] M. Prakash, I. Bombaci, M. Prakash, P. J. Ellis, J. M. Lattimer, R. Knorren, *Phys. Rep.* 280 (1997), 1.

[10] D. K. Campbell, R. F. Dashen and J. T. Manassah, *Phys. Rev.* D 12 (1975) 979, 1010.

[11] T. Muto, R. Tamagaki and T. Tatsumi, *Prog. Theor. Phys. Suppl.* 112 (1993) 159.

[12] T. Tatsumi, *Prog. Theor. Phys.* 80 (1988) 22.

[13] T. Muto, *Prog.. Theor. Phys.* 89 (1993) 415.

[14] J. Gasser, H. Leutwyler and M. E. Sainio, *Phys. Lett.* B 253 (1991), 252.

[15] T. Muto, T. Takatsuka, R. Tamagaki, T. Tatsumi, *Prog. Theor. Phys. Suppl.* 112 (1993) 221.

[16] T. Takatsuka, K. Tamiya, T. Tatsumi, R. Tamagaki, *Prog.. Theor. Phys.* 59 (1978), 1933.

[17] T. Takatsuka, T. Tatsumi, R. Tamagaki, *Prog. Theor. Phys. Suppl.* 112 (1993), 67 (Chap. III).

[18] K. Tamiya and R. Tamagaki, *Prog. Theor. Phys.* 66 (1981) 1361.

[19] T. Muto and T. Tatsumi, *Prog. Theor. Phys.* 78 (1987) 1405.

[20] T. E. O. Ericson and W. Weise, Pions and Nuclei (Oxford Univ. Press, Oxford, 1988).

[21] T. Wakasa et al., *Phys. Rev.* C 55 (1997), 2909.

[22] T. Suzuki and H. Sakai, *Phys. Lett.* B 455 (1999), 25.

[23] T. Suzuki, H. Sakai, T. Tatsumi, in Proc. RCNP Int. Symp. Nuclear Responses and Medium Effects (Tokyo, Universal Academy Press, 1999), 77.

[24] A. D. Martin, *Nucl. Phys.* B 179 (1981), 33.

[25] H. Fujii, T. Maruyama, T. Muto, T. Tatsumi, *Nucl. Phys.* A597 (1996) 645.

[26] T. Muto and T. Tatsumi, *Phys. Lett.* B 283 (1992), 165.

[27] G. E. Brown, K. Kubodera, M. Rho, V. Thorsson, *Phys. Lett.* 291 (1992), 355.

[28] T. Muto, T. Tatsumi, N. Iwamoto, *Phys. Rev.* D 61 (2000), 063001; ibid (2000), 083002 .

[29] H. Yabu, S. Nakamura, F. Myhrer, K. Kubodera, *Phys. Lett.* B 315 (1993),17.
H. Yabu, F. Myhrer, K. Kubodera, *Phys. Rev.* D 50 (1994), 3549.

[30] C. -H. Lee, H. Jung, D. -P. Min, M. Rho, *Phys. Lett.* B 326 (1994) 14.

[31] V. Thorsson and A. Wirzba, *Nucl. Phys.* A 589 (1995), 633.

[32] T. Waas,N. Kaiser and W. Weise, *Phys. Lett.* B365 (1996) 12 ; B379 (1996) 34 .
T. Waas and W. Weise, *Nucl.Phys.* A625 (1997) 287.

[33] A. Ramos and E. Oset, *Nucl. Phys.* A 671 (2000), 481.
J. Schaffner-Bielich, V. Koch, M. Effenberger, *Nucl. Phys.* A 669 (2000), 153.

[34] L. Tolós, A. Ramos, A. Polls, *Phys. Rev.* C 65 (2002), 054907.

[35] A. Sibirtsev and W. Cassing, *Nucl. Phys.* A 641 (1998), 476; nucl-th/9909024.

[36] C. J. Batty, E. Friedman and A. Gal, *Phys. Rep.* 287 (1997), 385.

[37] S. Hirenzaki, Y. Okumura, H. Toki, E. Oset, A. Ramos, *Phys. Rev.* C 61 (2000), 055205.

[38] A. Cieply, E. Friedman, A. Gal, J. Mareš, *Nucl. Phys.* A 696 (2001), 173.

[39] For a review, P. Senger and H. Ströbele, *J. Phys.* G 25 (1999), R59.

[40] Y. Akaishi and T. Yamazaki, *Phys. Rev.* C 65 (2002) 044005; A. Dote, H. Horiuchi, Y. Akaishi and T. Yamazaki, *Phys. Rev.* C 70 (2004) 044313.

[41] T. Kishimoto, *Phys. Rev. Lett.* 83 (1999), 4701.
M. Iwasaki et al., Nucl. Instrum. *Methods Phys. Res.* A 473 (2001) 286.

[42] B. F. Gibson, K. Imai, T. Motoba, T. Nagae and A. Ohnishi (Eds.), *Nucl. Phys.* A 835 (2010) 1.

[43] For example, J. M. Lattimer and M. Prakash, Astrophys. J. 550 (2001) 426; *Phys. Rep.* 442 (2007) 109.

[44] J. H. Taylor and J. M. Weisberg, *Astrophys. J.* 345 (1989), 434.

[45] L. S. Fin, *Phys. Rev. Lett.* 73 (1994), 1878.

[46] S. E. Thorsett and D. Chakrabarty, *Astrophys. J.* 512 (1999), 288.

[47] M. C. Miller, F. K. Lamb, D. Psaltis, Astrophys. J. 508 (1998), 791.

[48] M. Y. Fujimoto and R. E. Taam, *Astrophys. J.* 305 (1986), 246.
J. Cottam, F. Paerels, M. Mendez, Nature 420 (2002), 51.

[49] P. Demorest, T. Pennucci, S. Ransom, M. Roberts and J. W. T. Hessels, *Nature* 467 (2010) 1081.

[50] N. K. Glendenning, *Phys. Rep.* 342 (2001), 393.

[51] H. Heiselberg, C.J. Pethick, E.F. Staubo, *Phys. Rev. Lett.* 70 (1993), 1355.

[52] M. Christiansen, N. K. Glendenning, J. Schaffner-Bielich, *Phys. Rev.* C 62 (2000), 025804.

[53] T. Norsen and S. Reddy, *Phys. Rev.* C 63 (2001), 065804.

[54] T. Norsen, *Phys. Rev.* C 65 (2002), 045805.

[55] For a quark-hadron phase transition, D. N. Voskresensky, M. Yasuhira and T. Tatsumi, *Phys. Lett.* B 541 (2002) 93; *Nucl. Phys.* A 723 (2003) 291.

[56] T. Maruyama, T. Tatsumi, D. N. Voskresensky, T. Tanigawa, *Phys. Rev.* C 73 (2006) 035802.

[57] T. Maruyama, T. Tatsumi, T. Endo, S. Chiba, *Recent res. Devel. Phys.* 7 (2006) 1.

[58] J. J. Zach, *Phys. Rev.* C 65 (2002), 123002.

[59] G. E. Brown and H. A. Bethe, *Astrophys. J.* 423 (1994), 659.

[60] T. W. Baumgarte, S. L. Shapiro, S. Teukolsky, *Astrophys. J.* 443 (1995), 717; 458 (1996), 680.

[61] J. A. Pons et al., Phys. Rev. C 62 (2000), 035803.; *Astrophys. J.* 553 (2001), 382.

[62] T. Tatsumi and M. Yasuhira, *Nucl. Phys.* A 653 (1999), 133.
M. Yasuhira and T. Tatsumi, *Nucl. Phys.* A 690 (2001), 769.

[63] A. Li, X. R. Zhou, G. F. Burgio and H.-J. Schulze, *Phys. Rev.* C 81 (2010) 025806.

[64] M. A. Alpar, P. W. Anderson, D. Pines, J. Shaham, *Astrophys. J.* 276 (1984), 325.

[65] Y. Mochizuki and T. Izuyama, *Astrophys. J.* 440 (1995), 263.

[66] T. Takatsuka and R. Tamagaki, *Prog. Theor. Phys.* 79 (1988), 274; 82 (1989), 945.

[67] H. D. Politzer and M. B. Wise, *Phys. Lett.* B 273 (1991) 156.

[68] T. Muto, *Prog. Theor. Phys.* Supplement 153 (2004) 174.

[69] J. M. Lattimer, C. J. Pethick, M. Prakash, P. Haensel, *Phys. Rev. Lett.* 66(1991), 2701.

[70] H. Fujii, T. Muto, T. Tatsumi, R. Tamagaki, *Nucl. Phys.* A571 (1994), 758 ; *Phys. Rev.* C50 (1994), 3140.

[71] B. L. Friman and O. V. Maxwell, *Astrophys. J.* 232 (1979), 541.

[72] S. Tsuruta, M. A. Teter, T. Takatsuka, T. Tatsumi, R. Tamagaki, *Astrophys. J.* 571 (2002), L143.

[73] O. V. Maxwell et al., *Astrophys. J.* 216 (1977), 77.
T. Tatsumi, *Prog. Theor. Phys.* 68 (1982), 1231.
T. Muto and T. Tatsumi, *Prog. Theor. Phys.* 79 (1988), 461.

[74] G. E. Brown, K. Kubodera, D. Page, P. Pizzecherro, *Phys. Rev.* D 37 (1988), 2042.
T. Tatsumi, *Prog. Theor. Phys.* 80 (1988), 22.

[75] R. F. Sawyer and A. Soni, *Astrophys. J.* 216 (1977), 73.
T. Tatsumi, *Prog. Theor. Phys. Suppl.* 91 (1987), 299.

[76] T. Muto, T. Tatsumi, N. Iwamoto, *Phys. Rev.* D 67 (2003), 103002.

[77] H. Umeda, K. Nomoto, S. Tsuruta, T. Muto, T. Tatsumi, *Astrophys. J* 431 (1994), 309.

[78] S. Tsuruta, *Phys. Rep.* 292 (1988), 1.

[79] For a review, D. J. Dean and M. Hjorth-Jensen, *Rev. Mod. Phys.* 75 (2003) 607.

[80] T. Takatsuka, *Prog. Theor. Phys.* 48 (1972) 1517.
T. Takatsuka and R. Tamagaki, *Prog. Theor. Phys. Suppl.* 112 (1993) 27.

[81] T. Takatsuka and R. Tamagaki, *Prog. Theor. Phys.* 67 (1982) 1649; *Prog. Theor. Phys. Suppl.* 112 (1993) 107.

[82] N. K. Glendenning, *Astrophys. J.* 293 (1985) 470. ; N. K. Glendenning and S. A. Moszkowski, *Phys. Rev. Lett.* 67 (1991) 2414.

[83] P. J. Ellis, R. Knorren, and M. Prakash, *Phys. Lett.* B349, 11 (1995).
R. Knorren, M. Prakash, and P. J. Ellis, *Phys. Rev.* C52, 3470 (1995).

[84] J. Schaffner and I. N. Mishustin, *Phys. Rev.* C53, 1416 (1996).

[85] S. Pal, M. Hanauske, I. Zakout, H. Stöcker, and W. Greiner, *Phys. Rev.* C 60, 015802 (1999).

[86] M. Hanauske, D. Zschiesche, S. Pal, S. Schramm, H. Stöcker, and W. Greiner, *Astrophys. J.* 537, 958 (2000).

[87] P. K. Sahu, *Phys. Rev.* C 62, 045801 (2000).

[88] H. Huber, F. Weber, M. K. Weigel, and Ch. Schaab, *Int. J. Mod. Phys.* E 7, 301 (1998).

[89] M. Baldo, G. F. Burgio, and H. -J. Schulze, *Phys. Rev.* C 58, 3688 (1998) ; ibid C 61, 055801 (2000).

[90] I. Vidaña, A. Polls, A. Ramos, M. Hjorth-Jensen, and V. G. J. Stoks, *Phys. Rev.* C 61, 025802 (2000) ; I. Vidaña, A. Polls, A. Ramos, L. Engvik, and M. Hjorth-Jensen, *Phys. Rev.* C 62, 035801 (2000).

[91] S. Balberg and A. Gal, *Nucl. Phys.* A625, 435 (1997).

[92] S. Nishizaki, Y. Yamamoto, and T. Takatsuka, *Prog. Theor. Phys.* 108 (2002) 703.

[93] T. Takatsuka, *Prog. Theor. Phys.* Supplement 156, 84 (2004).

[94] S. T. Uechi and H. Uechi, *Adv. High Energy Phys.* 2009 (2009) 640919.

[95] S. Pal, D. Bandyopadhyay, and W. Greiner, *Nucl. Phys.* A 674, 553 (2000).

[96] S. Banik and D. Bandyopadhyay, *Phys. Rev.* C 63, 035802 (2001) ; ibid C 64, 055805 (2001).

[97] D. P. Menezes, P. K. Panda, and C. Providencia, *Phys. Rev.* C 72, 035802 (2005).

[98] C. Y. Ryu, C. H. Hyun, S. W. Hong, and B. T. Kim, *Phys. Rev.* C 75 (2007) 055804.

[99] T. Muto, *Nucl. Phys.* A 697 (2002) 225.

[100] E. E. Kolomeitsev and D. N. Voskresensky, *Phys. Rev.* C 68 (2003) 015803.

[101] T. Muto, *Phys. Rev.* C 77 (2003) 015810; *Nucl. Phys.* A 804 (2008) 322.

[102] A. Ohnishi et al. (Eds.), *Prog. Theor. Phys.* Supplement 186 (2010) 1.

In: Neutron Stars
Editors: Hiroshi Uechi et al, pp. 199-231
ISBN: 978-1-62081-123-8

Chapter 8

MIXED PHASES DURING THE PHASE TRANSITIONS

***Toshitaka Tatsumi*[1,*], *Nobutoshi Yasutake*[2,†] *and Toshiki Maruyama*[3,‡]**
[1]Department of Physics, Kyoto University, Kyoto, Japan
[2]Research Center for the Early Universe, School of Science,
University of Tokyo,Bunkyo-ku, Tokyo, Japan
[3]Advanced Science Research Center, Japan Atomic Energy Agency,
Tokai, Ibaraki, Japan

Abstract

Quest for a new form of matter inside compact stars compels us to examine the thermodynamical properties of the phase transitions. We closely consider the first-order phase transitions and the phase equilibrium on the basis of the Gibbs conditions, taking the liquid-gas phase transition in asymmetric nuclear matter as an example. Characteristic features of the mixed phase are figured out by solving the coupled equations for mean-fields and densities of constituent particles self-consistently within the Thomas-Fermi approximation. The mixed phase is inhomogeneous matter composed of two phases in equilibrium; it takes a crystalline structure with a unit of various geometrical shapes, inside of which one phase with a characteristic shape, called "pasta", is embedded in another phase by some volume fraction. This framework enables us to properly take into account the Coulomb interaction and the interface energy, and thereby sometimes we see the mechanical instability of the geometric structures of the mixed phase. Thermal effect on the liquid-gas phase transition is also elucidated.

Similarly hadron-quark deconfinement transition is studied in hyperonic matter, where the neutrino-trapping effect as well as the thermal effect is discussed in relation to the properties of the mixed phase. Specific features of the mixed phase are elucidated and the equation of state is presented.

1. Introduction

Equation of state (EOS) for matter inside compact stars is sometimes characterized by the phase transitions. There have been expected various first-order phase transitions in nuclear

*E-mail address: tatsumi@ruby.scphys.kyoto-u.ac.jp
†E-mail address: yasutake@resceu.s.u-tokyo.ac.jp
‡E-mail address: maruyama.toshiki@jaea.go.jp

matter, such as liquid-gas transition in the low densities, meson condensations and hadron-quark deconfinement transitions at high-density region [1, 2, 3, 4]. If such phase transitions occurs inside compact stars, they affect EOS as well as thermal, dynamical or magnetic properties [5, 6, 7, 8]. There has been a theoretical development and recent studies have shown that the inhomogeneous states called "pasta" may emerge as a mixed phase during the phase transitions, based on the Gibbs conditions [9]. Historically, the mixed phases have been mostly treated very naively by applying the Maxwell construction to get the EOS in thermodynamic equilibrium. Ravenhall et al. discussed the possible emergence of the pasta structures at low-density nuclear matter, since uniform nuclear matter is not stable at low densities [10]. Their pasta phase may be regarded as a mixed phase following liquid-gas phase transition. In 1990 Glendenning has pointed out that the phase equilibrium in multi-component system or in the case of plural chemical potential corresponding to the conserved quantities such as in neutron star matter must be carefully treated based on the more general Gibbs conditions [3, 11]. He emphasized that the phase transitions should be treated in a proper way in neutron-star matter, which is specified by two conserved quantities, electromagnetic charge as well as baryon number. In particular the local charge-neutrality condition used in the case of the Maxwell construction should be replaced by the global one in the mixed phase. Consequently the pressure becomes no more constant as in the Maxwell construction, but still increases in the mixed phase. He also demonstrated that the mixed phase is extended in the rather wide density region within a bulk calculation, where two sets of semi-infinite matter are considered in phase equilibrium, separated by a sharp boundary, and the pressure balance, chemical equilibrium and the global charge-neutrality conditions are imposed in the absence of the Coulomb interaction. It is important to see that particle fractions in each phase are no more the same to each other for given particle fractions as a whole. Nowadays we know that such phase transitions are rather common in other fields as well [12, 13, 14, 15, 16, 17] and we call them *non-congruent* phase transitions, following Iosilevskiy [17]. Generally any phase transition in a system with two or more conserved charges must be non-congruent.

In this calculation the Coulomb interaction is discarded and we can by no means estimate how important it is, or the surface energy can not be taken into account. If we consider a more realistic situation, we must consider inhomogeneous matter with various geometric structures called pasta, and take into account the finite-size effects such as the Coulomb interaction and the surface tension. Heiselberg et al. have demonstrated that such finite-size effects are important for the mixed phase in the context of hadron-quark transition [18]. Treating the pasta structure as in Ref. [10], they simply considered uniform particle densities in both phases. They showed a possibility of that the finite-size effects may largely restrict the pasta region. Subsequently, Voskresensky et al. have shown that the rearrangement of particle densities and the screening effect for the Coulomb interaction becomes important, and the large surface tension gives the mechanical instability of the geometrical structures in the mixed phase [19]. Following this idea, we have studied the pasta structures brought by the kaon condensation and hadron-quark deconfinement transition [20, 21, 22]. More recently we have extended our framework to deal with the finite temperature case or the neutrino-trapping matter, which may be relevant to supernova explosions or neutron star mergers [23].

In this chapter we first present thermodynamic concepts of the non-congruent transition by considering the liquid-gas phase transition in asymmetric nuclear matter (section 2). In Section 3 some results will be presented by a numerical calculation about the pasta structures in asymmetric nuclear matter. In Section 4 we will present the results about hadron-quark deconfinement transition, where the effects of finite temperature or neutrino trapping are also discussed. Section 5 is devoted to summary and concluding remarks.

2. Phase Equilibria in Thermodynamics

2.1. Maxwell Construction in the Pure System

Before beginning our discussion about the non-congruent transitions, let us briefly summarize more familiar case in the pure system, where the Maxwell rule can be applied to construct the equilibrium curve. Since there is only a single component, it is trivially congruent. Typical isotherms for liquid-gas phase transition are sketched for the van der Waals fluid with EOS,

$$P = \frac{RT}{V-b} - \frac{a}{V^2}, \tag{1}$$

in Figure 1, where we can see that the gas phase (low-density phase) and the liquid phase (high-density phase), both of which are thermodynamically stable due to $\partial P/\partial V < 0$ are clearly separated by the thermodynamically unstable region $\partial P/\partial V > 0$ at temperature T below the critical temperature T_c. Above T_c two phases are indistinguishable[1]. The Gibbs conditions for phase equilibrium read in this case as $P^G = P^L, \mu^G = \mu^L$ besides $T_G = T_L$ in terms of pressure P_i, chemical potential μ_i and temperature T_i in each phase $i = G$ or L. We can then construct the EOS by the graphical method, the Maxwell construction (see Figure 1).

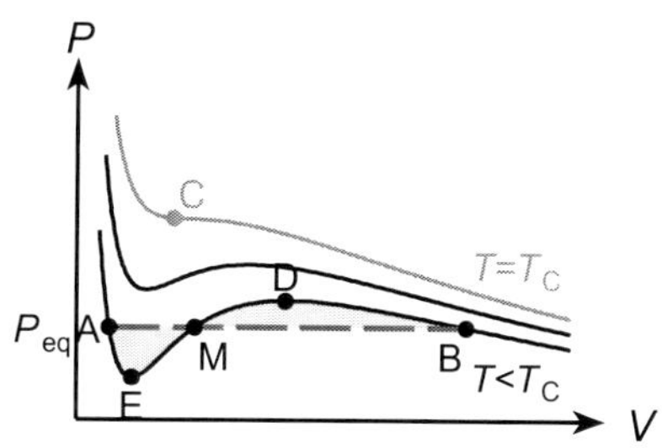

Figure 1. Isotherms of the van der Waals fluid on the P-V plane.

For a given T and fixed number N, points A and B denote the initial and final points of the phase coexistence, subjected to the equal-area condition,

$$\int_{AEM} V(P)dP = \int_{BDM} V(P)dP, \tag{2}$$

[1]Recently the supercritical fluid above the critical point has attracted much interest, especially in the context of various applications [24].

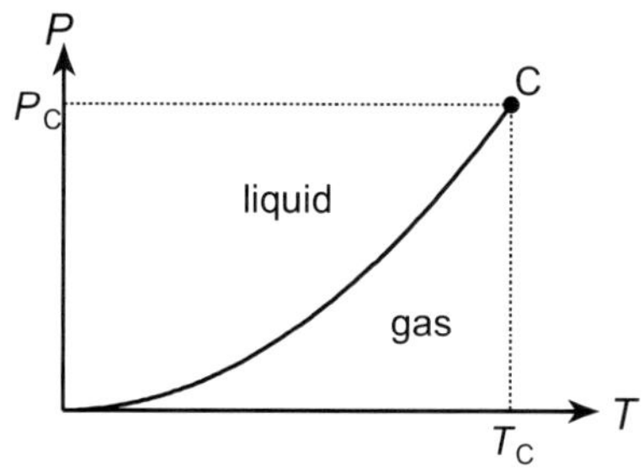

Figure 2. Saturation curve on the P-T plane.

which is equivalent with $\mu^L = \mu^G$ by way of the thermodynamic relation, $dG = Nd\mu = VdP$ with the Gibbs free energy G. Alternatively the double tangent method can be applied for the Helmholtz free energy $F = G - PV$. Anyway we can draw the *binodal curve* (coexistence curve) by connecting the equilibrium points, A and B, by changing T. At the same time we can also draw the *spinodal curve* by connecting the limit points of the stable region D and E where $\partial P/\partial V = 0$ by changing T. At the point E, where $\partial^2 P/\partial V^2 > 0$, the phase is intrinsically stable against the shrinkage of the volume, while it is unstable for the expansion of the volume. The similar argument can apply for the point D by exchanging the volume variations. Consequently, the segments AE and BD show the metastable regions.

We can draw the saturation curve, which is defined by the equilibrium pressure as a function of T, on the pressure-temperature plane, as in Figure 2. Then two phases are separated by a single curve terminated at the critical temperature T_c, which means P is constant in the mixed phase. In the following section we shall see that it spreads to involve a finite mixed-phase region in the general case with multi-component, since pressure still increases in the mixed phase. More interestingly the mixed phase emerges with peculiar geometrical structures.

2.2. Non-Congruent Transition and Structured Mixed Phase within the Bulk Calculation

First we briefly summarize the basic concepts of the non-congruent transition in a clear fashion by the bulk calculation without the Coulomb interaction or surface energy. We consider the liquid-gas transition at subnuclear densities in asymmetric nuclear matter (ASNM) as an example [25]. The proton-number fraction $Y_p \equiv N_p/N (0 \leq Y_p < 0.5)$ then specifies ASNM. Symmetric nuclear matter (SNM) with $Y_p = 0.5$ is the special case and the phase transition is congruent (see section 3.2)

It is well-known that EOS of nuclear matter exhibits a thermodynamically forbidden region, which separates "liquid" and "gas" phases at subnuclear densities. This phase transition looks like the first-order, so that we must treat the phase-equilibrium in the mixed phase. Since there are two conserved charges, electromagnetic charge and baryon number in ASNM, we must discuss the phase transition in the *binary* system. Different from the pure component system, thermodynamic behavior of the mixed phase becomes complicated in this case [24, 26]; two coexisting phases can vary the chemical composition with keeping the total charges.

Thermodynamical potential for ASNM is defined as

$$\Omega(T, V, \{\mu_i\}_{i=n,p}) = -\beta^{-1}\ln \mathrm{Tr}\exp\left[-\beta\left(\hat{H} - \sum_{i=n,p}\mu_i \hat{N}_i\right)\right]. \quad (3)$$

Then particle number and pressure are given as

$$\begin{aligned} N_i(T, V, \{\mu_j\}_{n,p}) &= -\left(\frac{\partial\Omega}{\partial\mu_j}\right)_{T,V,\mu_{j\neq i},} \\ P(T, V, \{\mu_i\}_{n.p}) &= -\left(\frac{\partial\Omega}{\partial V}\right)_{T,\{\mu_i\}_{n,p}.} \end{aligned} \quad (4)$$

The Helmholtz free energy can be expressed as

$$\begin{aligned} F(V, T, N_p, N_n) &= V\mathcal{F}(T, \{\rho_i\}_{n,p}) \\ &= \Omega(T, V, \{\mu_i\}_{n,p}) + V\sum_{i=n,p}\mu_i\rho_i(T, \{\mu_i\}_{n,p}), \end{aligned} \quad (5)$$

with the particle-number density ρ_i ,

$$N_i = V\rho_i(T, \{\mu_j\}_{n,p}) \quad (6)$$

for uniform matter. The Gibbs-Duhem relation gives $\Omega = -PV$.

Usually EOS (P vs V) renders the van der Waals type and thermodynamically unstable region ($dP/dV < 0$) appears at subnuclear density, which signals a phase transition (called the "liquid-gas" phase transition in nuclear matter) [25]. It looks like the first-order phase transition, so that there appears the mixed phase composing of two phases in equilibrium. Since there are two conserved charges, baryon number ($N_B = N_p + N_n = V\rho_B$) and isospin number ($I_3 = N_p - N_n = V\rho_I$), we must carefully treat the phase equilibrium in the mixed phase of this binary system[2].

In the following we assume the equilibrium between two phases (liquid(L) and gas(G)) in the mixed phase. The Gibbs conditions then read

$$T^G = T^L, \quad P^G = P^L, \quad \mu_i^G = \mu_i^L (i = n, p). \quad (7)$$

The phase diagram is then given in the T-P-Y_p space in accordance with the Gibbs phase rule [24]: phase-separation boundary is then given as a two dimensional surface. Note that the familiar Maxwell construction cannot be applied in this binary system, since it works only for single component system. When we draw the pressure surface as a function of μ_p and μ_n for given Y_p and T. we can graphically see a folded surface and the intersection gives the coexisting curve. The mixed phase develops along the coexisting

[2]In neutron-star matter electric charge is conserved instead of isospin at low temperature. Moreover, lepton number is also conserved in some situations as in supernovae, which makes a ternary system [36].

curve and obviously satisfies the Gibbs conditions. As a result, pressure is no more constant in the mixed phase. So this is not the first-order phase transition, but second-order one in the sense of Ehrenfest [29]. However, since it never accompanies the change of any symmetry, it is the first-order phase transition in the sense of Landau [26]. We shall see that the mixed phase consists of two distinct phases in equilibrium and there exist a metastable state and latent heat, which should characterize the first-order phase transition.

When we plot the beginning and terminating pressures in the P-T plane by changing T for a given Y_p, we have an extended coexisting domain. The phase-separation boundary is called *binodal curve*. Following Iosilevskiy, we call it *non-congruent* phase transition. Note that ρ_I as well as ρ_B takes a different value in each phase, keeping the volume average fixed. It is instructive to seek the coexisting region by daring to apply the Maxwell construction, as in the van der Waals fluid (Figure 1). One may see only a single curve on the same plane, which is called *congruent* phase transition and the water-vapor phase transition is a typical example [17]. When we apply the Maxwell construction to the ASNM, the first two conditions in (7) are obviously satisfied. However, the chemical equilibrium is partially violated since the proton fraction is forced to be the same in both phases: actually the baryon chemical potential, $\mu_B = \mu_p + \mu_n$, results in being equal at the end points, while the isospin chemical potential, $\mu_I = \mu_p - \mu_n$, is different.

Consider the manifold $\mathcal{F}$ in the parameter space spanned by $\{\rho_i\}_{n,p}$ with T fixed. If there is a common tangent plane connecting two distinct configurations with $\{\rho_i'\}_{n,p}$ in G and $\{\rho_i''\}_{n,p}$ in L, Gibbs' criteria are satisfied. Then the mixed phase can be constructed by combining two phases with volume fraction $\lambda = V^L/V(= V^L + V^G), 0 < \lambda < 1$. Total charge densities are given by $\rho_i = \lambda\rho_i'' + (1-\lambda)\rho_i'$ and Helmholtz free energy can be written as

$$\mathcal{F}(T, \{\rho_i\}_{n,p}) = \lambda\mathcal{F}(T, \{\rho_i''\}_{n,p}) + (1-\lambda)\mathcal{F}(T, \{\rho_i'\}_{n,p}). \quad (8)$$

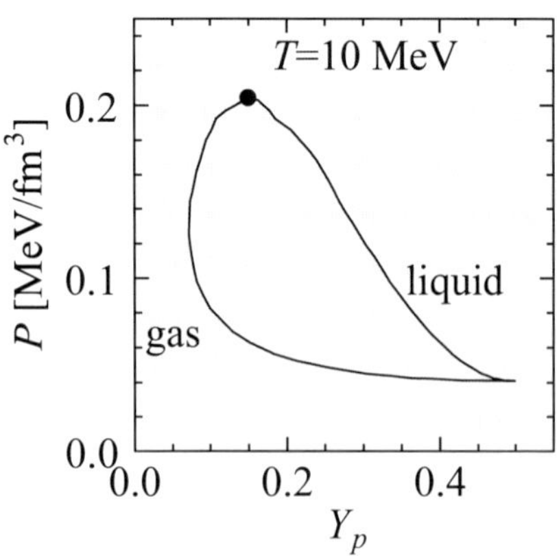

Figure 3. Binodal curve for ASNM at $T = 10$MeV (see section 3). The filled circle denotes the critical point.

The binodal points for given Y_p are then obtained by considering the two limits, $\lambda = 0, 1$ [3]. From Eq. (4),

$$P^G(\mu_i) = P^L(\mu_i), \tag{9}$$

and

$$\frac{\partial P^G}{\partial \mu_i} = \rho_i \tag{10}$$

for $\lambda = 0$, and

$$\frac{\partial P^L}{\partial \mu_i} = \rho_i \tag{11}$$

for $\lambda = 1$. Three equations in (7) or Eqs. (9)-(11) determine the unknown values of chemical potentials μ_i and the baryon densities for the binodal points. Finally the binodal curve is obtained by connecting the binodal points by changing the mixture Y_p (see Figure 3). Note that $P^L(\mu_i)$ in the former case and $P^G(\mu_i)$ in the latter case never correspond to the $\lambda = 0$ and $\lambda = 1$, respectively.

2.3. Stability Criteria

Spinodal instability has been often discussed in ASNM in relation to the nuclear multi-fragmentation in heavy-ion collisions or in the crust region of neutron stars [25, 30, 31, 32, 33, 35]. Since it should be also related to the formation of the pasta structures, we briefly summarize it. The stability criterion implies the convexity of the free energy density $\mathcal{F}$ of $\{\rho_i\}_{n,p}$: Hessian or the curvature matrix

$$C = \begin{pmatrix} \partial^2\mathcal{F}/\partial\rho_n^2 & \partial^2\mathcal{F}/\partial\rho_n\partial\rho_p \\ \partial^2\mathcal{F}/\partial\rho_p\partial\rho_n & \partial^2\mathcal{F}/\partial\rho_p^2 \end{pmatrix} = \begin{pmatrix} \partial\mu_n/\partial\rho_n & \partial\mu_n/\partial\rho_p \\ \partial\mu_p/\partial\rho_n & \partial\mu_p/\partial\rho_p \end{pmatrix} \tag{12}$$

has only positive eigenvalues [4],

$$\mathrm{tr}(C) \geq 0 \quad \text{and} \quad \det(C) \geq 0. \tag{13}$$

Since both eigenvalues never become negative in ASNM at subnuclear densities, one negative eigenvalue or equivalently $\det(C) < 0$ defines the spinodal region. The condition $\det(C) < 0$ is further equivalent with

$$\left(\frac{\partial P}{\partial \rho_B}\right)_{T,Y_p} < 0 \quad \text{or} \quad \left(\frac{\partial \mu_p}{\partial Y_p}\right)_{T,P} < 0, \tag{14}$$

where the first inequality is called "mechanical" instability and the second one "diffusive(chemical)" instability [31]. The mechanical instability is the familiar one and comes from the negative compressibility as in the van der Waals fluid (see Figure 1). Actually the spinodal condition is given only by this inequality for symmetric nuclear matter (SNM).

[3] Note that there is an exception: we shall see a possibility of the existence of $\lambda_{\max}$ in the special case, called retrograde condensation.

[4] In neutron-star matter, which is charge-neutral in the presence of electrons, the curvature matrix is modified by adding the term $\partial\mu_e/\partial\rho_e$ to the matrix element $\partial\mu_p/\partial\rho_p$ [30, 33].

However, the chemical instability first happens in ASNM, whereas curvature always remains positive and no spinodal instability is predicted in pure neutron matter [32]. So there should be a lower bound for Y_p, below which ASNM is always stable. The Gibbs free energy G is depicted in Figure 7, whose derivative with respect to proton number gives the proton chemical potential, $(\partial G/\partial N_p)_{T,P} = \mu_p$. Then the curve has a common tangent line at the binodal points $B_{1,2}$, indicating the equality of μ_i between two phases. In Figure 5 a chemical potential isobar is shown for proton. In the region between the binodal points $B_{1,2}$ the uniform matter is unstable and separated to two phases with different Y_p, $(0 \leq Y_p \leq 0.5)$ [5]; the spinodal region is specified by the spinodal points $S_{1,2}$ and two regions between B_1 and S_1 and B_2 and S_2 are the metastable regions. As ASNM enters the spinodal region, a unique mechanism of phase separation works and small fluctuations of Y_p grow to produce a non-uniform random pattern. This phenomenon is known as *spinodal decomposition* [24]. In the metastable regions ASNM is stable for small fluctuations of Y_p, but more likely becomes inhomogeneous by way of *nucleation* and growth. These features should be peculiar to the first-order phase transitions.

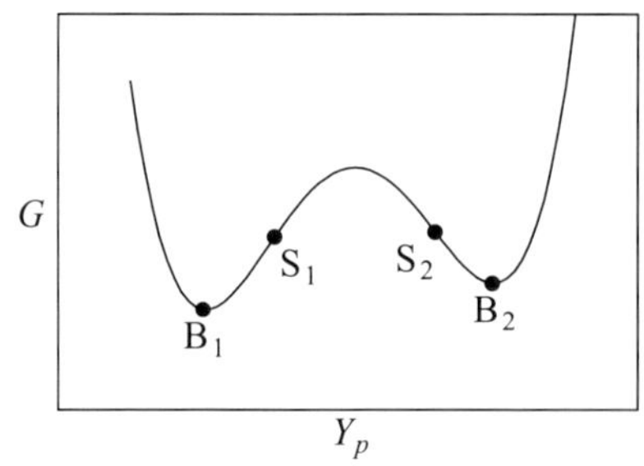

Figure 4. Schematic view of the Gibbs free energy G at fixed temperature as a function of Y_p. $B_{1.2}$ denote the binodal points and $S_{1,2}$ the spinodal points where $\left(\partial^2 G/\partial Y_p^2\right)_{T,P} = N_B(\partial \mu_p/\partial Y_p)_{T,P} = 0$.

2.4. Finite-Size Effects and "Pasta" Structures

We have only discussed the binary system to demonstrate the characteristic features of the non-congruent transition. Since there are usually two conserved quantities, baryon number and electromagnetic charge in the cold catalyzed matter, many cases are treated as the binary systems inside compact stars. Extension of the discussion to multi-component system is straightforward but very complicated. In the neutrino trapping case as in supernova explosions, we need another conserved quantity, lepton number. Thus we must consider the ternary matter. In the recent paper this problem has been generally addressed and classification of various situations has been presented [36].

We have presented general features of the non-congruent transition in a clear fashion with a help of the bulk calculation, discarding the Coulomb interaction and geometrical

[5]The region $0.5 \leq Y_p \leq 1$ is redundant due to symmetry of interchanging protons and neutrons.

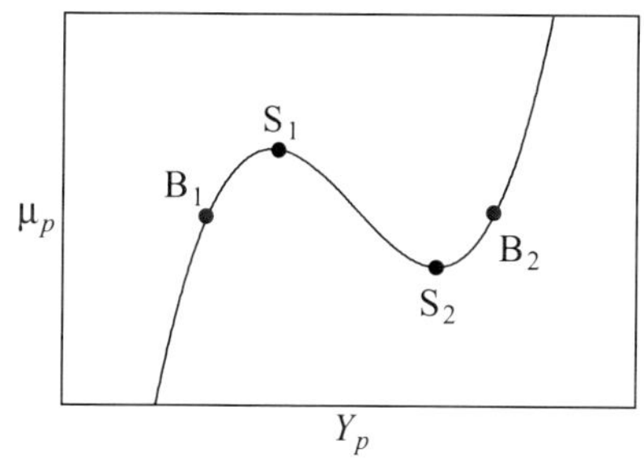

Figure 5. Proton chemical-potential isobar at fixed temperature as a function of Y_p. The segment between $S_{1,2}$ denotes the spinodal region, where $(\partial\mu_p/\partial Y_p)_{T,P} < 0$, and the segments between $B_{1,2}$ and $S_{1,2}$ correspond to the metastable regions.

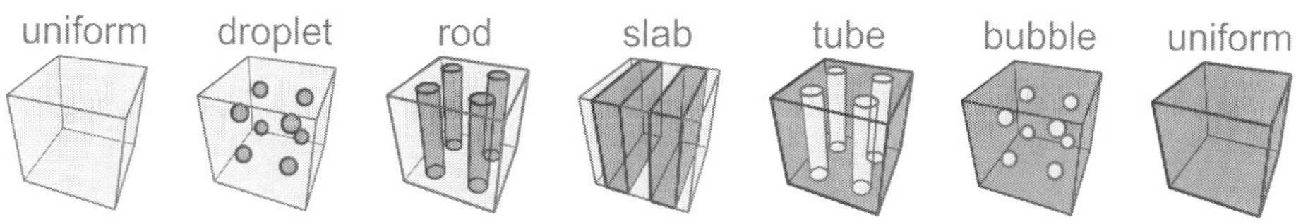

Figure 6. Schematic picture of "pasta" structures. When red phase appears in the blue phase, the red phase form droplets in the sea of blue. With increase of the volume fraction of red phase in the blue phase, the shape of the red phase changes from droplet to rod, slab, tube, and to bubble before pure red phase appears.

structures which are called the finite-size effects. However, the bulk calculation is insufficient to describe the essential aspects of the non-congruent transition. We can see the emergence of various geometrical shapes, called pasta (see Figure 6), by the cooperative effect of the Coulomb interaction and the surface energies, and the importance of the charge screening effect which is a typical many-body effect and makes pasta structures unstable, called mechanical instability.

A rough estimate about the finite-size effects can be done as a straight extension of the bulk calculation. Since the crystalline structure appears in this case we use the Wigner-Seitz cell approximation (see Figure 7). Consider a simple model with a droplet composed of the phase I embedded in the phase II in a Wigner-Seitz cell with the volume $V_W = 4\pi R^3/3$ at $T = 0$: particle density is assumed to be uniform in each phase, charge neutrality is achieved by electrons and their interface is represented by the sharp surface with a surface tension parameter τ_S. For a given volume fraction $\lambda = (R/R_W)^3$, the total energy then consists of three parts,

$$E = E_V + E_C + E_S, \tag{15}$$

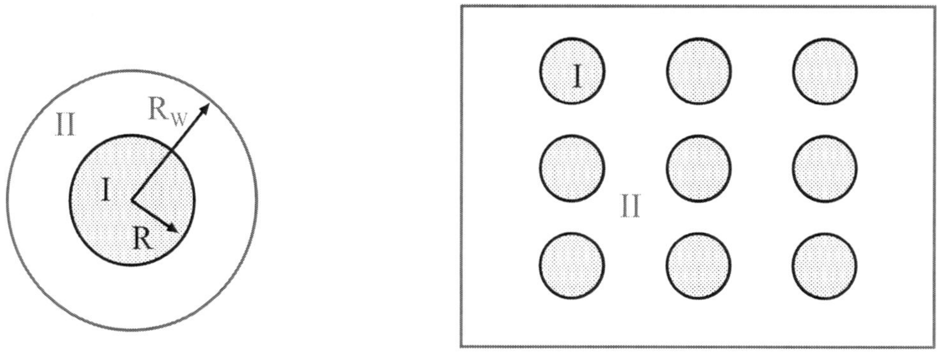

Figure 7. Wigner-Seitz cell for the droplet case (left panel) and crystalline structure composed of droplets (right panel).

where E_V is the volume energy computed with the strong-interaction Hamiltonian, and E_C and E_S are the Coulomb energy and the surface energy, respectively. The energy density E_V/V_W does not depend on R, while the Coulomb-energy density renders

$$E_C/V_W = \lambda \frac{16\pi^2}{15}(\rho^I_{\rm ch} - \rho^{II}_{\rm ch})^2 R^2, \tag{16}$$

and the surface-energy density

$$E_S/V_W = \lambda \frac{3\tau_S}{R}. \tag{17}$$

In Figure 8 we sketch the R dependence of these energy densities. The optimal value of R is then given by the minimum condition,

$$\frac{\partial(E/V_W)}{\partial R} = 0. \tag{18}$$

Using Eqs. (15), (16), (17), we can see the well-known relation $E_S = 2E_C$ holds at the minimum.

We shall see this simple consideration is too crude to describe the mixed phase in a realistic way. Once the Coulomb interaction is included, we must take into account the rearrangement of the particle densities at the same time, and in turn the screening effect for the Coulomb interaction. Consequently the Coulomb interaction becomes effectively short-range by the Debye mass $m_D^2 = \sum_i 4\pi\partial\rho^i_{\rm ch}/\partial\mu_i$. As shown in Figure 8 the Coulomb-energy density is no more an increasing function with respect to R, but is exponentially reduced at large R. The minimum point is shifted to larger R as τ_S increases, which gives rise to a mechanical instability of the droplet above the critical value of τ_S. The mechanical instability of the pasta structure also occurs in other geometrical shapes, which means that both R and R_W go to infinity with their ratio fixed. Thus we recover the picture of the phase equilibrium of two bulk matters, where the surface energy is irrelevant and the Coulomb energy vanishes due to the achievement of the local charge-neutrality. These features indicate that the EOS resembles the one given by the Maxwell construction.

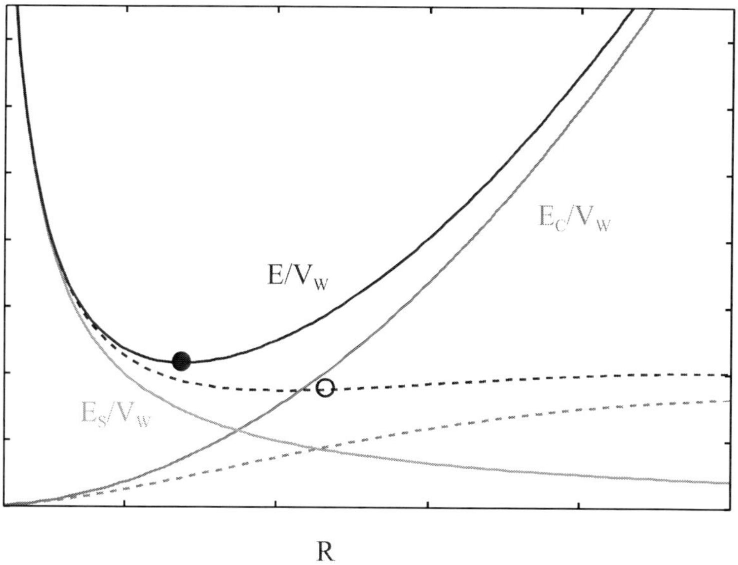

Figure 8. Energy densities as functions of the radius of droplets. Solid curves show the case without the screening effect, while dotted curves with screening.

The rearrangement of the particle densities also induces a change in the volume energy E_V, which is composed of kinetic energy and strong-interaction energy. In the above calculation E_V itself is intrinsically large but independent on R, since we have assumed uniform densities $\rho_i^{I,II}$ in each phase. Once the Coulomb interaction is properly taken into account, particle densities are no more uniform to produce a R dependence in E_V. The difference in E_V is called the correlation energy in Ref. [19], defined by

$$E_{\rm corr} \equiv E_V(\rho_i(\mathbf{r})) - E_V(\rho_i^{I,II}). \tag{19}$$

The detailed analysis showed that $E_{\rm corr}$ gives negative values and behaves R^{-1} for large R. Sometimes $E_{\rm corr}$ gives a significant contribution as later seen in hadron-quark deconfinement transition (Sec. 4).

3. Liquid-Gas Transition in Asymmetric Nuclear Matter

The properties of inhomogeneous low-density matter was already studied in the early 70th by Baym et al. [37]. They have presented a concept of structured mixed phase which includes nuclear droplet in the neutron gas as well as its reversal structure (droplet of gas in the nuclear sea).

About a decade later, Ravenhall et al. [10] have considered not only the three-dimensional structures but also those of one and two dimension. They have introduced a dimensionality d with a fractional number as well as other parameters such as volume fraction, density in the nuclear phase, and so on. With given average density and proton fraction, they looked for the parameters which give a minimum free energy. With increasing the average density, the structure of nucleus changes from droplet to rod, slab, tube, bubble, and finally uniform. Afterward, such a series of structures was called "pasta" (see Figure 6). The corresponding d changes from 3 for droplet to 2 for rod, 1 for slab, 2 for tube, and 3 for bubble. But d changes smoothly, which suggests the existence of intermediate shapes besides the typical pasta structures.

The next year, Hashimoto et al. [38] used a liquid-drop model with the Wigner-Seitz (WS) cell approximation, and showed that the stable nuclear shape changes as in the same way as has shown by the above Ref. [10]. After that there appeared studies of pasta structures by several groups using different frameworks which are, however, basically mean-field models [39, 40, 41, 42, 43, 44].

Besides the mean-field approaches, molecular dynamics (MD) simulations [45, 46] were applied to the inhomogeneous low-density matter [7, 46, 47, 48]. The advantage of MD simulation is that one needs to assume no structure, or capability in simulating the dynamical processes [7]. Some mean-field approaches are also possible without the assumption of the structure [49, 50].

3.1. Framework Based on Relativistic Mean-Field Theory

Here we use the relativistic mean-field (RMF) model to describe asymmetric nuclear matter. Exploiting the idea of the density functional theory (DFT) within RMF model [51], we can calculate the particle-density profiles as well as thermodynamic quantities. The Lagrangian density of the system is written as

$$\mathcal{L} = \mathcal{L}_N + \mathcal{L}_M + \mathcal{L}_e \tag{20}$$

$$\mathcal{L}_N = \bar{\Psi}\left[i\gamma^\mu\partial_\mu - m_N^* - g_{\omega N}\gamma^\mu\omega_\mu - g_{\rho N}\gamma^\mu\vec{\tau}\vec{b}_\mu - e\frac{1+\tau_3}{2}\gamma^\mu V_\mu\right]\Psi \tag{21}$$

$$\begin{aligned}\mathcal{L}_M = &+\frac{1}{2}(\partial_\mu\sigma)^2 - \frac{1}{2}m_\sigma^2\sigma^2 - U(\sigma)\\ &-\frac{1}{4}\omega_{\mu\nu}\omega^{\mu\nu} + \frac{1}{2}m_\omega^2\omega_\mu\omega^\mu\\ &-\frac{1}{4}\vec{R}_{\mu\nu}\vec{R}^{\mu\nu} + \frac{1}{2}m_\rho^2\vec{R}_\mu\vec{R}^\mu\end{aligned} \tag{22}$$

$$\mathcal{L}_e = -\frac{1}{4}V_{\mu\nu}V^{\mu\nu} + \bar{\Psi}_e\left[i\gamma^\mu\partial_\mu - m_e + e\gamma^\mu V_\mu\right]\Psi_e, \tag{23}$$

$$m_N^* = m_N - g_{\sigma N}\sigma \tag{24}$$

$$U(\sigma) = \frac{1}{3}bm_N(g_{\sigma N}\sigma)^3 - \frac{1}{4}c(g_{\sigma N}\sigma)^4 \tag{25}$$

where $F_{\mu\nu}$ stands for $\partial_\mu F_\nu - \partial_\nu F_\mu$. By the Euler-Lagrange equation

$$\partial_\mu\left[\frac{\partial\mathcal{L}}{\partial(\partial_\mu\phi)}\right] - \frac{\partial\mathcal{L}}{\partial\phi} = 0, \quad (\phi = \sigma, \omega_\mu, R_\mu, V_\mu, \Psi) \tag{26}$$

we get the coupled equations of motion for the scalar (σ), isoscalar-vector (ω), isovector-vector (ρ), electromagnetic and nucleon fields. Since the spatial components of vector fields become zero due to the isotropy of the system, the meson mean-fields (σ, ω_0, R_0) and electric field (V_0) obey the following equations of motion

$$
\begin{aligned}
-\nabla^2\sigma(\mathbf{r}) + m_\sigma^2\sigma(\mathbf{r}) &= -\frac{dU}{d\sigma}(\mathbf{r}) + g_{\sigma N}(\rho_n^{(s)}(\mathbf{r}) + \rho_p^{(s)}(\mathbf{r})), && (27)\\
-\nabla^2\omega_0(\mathbf{r}) + m_\omega^2\omega_0(\mathbf{r}) &= g_{\omega N}(\rho_p(\mathbf{r}) + \rho_n(\mathbf{r})), && (28)\\
-\nabla^2 R_0(\mathbf{r}) + m_\rho^2 R_0(\mathbf{r}) &= g_{\rho N}(\rho_p(\mathbf{r}) - \rho_n(\mathbf{r})), && (29)\\
\nabla^2 V_0(\mathbf{r}) &= 4\pi e^2\rho_{\rm ch}(\mathbf{r}) \quad (\rho_{\rm ch}(\mathbf{r}) = \rho_p(\mathbf{r}) - \rho_e(\mathbf{r})). && (30)
\end{aligned}
$$

For nucleons and electrons we employ the Thomas-Fermi approximation at finite temperature T and the distribution function f of proton, neutron and electron can be written as

$$
\begin{aligned}
f_{i=n,p}(\mathbf{r};\mathbf{p},\mu_i) &= \left(1 + \exp\left[\left(\sqrt{p^2 + m_N^*(\mathbf{r})^2} - \nu_i(\mathbf{r})\right)/T\right]\right)^{-1}, && (31)\\
f_e(\mathbf{r};\mathbf{p},\mu_e) &= \left(1 + \exp\left[\left(p - (\mu_e - V_0(\mathbf{r}))\right)/T\right]\right)^{-1}. && (32)\\
& && (33)
\end{aligned}
$$

For simplicity we neglect contributions of anti-particles, which is legitimate for high chemical potentials μ_i and low T. Using above distribution functions f_a, $(a = n, p, e)$, relations between chemical potentials μ_a and densities ρ_a are written as

$$
\begin{aligned}
\rho_e(\mathbf{r}) &= 2\int_0^\infty \frac{d^3p}{(2\pi)^3} f_e(\mathbf{r};\mathbf{p},\mu_e), && (34)\\
\rho_{i=p,n}(\mathbf{r}) &= 2\int_0^\infty \frac{d^3p}{(2\pi)^3} f_i(\mathbf{r};\mathbf{p},\mu_i), && (35)\\
\mu_n &= \nu_n(\mathbf{r}) + g_{\omega N}\omega_0(\mathbf{r}) - g_{\rho N}R_0(\mathbf{r}), && (36)\\
\mu_p &= \nu_p(\mathbf{r}) + g_{\omega N}\omega_0(\mathbf{r}) + g_{\rho N}R_0(\mathbf{r}) - V_0(\mathbf{r}), && (37)
\end{aligned}
$$

We demand that the chemical potentials μ_a $(a = p, n, e)$ become independent of the position $\mathbf{r}$. Note that the condition $\mu_p = \mu_n - \mu_e$ among chemical potentials is applied only in the case of beta-equilibrium. Here we'd like to make a comment about the inclusion of the Coulomb potential. The electron chemical potential as well as the proton chemical potential is the gauge-variant quantity and only the charge density is gauge-invariant. Hence, the electron density is not uniform in the presence of the Coulomb potential even when the condition μ_e =const. holds everywhere [19].

Accordingly we list in addition some necessary quantities, the scalar density $\rho^{(s)}$ used in Eq. (27), energy density ε and the entropy density s below:

$$\begin{aligned}
\rho^{(s)}_{i=p,n}(\mathbf{r}) &= \frac{m^*_N(\mathbf{r})}{\pi^2} H_3(\mathbf{r};\nu_i, m^*_N(\mathbf{r})) \quad (38)\\
\varepsilon(\mathbf{r}) &= \frac{1}{\pi^2}\sum_{i=n,p}\left[H_5(\mathbf{r};\nu_i, m^*_N(\mathbf{r})) + m^{*2}_N H_3(\mathbf{r};\nu_i, m^*_N(\mathbf{r}))\right]\\
&+\frac{1}{2}\left(\nabla\omega_0(\mathbf{r})\right)^2 + \omega_0(\mathbf{r})(\rho_p+\rho_n) - \frac{m_\omega^2}{2g_{\omega N}}\omega_0^2(\mathbf{r})\\
&+\frac{1}{2}\left(\nabla R_0(\mathbf{r})\right)^2 + g_{\rho N}R_0(\mathbf{r})(\rho_p(\mathbf{r})-\rho_n(\mathbf{r})) - \frac{m_\rho^2}{8g_{\rho N}^2}R_0^2(\mathbf{r})\\
&+\frac{1}{2}\left(\nabla\sigma(\mathbf{r})\right)^2 + \frac{m_\sigma^2}{2g_{\sigma N}^2}\sigma^2(\mathbf{r}) + \frac{bm_N}{3}\sigma^3(\mathbf{r}) + \frac{c}{4}\sigma^4(\mathbf{r})\\
&+\frac{1}{\pi^2}\int_0^\infty p^3 dp f_e(\mathbf{r};\mathbf{p},\mu_e) + \varepsilon_C(\mathbf{r})\\
s_{a=n,p,e}(\mathbf{r};\mu_a) &= 2\int_0^\infty \frac{d^3p}{(2\pi)^3}\left[f_a(\mathbf{r};\mathbf{p},\mu_a)\log\left(f_a(\mathbf{r};\mathbf{p},\mu_a)\right)\right.\\
&\left.+(1-f_a(\mathbf{r};\mathbf{p},\mu_a))\log\left(1-f_a(\mathbf{r};\mathbf{p},\mu_a)\right)\right], \quad (39)
\end{aligned}$$

where ε_C is the contribution from the Coulomb interaction[6],

$$\varepsilon_C(\mathbf{r}) = \frac{e^2}{2}\int_V d^3r' \frac{\rho_{\rm ch}(\mathbf{r})\rho_{\rm ch}(\mathbf{r}')}{|\mathbf{r}-\mathbf{r}'|}, \quad (40)$$

and the functions $H_n, (n=3,5)$ are defined by

$$\begin{aligned}
H_n(\mathbf{r};\nu_i,m) &\equiv \int_0^\infty \frac{p^{n-1}dp}{(p^2+m^2)^{1/2}}\left(f_i(\mathbf{r};\mathbf{p},\mu) - f_i(\mathbf{r};\mathbf{p},-\mu)\right)\\
&\simeq \int_0^\infty \frac{p^{n-1}dp}{(p^2+m^2)^{1/2}} f_i(\mathbf{r};\mathbf{p},\mu). \quad (41)
\end{aligned}$$

To solve the coupled equations of motion (27)-(37), we employ the Wigner-Seitz approximation , where whole the space is divided into equivalent cells with a geometrical symmetry in one, two or three dimension. The cell is divided into grid points and the coupled equations of motion are solved numerically. The size of the WS cell is optimized so that the Helmholtz free energy,

$$F_{WS} = \int_{V_{\rm WS}} d^3r\left(\varepsilon - T\sum_{a=n,p,e} s_a\right) \quad (42)$$

becomes minimum, where $V_{\rm WS}$ denotes the volume of the WS cell. In the following we consider two cases: one is a given proton fraction case, $Y_p \equiv \int_{V_{WS}}\rho_p(\mathbf{r})/\int_{V_{WS}}(\rho_p(\mathbf{r})+\rho_n(\mathbf{r}))$ and the other is β-equilibrium case with additional condition, $\mu_n = \mu_p + \mu_e$ The value of μ_e

[6]We here discarded the exchange term, which is evaluated in the local-density approximation as $\varepsilon_C^{\rm ex} = -3e^2/4(3/\pi)^{1/3}\rho_{\rm ch}(\mathbf{r})^{4/3}$ [52].

Table 1. Parameter set used in RMF in our calculation. With these parameters, the saturation property of nuclear matter is reproduced: minimum energy per baryon -16.3 MeV at $\rho = 0.153$ fm^{-3} with the incompressibility $K = 240$ MeV

$g_{\sigma N}$	$g_{\omega N}$	$g_{\rho N}$	b	c	m_σ [MeV]	m_ω [MeV]	m_ρ [MeV]
6.3935	8.7207	4.2696	0.008659	−0.002421	400	783	769

is adjusted at any time step to maintain the global charge neutrality, $\int_{V_{WS}} d^3r\rho_{\rm ch}(\mathbf{r}) = 0$: we decrease μ_e when the total charge in a cell is positive and vice versa. All the above relaxation procedures are carried out simultaneously. The parameters used in our calculation are fitted to give the saturation properties of nuclear matter at $T = 0$ and properties of finite nuclei in the ground states [53], which are listed in Table I.

3.2. Bulk Gibbs Calculation

In this section we discuss about bulk properties of nuclear matter without taking into account of electrons and the finite-size effects. We give some isothermal curves in the P_B-Y_p plane and isolepthal curves in the P_B-T plane by the bulk Gibbs calculation, where P_B denotes the baryon partial pressure. Dropping the electric field V and assuming uniform densities $\rho_i, (i = n, p)$ as well as the meson mean-fields, the thermodynamic potential density and the energy density render [31],

$$\Omega_B/V \equiv -P_B$$

$$= -\frac{1}{3\pi^2}\sum_{i=n,p} H_5(\nu_i, m_N^*) - \frac{m_\omega^2}{2g_{\omega N}}\omega_0^2 - \frac{m_\rho^2}{8g_{\rho N}^2}R_0^2 + \frac{m_\sigma^2}{2g_{\sigma N}^2}\sigma^2 + \frac{bm_N}{3}\sigma^3 - \frac{c}{4}\sigma^4$$

$$\varepsilon_B = \frac{1}{\pi^2}\sum_{i=n,p}\left[H_5(\nu_i, m_N^*) + m_N^{*2}H_3(\nu_i, m_N^*)\right]$$

$$+\omega_0(\rho_p+\rho_n) - \frac{m_\omega^2}{2g_{\omega N}}\omega_0^2 + g_{\rho N}R_0(\rho_p-\rho_n) - \frac{m_\rho^2}{8g_{\rho N}^2}R_0^2 + \frac{m_\sigma^2}{2g_{\sigma N}^2}\sigma^2 + \frac{bm_N}{3}\sigma^3 - \frac{c}{4}\sigma^4, \quad (43)$$

with the meson mean-fields,

$$\begin{aligned} \sigma &= \frac{1}{m_\sigma^2}\left[g_{\sigma N}(\rho_p^{(s)} + \rho_n^{(s)}) - \frac{dU}{d\sigma}\right] \\ \omega &= \frac{g_{\omega N}}{m_\omega^2}(\rho_p + \rho_n) \\ \rho &= \frac{g_{\rho N}}{m_\rho^2}(\rho_p - \rho_n), \end{aligned} \quad (44)$$

where we subtracted a rather trivial electron contribution. The entropy density is given by Eq. (43) or follows from the Gibbs-Duhem relation,

$$s_B \equiv (P_B + \varepsilon_B - \sum_{i=n,p}\mu_i\rho_i)/T, \quad (45)$$

so that we have the Helmholtz free energy density,

$$\mathcal{F}_B = \varepsilon_B - T s_B. \quad (46)$$

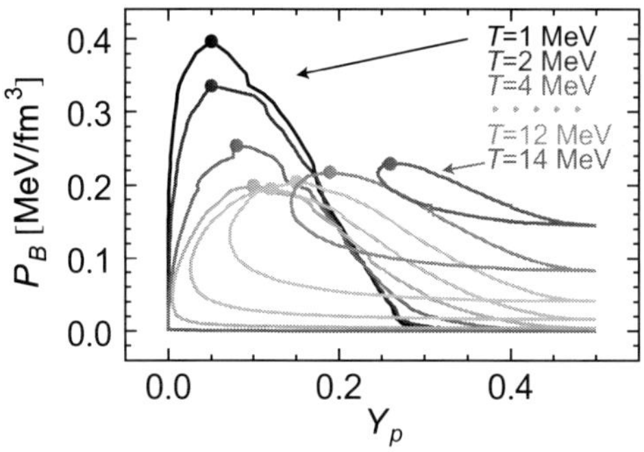

Figure 9. Temperature-dependence of phase coexistence curves in Y_p-P_B plane. Filled circles show the critical points.

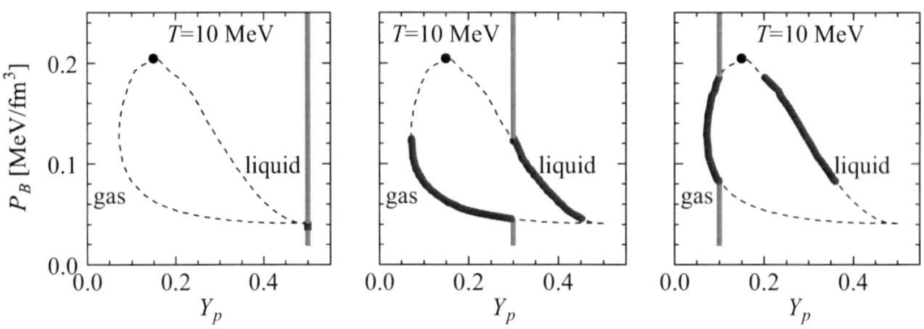

Figure 10. Same as Figure 9 for $T = 10$ MeV.

In Figure 9, binodal curves with several temperatures are presented in the Y_p-P_B plane. For a given Y_p and temperature T, e.g. $Y_p = 0.3$ and $T = 10$ MeV (Figure 10, ASNM is in the gas phase when pressure is small enough. As pressure increases the system crosses the binodal curve, where the phase separation occurs and mixed phase begins ($\lambda = 0$). The gas and liquid phases coexist in the mixed phase and the proton fraction as well as baryon number density in each phase are different. As pressure further increases, the volume fraction of each phase λ get larger, and eventually total system becomes the pure liquid phase ($\lambda = 1$). The critical point is given by the condition, $dP_B/dY_p|_{\rm CP} = 0$, since the two phases cannot be distinguished there, $P_B(T, Y_p^{\rm CP}) = P_B(T, Y_p^{\rm CP} + \delta Y_p)$. On the other hand, if we dare to apply the Maxwell construction (forced congruent transition), we have the congruent curve, where the critical point is the terminal point.

Note that the phase transition is congruent for symmetric nuclear matter ($Y_p = 0.5$). This is because the free energy should be smallest at $Y_p = 0.5$ for every density. In this case the trajectory of the system and the coexistence curve meet at a single point on the Y_p-P_B plane and the mixed phase of this system consists of gas and liquid with the same value of $Y_p = 0.5$. Accordingly $\partial P_B/\partial Y_p = 0$ at this point (Gibbs-Konovalov theorem [24]).

In the case of $Y_p = 0.1$, the system starts as a gas phase in the panel at the far right of Figure 10. With increase of density, the pressure goes up till $P_B \approx 0.08$ MeV $\cdot$ fm^{-3}. Then the LG mixed phase appears as in the central panel. With further increase of density, the pressure goes up until $P_B \approx 0.18$ MeV $\cdot$ fm^{-3}. At this point Y_p of the gas phase becomes that of the total system, 0.1, which means that the system becomes a gas again. However such "supercritical" gas has very different properties from the usual gas at low pressure. Such a transition, (gas) $\rightarrow$ (mixed phase) $\rightarrow$ (supercritical gas), is peculiar to systems with two or more chemical components and called "retrograde condensation " [26].

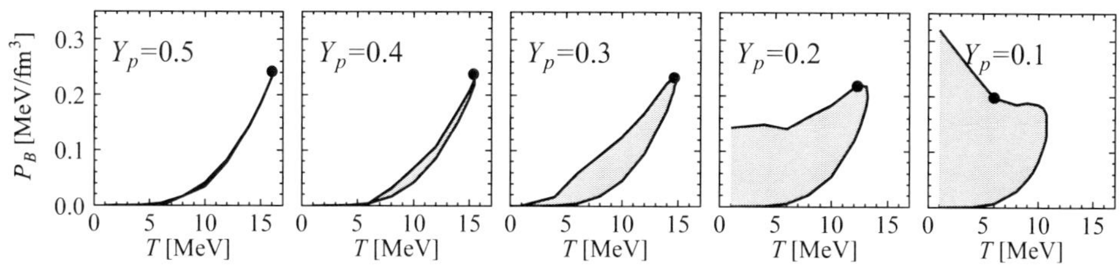

Figure 11. Phase coexistence curve on T-P_B plane. Gray area shows the region of LG mixed phase. Dots on the curves indicate the critical points. The panel at the far left exhibits the congruent curve (see Figure2), while others the non-congruent ones.

This situation is more clearly seen in Figure 11, where the binodal curves are shown in the P_B-T plane for given Y_p. The critical point ($P_{\rm CP}, T_{\rm CP}$) is located on the binodal curve, but it is not the special point on the curve. The supercritical region ($T > T_{\rm CP}$) also appears in this plane for $Y_p = 0.1$ and 0.2. There is no area inside the coexistence curve for $Y_p = 0.5$, since the system behaves congruently. For smaller Y_p, on the other hand, the area become wider. The gas phase for $Y_p \leq 0.2$ includes dripped neutrons even at $T = 0$. therefore, the curve is not closed at $T = 0$.

3.3. Pasta and EOS

In this section we discuss the non-uniform nuclear matter by taking into account the finite-size effects. We include electrons to make the total system to be charge-neutral. Electrons are distributed quasi uniformly even in the mixed phase and the finite-size effects by electrons is small. We can see that pasta structures with various geometrical shapes appear in the mixed phase. The properties of asymmetric nuclear matter with pasta structures will be elucidated by comparing them with those given by the bulk Gibbs calculation in the previous section.

We show in Figure 12 some typical density profiles in the cell. The left and the middle panels show the cases of proton fraction $Y_p = 0.1$ and $Y_p = 0.5$. Apparently, dense nuclear phase (liquid) and dilute nuclear/electron phase (gas) are separated in space and they form pasta structures depending on density. One should notice that coexisting two phases are non-congruent and have different components, i.e. nuclear matter and electron gas. Therefore the EOS of the whole system cannot be obtained by the Maxwell construction. Since

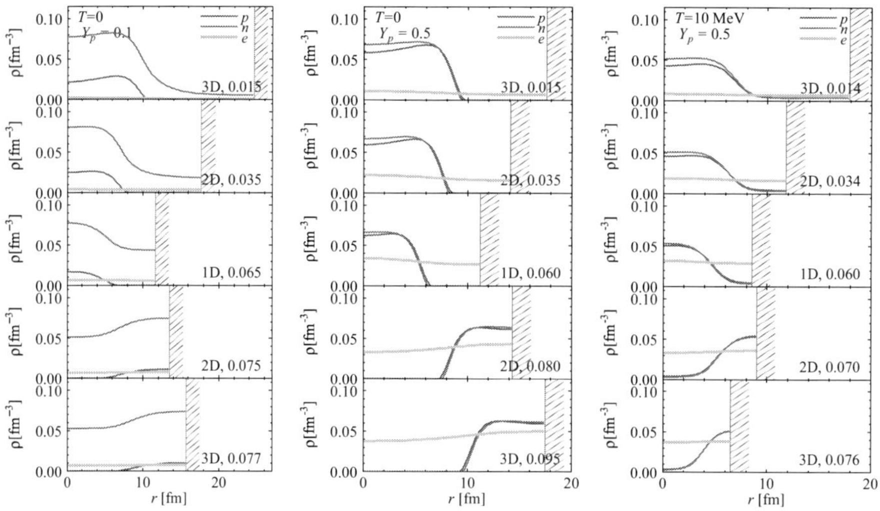

Figure 12. Examples of the density profiles in the Wigner-Seitz cells with $Y_p = 0.1$ (left), $Y_p = 0.5$, $T = 0$ (middle) and $T = 10$ MeV (right).

electron density is almost uniform and independent of baryon density distribution, we can separately discuss the properties of the baryon partial system.

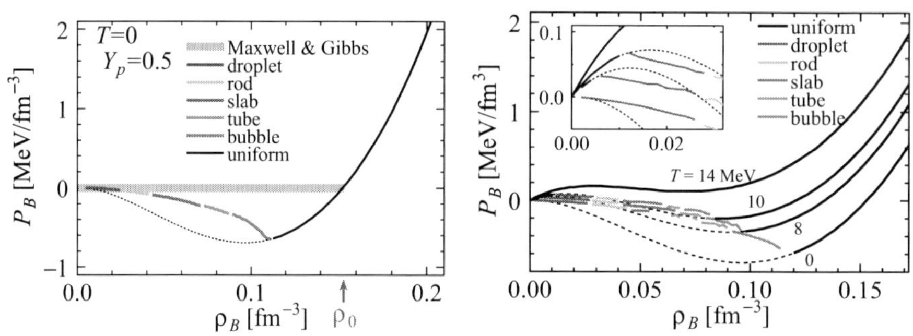

Figure 13. Baryon partial pressure as a function of density ρ_B for symmetric nuclear matter $Y_p = 0.5$, which is given by subtracting electrons pressure from the total one.

In the case of $Y_p = 0.5$, as discussed in Sec. 3.2, the local proton fraction stays almost constant. Therefore the system behaves like a system with single component. This means that one can apply the Maxwell construction to get the baryon partial pressure P_B as in the left panel of Figure 13: uniform low-density matter with a negative partial pressure is not favored and the Maxwell construction gives $P_B = 0$ for the mixed phase. By the finite-size effects, i.e. the Coulomb repulsion and the surface tension, the structured mixed phase becomes unstable in the density region just below ρ_0, and consequently uniform matter with a negative partial pressure is allowed. Note again that the total pressure including electrons is always positive even in this case, so that the system is thermodynamically stable.

In the case of asymmetric nuclear matter, e.g. $Y_p = 0.1$ in the left panel of Figure 12, the proton fraction in the dilute and dense phases are different, especially for low Y_p. Matter behaves as a system with multi chemical components and the Maxwell construction does not satisfy the Gibbs conditions.

3.4. Thermal Effects

Let us discuss the thermal effects on the LG mixed phase of low density nuclear matter. By comparing the density profiles in the middle and right panels of Figure 12, we easily notice that the dilute phase at finite temperature always contains baryons while they are absent in the dilute phase at zero temperature if $Y_p \approx 0.5$. This is due to the Fermi-Dirac distribution at finite temperature, where density as a function of chemical potential is always positive.

We also notice that the size of the pasta structure is smaller in the case of finite temperature. This comes from a reduction of the surface tension between two phases at finite temperature since the difference of baryon densities between two phases get smaller.

The isothermal EOS's (baryon partial pressure as a function of baryon number density) of symmetric nuclear matter at various temperatures are shown in the left panel of Figure 13. Dotted and thick solid curves show the cases of uniform matter, while thin solid curves are the cases where non-uniform pasta structures are present. As shown in the right panel of Figure 13, pasta structures appear at finite temperatures as well as the case of $T = 0$. But there appears uniform matter (gas phase) at the lowest-density region [27, 28] since the baryon partial pressure of uniform matter has a positive gradient against density. On the other hand, the uniform matter is unstable where the pressure gradient is negative even if the pressure itself is positive. At $T = 14$ MeV, we obtain no pasta structure since the baryon partial pressure of uniform matter becomes a monotonic function of density above this temperature. Study of the instability of uniform matter and the appearance of the pasta structures in connection with the spinodal region is in progress.

In Sec. 3.2 we have shown a retrograde condensation at $T = 10$ MeV and $Y_p = 0.1$ by the bulk calculation. In the full calculation with pasta structures, however, there is no evidence to have such phenomenon so far. Probably it is washed away by the finite-size effects.

3.5. Discontinuity in Pressure

If one applies bulk Gibbs conditions to the low-density nuclear matter, the pressure changes continuously with the density. Mechanically unstable region with negative gradient of pressure $dP/d\rho_B < 0$ never appears. However, with inclusion of the finite-size effects for inhomogeneous nuclear pasta, some part of density region becomes unstable for the mixed phase. In this density region, uniform matter with negative baryon pressure is allowed (see Figure 13). The baryon pressure of the mixed phase also can be negative. However, the total pressure including that of electron and its gradient are always positive.

Another thing to note is that there are discontinuities in pressure at the densities where the geometrical structure changes. The free-energy density as a function of density is continuous since it is the criterion of choosing the optimal structure. On the other hand the pressure $P_B = \sum_i \mu_i \rho_i - \varepsilon$ is not necessarily continuous since the chemical potential μ_i should be dependent on the structure. So far we employ the Wigner-Seitz cell approximation for matter with the pasta structures. This is rather a strong constraint on the structure. In the work of Ravenhall et al. [10] dimension parameter is optimized to get the ground state structure. It behaves smoothly at the transient region except for a jump between uniform and bubble. If we could consider more than the typical pasta structures [50], the discontinuity of pressure may diminish.

4. Hadron-Quark Deconfinement Transition

In this section we discuss the emergence of the pasta structure associated with the hadron-quark deconfinement transition[7].

Nowadays the hadron-quark deconfinement transition and its implications has been extensively studied from various viewpoints: a great deal of effort has been paid for quest of signals in relativistic heavy-ion collisions [56], compact-star phenomena [5] or remnants of primordial QCD transition [57]. Theoretically, numerical simulations based on the lattice gauge theory have been done to estimate the critical temperature and figure out the properties of the phase transition. However, for the present, any numerical method is useless for the deconfinement in compact stars, since it suffers from the sign problem at finite density [58]. Instead, the studies have been done by using the effective models of QCD such as the MIT bag model or the NJL model.

Although the order of the phase transition is not clear, theoretical arguments have suggested it should be of the first order in the low temperature and high-density regime [56]. So, let us assume here it is of the first order. Then we can expect the emergence of the structured mixed phase as in liquid-gas transition. Actually Glendenning first discussed it to demonstrate how the Maxwell construction is insufficient in neutron-star matter. Before the appearance of his paper, when the first-order phase transition occurs and new phase appears, EOS has been constructed by simply applying the Maxwell construction. Using the bulk calculation without the Coulomb interaction or the surface energy, called the finite-size effects, he could show a wide density region of the mixed phase for hadron-quark deconfinement transition, the pressure of which is by no means constant.

Once the finite-size effects are properly taken into account, inhomogeneous matter called pasta emerges as is already shown in the previous section. Heiselberg et al. immediately pointed out the importance of the finite-size effects and demonstrated how the surface energy parametrized by the surface-tension parameter makes the region of the structured mixed phase narrower [18]: they showed the disappearance of the structured mixed phase for a critical value of the surface tension. They considered only droplets for hadron-quark deconfinement transition, while more general pasta structures or the crystalline structures have been subsequently discussed [55]. Voskresensky et al. discussed the charge screening effect on the stability of the structured mixed phase during the deconfinement transition, and showed its relevance for a given surface tension [19]. They emphasized that charge screening and rearrangement of particle densities become important for the description of the mixed phase and EOS. Using the effective model for hadron matter and the MIT bag model for quark matter, Endo et al. explicitly demonstrated these features [21].

The surface tension should be calculated self-consistently with models of hadron and quark matter. Unfortunately, there is a large ambiguity about its value, whereas a large value of $100\mathrm{MeV}\cdot\mathrm{fm}^{-2}$ might be more realistic [54].

[7]We, hereafter, dismiss the possibility of color superconductivity [54].

4.1. Framework Using Brueckner-Hartree-Fock Calculation and the MIT Bag Model

There are many models to describe the EOS with hyperons. However, nowadays, the potentials of the baryon-baryon interactions have been directly calculated by the great successes in the lattice QCD calculations [59]. Hence, it is very hot topic to calculate the EOS from the baryon-baryon interaction directly. Here, we show the EOS by the theoretical framework of the nonrelativistic Brueckner-Hartree-Fock approach [60] based on the microscopic nucleon-nucleon (NN), nucleon-hyperon (NY). The Brueckner-Hartree-Fock calculation is a reliable and well-controlled theoretical approach for the study of dense baryonic matter. Detailed procedure can be found in Refs. [60, 61, 62, 63].

We adopt the Argonne V_{18} potential [64] for NN potentials, and semi-phenomenological Urbana UIX nucleonic three-body forces [65] and the Nijmegen soft-core NSC89 NY potentials [66]. Unfortunately, there are not reliable potentials for the hyperon-hyperon interaction now. Therefore, we neglect them here. We strongly hope that the lattice QCD calculations or the experiments, such as J-PARC, will reveal the these interactions in the future.

Moreover, we discuss the hadron-quark mixed phase with the finite-size effects at finite temperature with/without trapped neutrinos. To take into account the effects of finite temperature, we adopt *Frozen Correlations Approximation* [67, 68, 69]. In this approximation, the correlations at finite temperature are assumed to be the same with the ones at zero temperature. It is found to be good accuracy at finite temperature by past studies [67, 68, 69]. Accordingly, we focus on the hadron-quark mixed phase at finite temperature and adopt this approximation.

In the following, we briefly describe our framework. First we get the chemical potential μ_i from the number density n_i,

$$\rho_i = \frac{g}{(2\pi)^3}\int_0^\infty f_i(p)\, 4\pi p^2 dp, \tag{47}$$

$$f_i(p) = \frac{1}{\exp\{(\varepsilon_i - \mu_i)/T\}+1}, \tag{48}$$

where ε_i and $f_i(p)$ are the single-particle energy and the Fermi-Dirac distribution function, respectively, whereas subscript i denotes the particle species, $i = n, p, \Lambda, \Sigma^-$. We set each degeneracy factor $g = 2$, and adopt each mass as $m_n = m_p = 939$ MeV, $m_\Lambda = 1115.7$ MeV, and $m_{\Sigma^-} = 1197.4$ MeV. We note that ε_i includes the interaction energy U_i as well as the kinetic energy [67, 70],

$$\varepsilon_i = \sqrt{m_i^2 + p^2} + U_i. \tag{49}$$

Finally, we get the free-energy density $\mathcal{F}$ as following

$$\mathcal{F}_H = \sum_i \left\{ \frac{g}{(2\pi)^3}\int_0^\infty \sqrt{m_i^2+p^2} f_i(p)\, 4\pi p^2 dp + \frac{1}{2}U_i\rho_i \right\} - Ts_H, \tag{50}$$

where s_H is the entropy density calculated from

$$s_H = -\sum_i \frac{g}{(2\pi)^3}\int_0^\infty \{f_i(p)\ln f_i(p) + (1-f_i(p))\ln(1-f_i(p))\}\, 4\pi p^2 dp. \tag{51}$$

The total pressure for uniform hadron phase is given by

$$P_H = \sum_i \mu_i \rho_i - \mathcal{F}_H. \tag{52}$$

Next, we show the quark matter based on the thermodynamic bag model. In this model, we get the number density n_Q, the pressure P_Q, and the energy density ϵ_Q,

$$\rho_Q = \sum_q \frac{g}{(2\pi)^3} \int_0^\infty f_q(p) 4\pi p^2 dp, \tag{53}$$

$$\epsilon_Q = \sum_q \frac{g}{(2\pi)^3} \int_0^\infty \sqrt{m_q^2 + p^2} f_q(p) 4\pi p^2 dp + B, \tag{54}$$

$$s_Q = -\sum_q \frac{g}{(2\pi)^3}, \int_0^\infty \{f_q(p)\ln f_q(p) + (1 - f_q(p))\ln(1 - f_q(p))\} 4\pi p^2 dp, \tag{55}$$

$$\mathcal{F}_Q = \epsilon_Q - T s_Q, \tag{56}$$

$$P_Q = \sum_q \mu_q \rho_q - \mathcal{F}_Q. \tag{57}$$

for uniform quark matter, where $f_q(p)$ is the Fermi-Dirac distribution function of the quark q $(= u, d, s)$, m_q its current mass, and B the energy-density difference between the perturbative vacuum and the true vacuum, the bag constant B set to be 100 MeV$\cdot$fm^{-3}.

However, there are many uncertainties for B. Here, we also adopt the another type of B by the density dependent bag model. In this model, we set the vacuum energy $B(\rho_B)$ as

$$B(\rho_B) = B_\infty + (B_0 - B_\infty) \exp[-\beta \frac{\rho_B}{\rho_0}]$$

with $B_\infty = 50$ MeV$\cdot$fm^{-3}, $B_0 = 400$ MeV$\cdot$fm^{-3}, and $\beta = 0.17$. Here, we set the saturation density as $\rho_0 = 0.17$ fm^{-3}. This approach has been proposed in Ref. [69], on the basis of experimental results obtained at CERN SPS on the formation of a quark-gluon plasma. The just mentioned choice of the parameters allows symmetric nuclear matter to be in the pure hadronic phase at low densities, and in the quark phase at large densities, while the transition density is taken as a parameter.

For the mixed phase, we must numerically search the minimum of the free energy

$$F = \int_{V_H} dr^3 \mathcal{F}_H[n_i] + \int_{V_Q} dr^3 \mathcal{F}_Q[n_q] + F_e + F_{\nu_e} + E_C + E_S \tag{58}$$

changing the volume ratio of quark matter to hadron matter, the geometrical types of structures, and the size of Wigner-Seitz cell R_W. Here, the free energies of electrons, F_e, and neutrinos, F_{ν_e}, are calculated by the use of the Fermi-Dirac distribution functions. We do not consider all kinds of anti-particles for simplicity.

In the above equation, E_S stands for the surface energy which comes from a sharp boundary between hadron and quark phases with a fixed surface tension τ_S. We demand

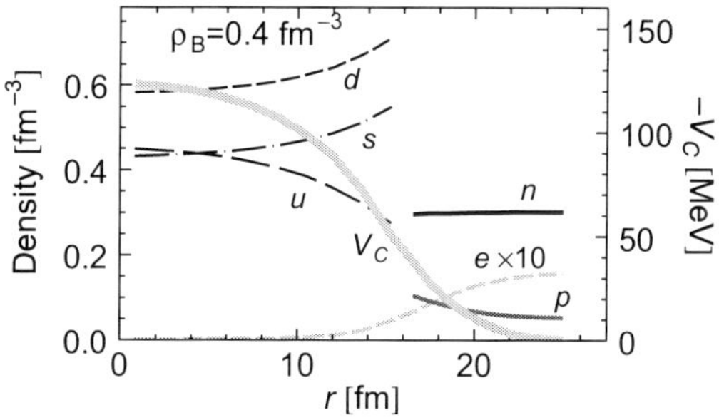

Figure 14. Density profiles and Coulomb potential V_C within a 3D (quark droplet) Wigner-Seitz cell of the mixed phase at $\rho_B = 0.4\ \mathrm{fm}^{-3}$. The cell radius and the droplet radius are $R_W = 26.7$ fm and $R = 17.3$ fm, respectively.

the global charge neutrality and the chemical equilibrium as,

$$\begin{aligned} &\mu_u + \mu_e - \mu_{\nu_e} = \mu_d = \mu_s, \\ &\mu_p + \mu_e - \mu_{\nu_e} = \mu_n = \mu_\Lambda = \mu_u + 2\mu_d, \\ &\mu_{\Sigma^-} + \mu_p = 2\mu_n. \end{aligned} \tag{59}$$

Although the surface tension of the hadron-quark interface is poorly known, some theoretical estimates based on the MIT bag model for strangelets [71] and lattice gauge simulations at finite temperature [72] suggest a range of $\tau_S \approx 10\text{–}100\ \mathrm{MeV}\cdot\mathrm{fm}^{-2}$. We will discuss the effects of surface tension on the pasta structures in subsection 1.4.4.

Figure 14 shows an example of the resulting density profile in the droplet phase for $\rho_B = 0.4\ \mathrm{fm}^{-3}$ at $T = 0$. One can see the non-uniform density distributions of particle species together with the non-vanishing Coulomb potential. The quark phase is negatively charged, so that d and s quarks are repelled to the phase boundary, while u quarks gather at the center. The protons in the hadron phase are attracted by the negatively charged quark phase, while the electrons are repelled.

4.2. Hadron-Quark Mixed Phase and Compact Stars

Once obtained the EOS of matter, we can know the density profile (density as a function of radial distance from the center) and consequently the mass and the radius of compact stars. This is done by numerically solving the Tolman-Oppenheimer-Volkoff (TOV) equations [1] as

$$\frac{dp}{dr} = -\frac{Gm\epsilon}{r^2}\frac{(1+p/\epsilon)\left(1+4\pi r^3 p/m\right)}{1-2Gm/r}, \tag{60}$$

$$\frac{dm}{dr} = 4\pi r^2 \epsilon, \tag{61}$$

being G the gravitational constant. Starting with a central mass density $\epsilon(r=0) \equiv \epsilon_c$, one integrates out Eqs. (60) and (61) until the surface pressure becomes to zero. This gives the stellar radius R and its gravitational mass $M = m(R)$.

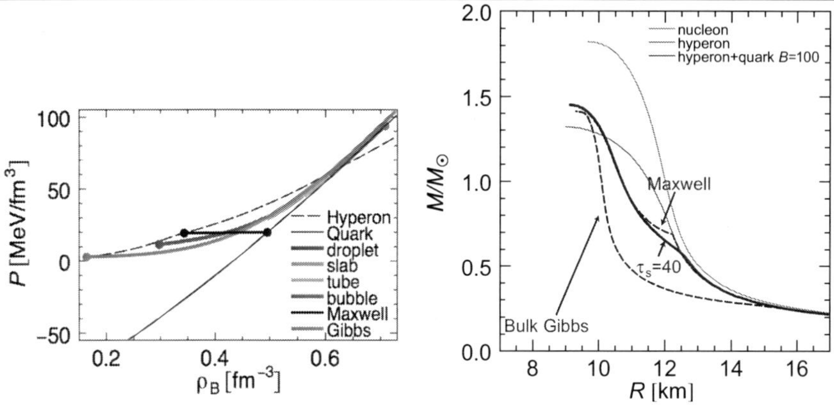

Figure 15. Left panel shows the EOS of the mixed phase (thick dots) in comparison with pure hadron and quark phases (thin curves) at zero temperature. We also show, for comparison, the mixed phase by the Maxwell construction by thin solid line. Note that the Maxwell construction is incorrect in the present case, but it is still useful as an eye guideline. Right panel show mass-radius relation of calculated neutron stars.

We show the relation between EOSs and structures of compact stars in Figure 15. The left panel is the EOSs and the right panel shows the mass-radius relation of calculated neutron stars. As for the EOSs, we shows three types of quark-hadron phase transitions; c.g. the EOSs by the bulk Gibbs condition, the Maxwell construction, and full calculation considering the finite size effects. Here, the surface tension is set as $\tau_S = 40$ MeV·fm^{-2}. We also show the uniform hadron and quark matters for comparison. Using these EOSs and the TOV equations, we can calculate the mass-radius relations shown in the right panel. It is well-known that the observed masses of neutron stars are distributed in a narrow band close to 1.4 times the solar mass $M_\odot$ [73]. The maximum mass of neutron star is enough larger than the observed masses. With inclusion of hyperon degree of freedom, however, the softening of matter reduces the possible mass of neutron stars as low as 1.3 $M_\odot$, which contradicts the observation. The blue curves show the cases with hadron-quark mixed phase. The upper most one and the bottom are the results with the EOSs obtained by the Maxwell construction and the bulk Gibbs calculation, respectively. The result with inclusion of the full pasta structures with $\tau_S = 40$ MeV $\cdot$ fm^{-2} lies just below the Maxwell construction case.

In any case with the hadron-quark mixed phase, the calculated maximum mass is slightly above $1.4M_\odot$, which might be tolerable in confrontation with the observation. What is important is that the mixture of hyperons in nuclear matter softens the EOS, which causes the contradiction with the observation, i.e. the "maximum-mass problem". By considering the deconfinement transition to quark matter, the EOS gets stiffer at higher density region. This stiffness at high density sustains the neutron stars with $M \sim 1.4M_\odot$.

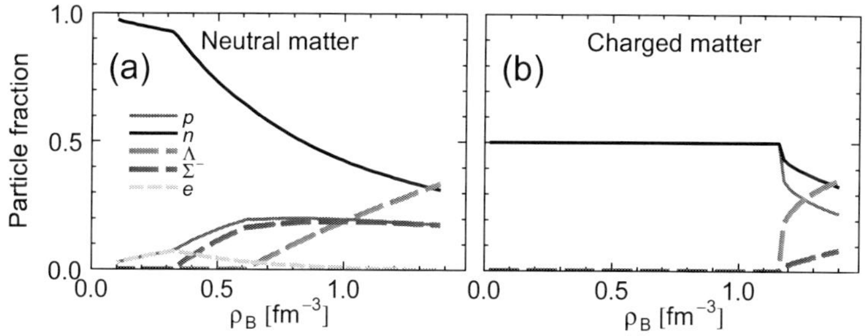

Figure 16. (a) Particle fractions of neutral matter with electrons. (b) The same quantity for charged matter without electrons, the low-density part of which corresponds to symmetric nuclear matter.

4.3. Hyperon Suppression in the Mixed Phase

Though we allow the freedom of hyperon mixture, hyperons don't actually come into the mixed phase. The suppression of hyperon mixture in the mixed phase is due to the fact that the hadron phase is positively charged. As shown in Figure 16, hyperons (Σ^-) appear in charge-neutral hadronic matter at low density to reduce the Fermi energies of electron and neutron. Without the charge-neutrality condition, on the other hand, there is symmetric nuclear matter at lower density and hyperons will be mixed there at higher density due to the large hyperon masses. Generally speaking hyperons are hardly mixed in the positively charged matter. Thus, the mixture of hyperons is suppressed in the mixed phase where the hadron phase is positively charged.

4.4. Effects of Surface Tension

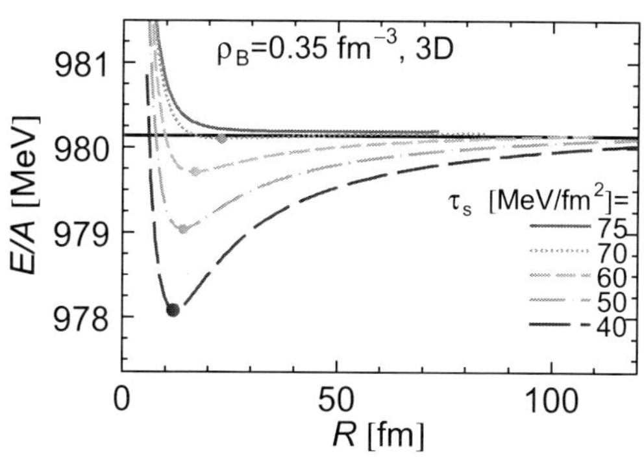

Figure 17. Droplet radius (R) dependence of the energy per baryon for fixed baryon density $\rho_B = 0.35$ fm^{-3} and different surface tensions. The temperature T is zero for all cases. The quark volume fraction $(R/R_W)^3$ is fixed for each curve. Dots on the curves show the local energy minima. The black line shows the energy of the Maxwell construction case.

Let us consider the role of the surface tension on the mixed phase. As already mentioned, the surface tension between hadron and quark phases is unknown. If one uses a smaller surface tension parameter τ_S, the energy gets lower and the density range of the mixed phase gets wider. The limit of $\tau_S = 0$ leads to a bulk application of the Gibbs conditions without the Coulomb and surface effects. On the other hand, using a larger value of τ_S, the geometrical structures increase in size and the EOS gets closer to that of the Maxwell construction case. Above a limiting value of $\tau_S \approx 65\ \mathrm{MeV}\cdot\mathrm{fm}^{-2}$ the structure of the mixed phase becomes mechanically unstable [19]: for a fixed volume fraction $(R/R_W)^3$ the optimal values of R and R_W diverge and local charge neutrality is recovered in the mixed phase, where the energy density equals that of the Maxwell construction (see Figure 17).

4.5. Thermal and Neutrino Trapping Effects

In following we discuss the thermal and neutrino trapping effects on the quark-hadron phase transition. In the left panel of Figure 18 we show the dependencies of the isothermal free-energy on the cell size for several temperatures in Eq. (59) at $\rho_B = 2\ \rho_0$, using the constant bag constant and the surface tension as $B = 100$ MeV·fm^{-3} and $\tau_S = 40$ MeV·fm^{-2}, respectively.

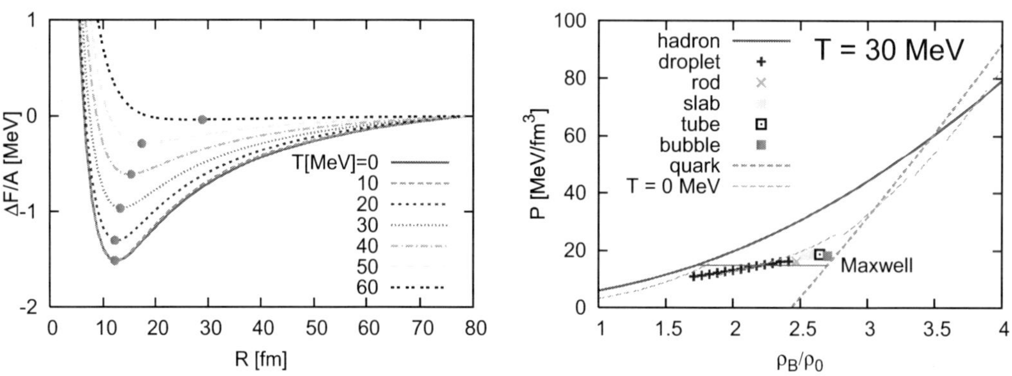

Figure 18. Left panel shows that the radius R dependence of the free-energy per baryon for the droplet case at $\rho_B = 2\ \rho_0$ and different temperatures. The free energy is normalized by the value at $R \to \infty$. The filled circles on each curve shows the energy minimum. The results are for $B = 100$ MeV·fm^{-3}, $\tau_S = 40$ MeV·fm^{-2}. The right panel shows the EOS of the mixed phase (thick dots) in comparison with pure hadron and quark phases (thin curves) at $T = 30$ MeV. We also show, for comparison, the mixed phase by the Maxwell construction by thin solid line.

Here we assume the non-uniform structure is the droplet and do not take into account neutrinos to see the thermal effects. The quark volume fraction $(R/R_W)^3$ is fixed to be the optimal value at $T = 0$ MeV for each curve. From the this panel, we can see that the minimum point is shifted to a larger value of R as temperature is increased, and eventually disappears for $T > 60$ MeV. The temperature dependence of the free energy comes from the Coulomb energy, the surface energy and the correlation energy. By comparing the temperature dependence of these contributions, we can see that the correlation energy is primarily responsible to the behavior of the minimum point. It means that the mixed phase becomes less stable as temperature is increased [23].

After the minimization of the free energy, we can get the thermodynamic quantities, such as pressure, entropy, etc. In the right panel of Figure 18, we show the EOS (baryon density vs. pressure for the neutrino-free case at $T = 30$ MeV. Clearly, the pressure becomes to close to the one given by the Maxwell construction at finite temperature, since the mixed phase becomes unstable and its density regime is thereby largely restricted. In Figure 19, we show the dependencies of neutrino trappings on the free energy of the mixed phase. Here, we adopt the density-dependent bag constant, and fixed the quark volume fraction $(R/R_W)^3$ at $Y_{\nu_e} = 0.01$ for each curve. For $Y_{\nu_e} > 0.1$, the minimum points disappear. The right panel shows the each component of the free energy. From this figure, we can see that both the correlation energy and the Coulomb energy mainly contribute to the behavior of the minimum point in the presence of neutrinos, while it is a only the correlation energy for the temperature dependence.

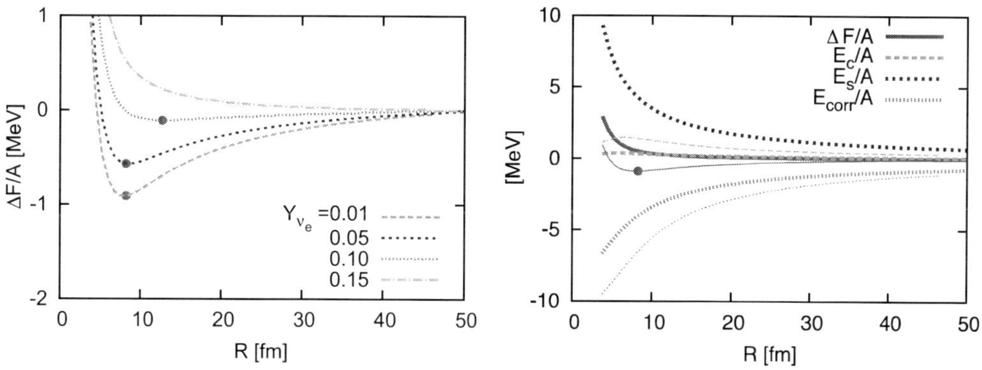

Figure 19. Left panel is the same as the left panel of Figure 18, but for different neutrino fractions Y_{ν_e} at a given temperature, $T = 10$ MeV. In this case, we set baryon density as $\rho_B = 2.5\ \rho_0$. The thick lines of the right panel are the case of $Y_{\nu_e} = 0.15$, the thin lines $Y_{\nu_e} = 0.01$. The right panel shows each component of the free energy ($\Delta F/A$); e.g. the free energy, the Coulomb energy (E_C/A), the surface energy (E_S/A), and the correlation energy (E_{corr}/A). Here they are shown per baryon.

To elucidate the neutrino-trapping effect in the mixed phase, we show the density profiles and the Coulomb potential for the slab case in Figure 20. For comparison, we use the same cell-size here. High neutrino fraction enhances the electrons to satisfy the chemical equilibrium (see $\mu_e - \mu_\nu$ in Eq. (59)). As a result, the Coulomb potential changes drastically at the trapped-neutrino case as shown in this figure.

We do not present the pressure or EOS at the neutrino-trapped case as in the case of finite temperature because, at the realistic situations in proto-neutron stars as T =40 MeV, $Y_l(\equiv Y_e + Y_{\nu_e}) = 0.4$, the mixed phase becomes unstable. Related to this discussion, the recent study suggested a new possible mixed phase for the neutrino trapped matter [36]. They imposed the local charge neutrality a priori instead of directly solving the Poisson equation, infered from our previous result in Figure 19. In such case, the surface tension and the Coulomb energy do not work to make any pasta structure, and the mixed phase may resemble the amorphous phase of quarks and hadrons. Our results look to show an establishment of the local charge neutrality, but details will be discussed elsewhere.

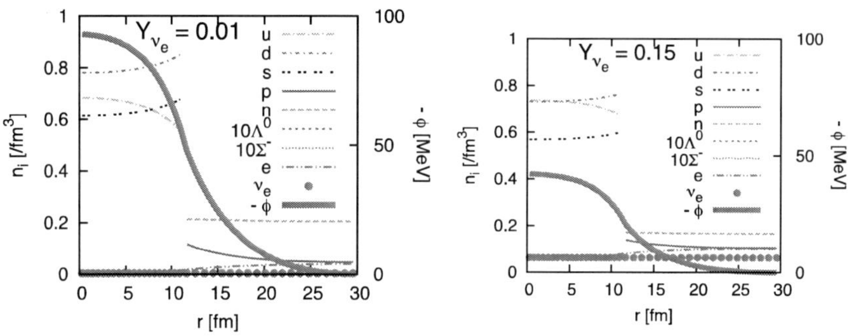

Figure 20. Density profiles and Coulomb potential ϕ within 1D (slab) for $\rho_B = 2.5\ \rho_0$ at $T = 10$ MeV. Here, the neutrino fractions are set to $Y_{\nu_e} = 0.01$ (left panel) and $Y_{\nu_e} = 0.15$ (right panel). The cell sizes are $R_W = 30$ fm in these figures. The slab size are $R = 10.9$ fm. $(R/R_W)^3$ is fixed to be the optimal value at $Y_{\nu_e} = 0.01$.

5. Summary and Concluding Remarks

We have discussed general features of the phase transitions in compact stars, where multi-component substance are in equilibrium. These phase transitions are non-congruent, and the usual Maxwell construction cannot be applied any more. Using the bulk calculation, we have presented basic concepts of the non-congruent transition. Although these concepts are helpful in discussing and analyzing the properties of the phase transitions in compact stars, we must bear in mind that they are obtained in an idealized situation and must be carefully applied in realistic calculations by taking into account rather complicated effects such as the finite-size effects. Actually we found that uniform nuclear matter can exist in the mechanically unstable region.

When the finite-size effects are taken into account, the structured mixed phase (pasta phase) emerges to form the crystalline structures. Charge screening effect and the rearrangement effect of particle densities in the presence of the Coulomb interaction is also emphasized; they sometimes cause an instability of the geometrical structures in the mixed phase. In the extreme case the EOS resembles the one given by the Maxwell construction.

Liquid-gas transition in asymmetric nuclear matter and hadron-quark deconfinement transition are studied and their mixed phases have been calculated by using the Thomas-Fermi approximation for particle densities and the Wigner-Seitz (WS) cell approximation for the pasta structure. We have shown many interesting aspects of these phase transitions. However, we have also seen that the maximum mass of hybrid stars is rather low in the light of a recent observation [74], since contamination of hyperons considerably softens EOS. The transition to quark matter suppresses their appearance, but the maximum mass is still low due to the rather soft EOS of quark matter. So, we need another idea about EOS's of hadrons and quarks to circumvent this situation.

So far, most works on the pasta structures have used the WS cell with ansatz about the geometrical structures like droplet, rod, slab and so on. If we want to see the possibility of the intermediate shapes, or study the mixed phase in a more realistic way, a study is desirable without the WS cell approximation or ansatz about the geometrical structures. Recently we have performed a three-dimensional calculation of non-uniform nuclear matter based on the relativistic mean-field model and the Thomas-Fermi approximation [50]. Introducing a cubic cell with periodic boundary conditions and discretizing it into grid, we numerically solve the coupled field equations for meson mean-fields and the Coulomb potential. Randomly distributing fermions (n, p, e) on the grid points as an initial condition, we relax their density distributions to attain the uniformity of the chemical potentials. As a preliminary result we demonstrate that pasta structures, previously given by the WS cell approximation, also appear in the three-dimensional calculation.

Formation of nuclear pasta has been discussed in supernovae [7] or possible observation of the signals of the mixed phase has been suggested in the spectra of the gravitational waves [8]. The phenomenologically the pasta structures should involve various implications and deserve more elaborate studies; their thermodynamical or dynamical property is interesting; e.g. the neutrino opacity or the thermal conductivity may be affected by the pasta structures, and their elasticity may cause a discontinuous change of internal structure of compact stars; the pasta structures may be regarded as "liquid crystals" in the latter context. [75] In particular the crystalline structure in the core region may bring about novel dynamical effects on compact-star phenomena by way of deformation or oscillation.

References

[1] S. L. Shapiro and S. A. Teukolsky, *Black Holes, White Dwarfs, and Neutron Stars*, John Wiley & Sons Inc, 1983.

[2] T. Kunihiro, T. Muto, T. Takatsuka, R. Tamagaki, T. Tatsumi, *Prog. Theor. Phys. Suppl.* 112 (1993).

[3] N.K. Glendenning, *Compact Stars*, Springer, 2000.

[4] P. Haensel, A.Y. Pothekhin and D.G. Yakovlev, Neutron Stars 1: Equation of State and Structure, Springer, 2007.

[5] W. Becker ed., *Neutron Stars and Pulsars*, Springer, 2009.

[6] I. Sagart et al., *Phys. Rev. Lett.* 102 (2009) 081101.

[7] G. Watanabe, T. Maruyama, K. Sato, K. Yasuoka and T. Ebisuzaki, *Phys. Rev. Lett.* 94(2005) 031101.
G. Watanabe, H. Sonoda, T. Maruyama, K. Sato, K. Yasuoka, and T. Ebisuzaki, *Phys. Rev. Lett.* 103 (2009) 121101.

[8] H. Sotani, N. Yasutake, T. Maruyama and T. Tatsumi, *Phys. Rev.* D83 (2011) 024014.

[9] T. Maruyama, T. Tatsumi, T. Endo, and S. Chiba, *Recent Res. Devel. in Physics* 7, 1 (2006).

[10] D. G. Ravenhall, C. J. Pethick and J. R. Wilson, *Phys. Rev. Lett.* 27 (1983) 2066.

[11] N.K. Glendenning, *Phys. Rev.* D46 (1992) 1274.
N.K. Glendenning, *Phys. Rep.* 342 (2001) 393.

[12] W.R. Magro, D.M. Ceperly, C. Pierleoni, and B. Bernu, *Phys. Rev. Lett.* 76 (1996) 1240.

[13] C.R. Ronchi, I.L. Iosilevski, E.S. Yakub, *Equation of State of Uranium Dioxide*, Springer, 2004.

[14] S.A. Khrapak et al., *Phys. Rev. Lett.* 96 (2006) 015001.

[15] V.E. Fortov et al., *Phys. Rev. Lett.* 99 (2007) 185001.

[16] N. Nettelmann, B. Holst, A. Kietzmann, M. French, R. Redmar and D. Blaschke, *Astrophys. J.* 683(2008) 1217.

[17] I. Iosilevskiy, Talk at Int. Congress on Plasma Physics, Fukuoka, Japan, Sept. 2008.

[18] H. Heiselberg, C.J. Pethick and E.F. Staubo, *Phys. Rev. Lett.* 70 (1993) 1355.

[19] D. N. Voskresensky, M. Yasuhira and T. Tatsumi, *Phys. Lett.* B541 (2002) 93; *Nucl. Phys.* A723 (2003) 291.

[20] T. Maruyama, T. Tatsumi, V. N. Voskresensky, T. Tanigawa, and S. Chiba, *Phys, Rev.* C73 (2006) 035802.

[21] T. Endo, T. Maruyama, S. Chiba, and T. Tatsumi, *Nucl. Phys.* A749 (2005) 333;
T. Endo, T. Maruyama, S. Chiba, and T. Tatsumi, *Prog. Theor. Phys.* 115 (2006) 337;

[22] T. Maruyama, S. Chiba, H.-J. Schulze and T. Tatsumi, *Phys. Rev.* D76 (2007) 123015; *Phys. Lett.* B659 (2008) 192.

[23] N. Yasutake,T. Maruyama, T. Tatsumi, *Phys. Rev.* D80 (2009) 123009; in preparation (2011).

[24] J.W.Tester and M. Mondell, *Thermodynamics and its applications*, Prentice Hall PTR, 1997.

[25] P. Chomaz, C. Colonna and J. Randrup, *Phys. Rep.* 389 (2004) 263.

[26] L.D. Landau and E.M. Lifshitz, Statistical Physics, Elsevier, 2007.

[27] S. S. Avancini, L. Biro, Ph. Chommaz, D. P. Menezes, and C. Providẽncia, *Phys. Rev.* C74 (2006) 24317.

[28] B. Friedman and V. R. Pandharipande, *Nucl. Phys.* A361 (1981) 502.

[29] H.E. Stanley, *Introduction to Phase Transitions and Critical Phenomena*, Oxford U. Press, 1971.

[30] C. Pethick, D.G. Ravenhall and C.P. Lorenz, *Nucl. Phys.* A584 (1995) 675.

[31] H. Müller and B.D. Serot, *Phys. Rev.* C52 (1995) 2072.

[32] J. Margueron and P. Chomaz, *Phys. Rev.* C67 (2003) 041602(R).

[33] C. Providência, L. Brito, S.S. Avancini, D.P. Menezes and Ph. Chomaz, *Phys. Rev.* C73 (2006) 025805.
S.S. Avancini, L. Brito, Ph. Chomaz, D.P. Menezes and C. Providência, *Phys. Rev.* C74 (2006) 024317.
L. Brito, C. Providência, A.M. Santos, S.S. Avancini, D.P. Menezes and P. Chomas, *Phys. Rev.* C74 (2006) 045801.

[34] S.S. Avancini, D.P. Menezes, M.D. Alloy, J.M. Marinelli, M.M.W. Moraes and C. Providência, *Phys. Rev.* C78 (2008) 015802.

[35] J. Randrup, *Phys. Rev.* C79 (2009) 054911.

[36] G. Pagliara, M. Hempel and J. Schaffner-Bielich, *Phys. Rev. Lett.* 103 (2009) 171102.
M. Hempel, G. Pagliara, J. Schaffner-Bielich, *Phys. Rev.* D80(2009) 125014.

[37] G. Baym, H. A. Bethe, and C. J. Pethick, *Nucl. Phys.* A175 (1971) 225.

[38] M. Hashimoto, H. Seki and M. Yamada, *Prog. Theor. Phys.* 71 (1984) 320.

[39] K. Oyamatsu, *Nucl. Phys.* A561 (1993) 431.

[40] C. P. Lorentz, D. G. Ravenhall and C. J. Pethick, *Phys. Rev. Lett.* 25 (1993) 379.

[41] K. S. Cheng, C. C. Yao and Z. G. Dai, *Phys. Rev.* C55 (1997) 2092.

[42] H. Shen, H. Toki, K. Oyamatsu and K. Sumiyoshi, *Nucl. Phys.* A637, 435 (1998).

[43] P. Gögelein, E.N.E. van Dalen, C. Fuchs and H. Müther, *Phys. Rev.* C76 (2007) 024312.
P. Gögelein and H. Müther, *Phys. Rev.* C77 (2008) 025802.

[44] K. Nakazato,K. Oyamatsu, S. Yamada, *Phys. Rev. Lett.* 103 (2009) 132501.

[45] J. Aichelin, *Phys. Rep.* 202 (1991) 233; and references cited therein.

[46] T. Maruyama, K. Niita, K. Oyamatsu, T. Maruyama, S. Chiba and A. Iwamoto, *Phys. Rev.* C57 (1998) 655. T. Kido, T. Maruyama, K. Niita and S. Chiba, *Nucl. Phys.* A663- 664 (2000) 877.

[47] G. Watanabe, K. Sato, K. Yasuoka and T. Ebisuzaki, *Phys. Rev.* C66 (2002) 012801.
G. Watanabe, *Prog. Thor. Phys. Suppl.* 186 (2010) 45.

[48] G. Watanabe, K. Sato, K. Yasuoka and T. Ebisuzaki, *Phys. Rev.* C69 (2004) 055805.

[49] R. D. Williams and S. E. Koonin, *Nucl. Phys.* A435 (1985) 844.

[50] M. Okamoto. T. Maruyama, K. Yabana and T. Tatsumi, arXiv:1106.3407.

[51] *Density Functional Theory*, ed. E. K. U. Gross and R. M. Dreizler, Plenum Press, 1995.

[52] R.G. Parr and W. Yang, *Density-Functional Theory of Atoms and Molecules*, Oxford U. Press, 1989.

[53] T. Maruyama, T. Tatsumi, D.N. Voskresensky, T. Tanigawa and S. Chiba, *Phys. Rev.* C72 (2005) 015802.

[54] M. Alford, K. Rajagopal, S. Reddy and F. Wilczek, *Phys.Rev.* D64 (2001) 074017.

[55] N.K. Glendenning and S. Pei, *Phys. Rev.* C52 (1995) 2250.

[56] T. Schaefer, arXiv:hep-ph/0509068.
P. Braun-Munzinger and J. Wambach, *Rev. Mod. Phys.* 81 (2009) 1031.
K. Fukushima and T. Hatsuda, arXiv:1005.4814

[57] D. Boyanovsky, H.J. de Vega, D.J. Schwarz, *Ann. Rev. Nucl. Part. Sci.* 56 (2006) 441.

[58] For reviews, O. Philipsen, *Eur. Phys.* J. ST 152 (2007) 29.
A. Bazavov et al., *Phys. Rev.* D80 (2009) 014504.

[59] H. Nemura, N. Ishii, S. Aoki, and T. Hatsuda. *Phys. Lett.* B673 (2009) 2.

[60] Nuclear Methods and the Nuclear Equation of State, M. Baldo, World Scientific, Singapore, (1999).

[61] M. Baldo, G. F. Burgio, and H.-J. Schulze. *Phys. Rev.* C58 (1998) 3688.

[62] H.-J. Schulze et al. *Phys. Lett.* B 355 (1995) 21.

[63] I. Vidaña et al. *Phys. Rev.* C 61 (2000) 025802.

[64] R. B. Wiringa, V. G. J. Stoks, and R. Schiavilla. *Phys. Rev.* C 51 (1995) 38.

[65] B. S. Pudliner et al. *Phys. Rev.* C 56 (1997) 1720.

[66] P. M. M. Maessen, T. A. Rijken, and J. J. de Swart, *Phys.Rev.* C 40 (1989) 2226.

[67] M. Baldo and L. S. Ferreira, *Phys. Rev.* C 59 (1999) 682.

[68] O. E. Nicotra, M. Baldo, G. F. Burgio, and H.-J. Schulze, *Astron. Astrophys.* 451 (2006) 213.

[69] O. E. Nicotra, M. Baldo, G. F. Burgio, and H.-J. Schulze, *Phys. Rev.* D 74 (2006) 123001.

[70] M. Baldo, L. S. Ferreira, and O. E. Nicotra, *Phys. Rev.* C 69 (2004) 034321.

[71] E. Farhi and R. L. Jaffe, *Phys. Rev.* D 30 (1984) 2379.

[72] K. Kajantie, L. Kaȓkaïnen, and K. Rummukainen, *Nucl. Phys.* B357 (1991) 693.

[73] For a recent review, see D. Page and S. Reddy, *Annu. Rev. Nucl. Part. Sci.* 56 (2006) 327.

[74] P.B. Demorest et al., *Nature* 467(2010) 1081.

[75] C.J. Pethick and A.Y. Potekhin, *Phys. Lett.* B427 (1998) 7.

In: Neutron Stars
Editors: Hiroshi Uechi et al, pp. 233-257
ISBN: 978-1-62081-123-8

Chapter 9

MAGNETIC FIELD AND QUARK MATTER IN THE CORE

Toshitaka Tatsumi*
Department of Physics, Kyoto University,
Kyoto 606-8502

Abstract

Magnetic properties of quark matter are discussed in the light of the observation of pulsars. Our works about spontaneous spin polarization and spin density wave are reviewed and their implications on compact-star phenomena are discussed. In particular, the former subject may be directly related to the origin of strong magnetic fields. An inhomogeneous state emerges following the chiral transition, where a kind of spin density wave develops.

1. Introduction

Pulsars are rotating neutron stars and have a strong magnetic field, by which they emit optical, radio , X-ray and γ-ray pulses.

Although it is not well understood yet where or how the pulsar emission takes place, the magnetic dipole model of pulsars simply states, from the viewpoint of energetics, how the rotation energy of neutron stars is converted into pulsar emission [1]. First of all, let us briefly see the physical mechanism. Consider the sphere with radius a composed of uniform magnetization $\mathbf{M}$ ($\mathbf{M}//\hat{z}$). Then the solution of the vector potential is

$$A_\phi = \frac{4\pi}{3} M a^2 \frac{r_<}{r_>} \sin\theta, \quad (1)$$

where $(r_>, r_<)$ are the larger and smaller of (r, R) [2]. The magnetic field $\mathbf{B}$ is then given

*E-mail address: tatsumi@ruby.scphys.kyoto-u.ac.jp

by

$$
\begin{aligned}
B_r &= \frac{1}{r \sin\theta} \frac{\partial}{\partial \theta} (\sin\theta A_\phi) \\
B_\theta &= -\frac{1}{r} \frac{\partial}{\partial r} (r A_\phi) \\
B_\phi &= 0.
\end{aligned} \tag{2}
$$

The magnetic field outside the sphere is

$$
\begin{aligned}
B_\theta^{(\mathrm{out})} &= \frac{|\mathbf{m}| \sin\theta}{r^3} \\
B_r^{(\mathrm{out})} &= \frac{2|\mathbf{m}| \cos\theta}{r^3}.
\end{aligned} \tag{3}
$$

Hence it takes the maximum magnitude at $\theta = 0$,

$$
B_{\max}^{(\mathrm{out})} = 2|\mathbf{m}| r^{-3}, \tag{4}
$$

with the total magnetic moment $\mathbf{m} = 4\pi/3a^3\mathbf{M}$. When we consider a star with radius R $(a = R)$, $|\mathbf{m}|$ can be related to the magnetic field at the magnetic pole of the star, $B_p \equiv 2|\mathbf{m}|R^{-3}$, by way of Eq. (4). Accordingly the magnetic-field lines can be drawn by solving the differential equation,

$$
\frac{dr}{B_r^{(\mathrm{out})}} = \frac{r d\theta}{B_\theta^{(\mathrm{out})}} \tag{5}
$$

to give

$$
r = R(\sin\theta / \sin\theta_0)^2, \tag{6}
$$

outside the star (Fig. 1).

The magnetic field is uniform inside the star,

$$
\boldsymbol{B}^{(\mathrm{in})} = \frac{2\mathbf{m}}{R^3}. \tag{7}
$$

The *light cylinder* is defined as a cylinder with an axis along the pulsar rotation axis and with the radius $r_c = c/\omega$, where the velocity of the co-rotating frame with the pulsar attains the speed of light. Then the emission activity of the pulsar has been considered to originate from the magnetosphere with the radius r_c, while its mechanism is not clear yet [1].

Denoting the inclined angle of the magnetic moment $\mathbf{m}$ to the rotation axis $\mathbf{e}_3$ by α, we can evaluate the luminosity (energy radiation per unit time) L_m. Then we can immediately see that the component perpendicular to $\mathbf{e}_3$, $\mathbf{m}_\perp$, is responsible to the luminosity, which is rotating in the plane perpendicular to the rotation axis, $\mathbf{m}_\perp = |\mathbf{m} \sin\alpha| [\mathbf{e}_1 \cos(\omega t) + \mathbf{e}_2 \sin(\omega t)]$, with $\mathbf{e}_i \cdot \mathbf{e}_j = \delta_{ij}$. Thus the luminosity becomes equivalent with the one brought about by two oscillating magnetic dipoles (see Fig. 2). A standard formula gives

$$
\frac{dL_m}{d\Omega_\omega} = \frac{\omega^4}{4\pi c^3} |\mathbf{m} \sin\alpha|^2 \frac{1}{2} (1 + \cos^2\theta_\omega) \tag{8}
$$

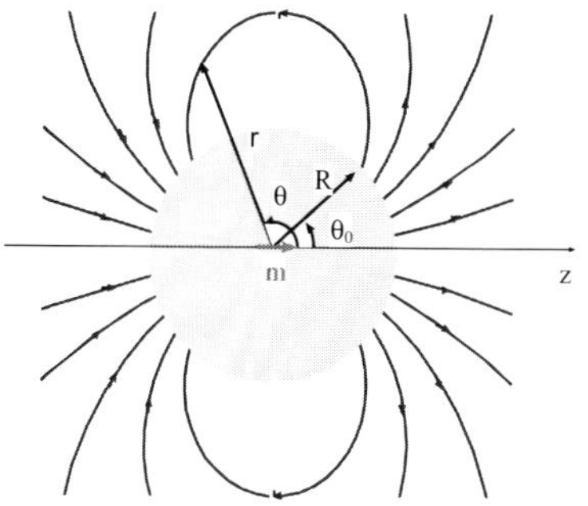

Figure 1. Dipole magnetic field-lines outside the star.

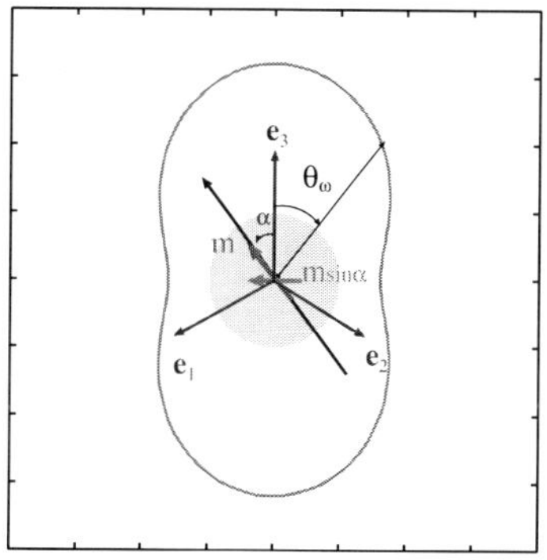

Figure 2. Angular distribution of the luminosity.

for differential luminosity , where θ_ω is an angle relative to the rotation axis [2].

Finally we have

$$\begin{aligned} L_m &= \int d\Omega_\omega \frac{dL_m}{d\Omega_\omega} \\ &= \frac{2\omega^4}{3c^3}|\mathbf{m}\sin\alpha|^2 \\ &= \frac{B_p^2 R^6 \omega^4 \sin^2\alpha}{6c^3}, \end{aligned} \tag{9}$$

for total luminosity. It would be interesting here to see that the luminosity (9) can be recast into the form,

$$L_m \simeq c\left(\frac{B_p R^3}{2r_c^3}\right)^2 r_c^2 \sim c\frac{B(r_c)^2}{8\pi} \cdot 4\pi r_c^2, \tag{10}$$

which implies that the energy density of the magnetic field ($\sim B(r_c)^2/8\pi$) flows outward through the surface of $4\pi r_c^2$ with the velocity of light c.

Assuming that this energy release originates from the rotational energy of neutron star,

$$\frac{dE_{\rm rot}}{dt} = -L_m, \tag{11}$$

where the rotation energy is given as $E_{\rm rot} = \frac{1}{2}I\omega^2$ with the moment of inertia I, assuming the rigid body. From Eqs, (9), (11), we have

$$P\dot{P} = \frac{2\pi^2 R^6 \sin^2\alpha}{3Ic^3} B_p^2, \tag{12}$$

where $P = 2\pi/\omega$ is the period of the rotation. The *characteristic age* τ_c is defined by

$$\tau_c \equiv P/2\dot{P}, \tag{13}$$

which gives a rough estimate of the age of pulsars.

The spin-down rate $\dot{\omega}$ is given as $\dot{\omega} \propto -\omega^3$ from Eqs. (9) and (11), so that the *braking index* $n \equiv -\omega\ddot{\omega}/(\dot{\omega})^2$ is three in this model. Note that the observed values of the braking index are less than three, e.g. $n = 2.515 \pm 0.005$ for the Crab, and it is usually considered that the significant fraction of the rotation energy is carried away by a pulsar wind [4]. Thus the magnetic dipole model is not sufficient to understand the emission processes in pulsars, while it can provide plausible estimates for the magnetic field and the characteristic ages.

Nowadays observation of $P - \dot{P}$ for pulsars are summarized on the plane [4], where we can see three clusters of pulsars: ordinary radio pulsars, the millisecond pulsars and *magnetars* . We can easily see that most of the radio pulsars are centered around 10^{12}G. Magnetars are compact stars with huge magnetic field ($B = O(10^{14-15})$G), including soft γ-ray repeaters (SGRs) or anomalous X-ray pulsars (AXPs) [3, 5, 6]. It exceeds the critical magnetic field defined as

$$B_{\rm QED} = \frac{m_e^2 c^3}{e\hbar} \simeq 4.4 \times 10^{13}{\rm G}, \tag{14}$$

which is obtained by equating the spacing of the Landau levels, $\delta E_e = \hbar eB/(m_e c)$, with the mass, $m_e c^2$. So it presents a criterion about how the relativistic effects are important and the quantum effects such as e^+e^- creation or photon splitting becomes significant. It should be also interesting to recall the relation, $1{\rm MeV}^2 \simeq 10^{13}$G in the natural unit. So, $B = 10^{15}$G expected in magnetars may correspond to 10MeV, which gets to the strong-interaction energy scale or the Fermi energy of nuclear matter or quark matter at nuclear density.

We can easily asses how the magnetic field affects EOS inside compact stars. As shown in Eq. (7) we can consider the uniform magnetic field inside the stars. Then δE_p is small for protons, $\delta E_p \ll m_p$, while it is comparable with light quark mass, $\delta E_q \sim m_q$, for $B = O(10^{15})$G. On the other hand, the magnetic-interaction energy can be estimated by a simple formula, $E_{\rm mag} = \mu_i B$, with the Dirac magnetic moment, $\mu_i = e_i/(2m_i)$, for i-th particle having mass m_i and electric charge e_i. For electrons it gives $O({\rm keV})$ for the canonical value of $B = O(10^{12})$G, which is comparable with the atomic-energy scale. So we may easily expect that thermodynamic quantities or EOS of the neutron-star envelope or the crust region should be much affected by the magnetic field [7]. On the other hand, it becomes very tiny for protons ($E_{\rm mag} \sim$ keV) even for $B = O(10^{15})$G, which implies that the magnetic field little affect the EOS of nuclear matter. Since it amounts to $E_{\rm mag} \sim$ MeV for light quarks, the magnetic field looks to modify the EOS of quark matter. However, the Fermi energy should be very large ($O(100){\rm MeV} \gg \delta E_q$), and a large number of the

Landau levels are occupied, which may be well approximated by the usual Fermi sphere. Thus we can conclude the effect of the magnetic field little modifies the EOS in the core region, even for $B \sim 10^{15}$G.

Other high-field radio emitting pulsars with $B \geq 10^{14}$G or rotation-powered pulsar/magnetar transition objects have been also observed [8]. These observations may give a hint about the relationship between magnetic activity and neutron star spin-down.

The origin of such strong magnetic field has been a basic but a long-standing problem since the first discovery of pulsars, while many people believe that it originates from dynamo scenario due to the charged current or inheritance of the magnetic field from the progenitor main-sequence stars (fossil-field hypothesis) [9]. This problem becomes a current issue, stimulated by the discoveries of magnetars. Due to the dynamo scenario, magnetic field is produced or maintained by the rotation or convection of charged fluid, but seems irrelevant for generation of high magnetic field in magnetars. It requires that magnetars be born with very short rotation periods of the order of 1-2 ms, and may not be supported by observations by two reasons [6]. First, the combination of high magnetic field and very rapid rotation is expected to impart a high velocity to the neutron star. However, up to now, the observational evidence for large spatial velocities in SGRs and AXPs is poor. Secondly, a large fraction of the rotational energy of a newly born magnetar, a few 10^{52}erg, is lost due to the strong magnetic braking. However, an estimate of the explosion energy of the remnants containing magnetars yields values close to the canonical supernova explosion energy of 10^{51}erg, implying initial periods longer than 5 ms.

Fossil-field hypothesis assumes the conservation of magnetic flux during the evolution of main-sequence progenitor star to a compact star: $B = (R_M/R)^2 B_M$ with R_M, B_M being the radius and magnetic field of the progenitor, respectively. It should be interesting to compare the radius R with R_M. Taking the sun as a typical main-sequence star, $(B_\odot)_{\rm Max} \simeq$ several thousands G and $R_\odot \simeq 7 \times 10^{10}$cm. Squeezing the magnetic flux from $R_\odot$ to R, we have $R = R_\odot (B_\odot/B)^{1/2} \simeq$ 10km for usual pulsars with $B = 10^{12}$G, which should be consistent with standard neutron stars. However, $R <$ several $\times\, 10^{4-5}$cm for $B \simeq 10^{14-15}$G, which should be compared with the Schwartzschild radius, $R_S = 2GM/c^2 = 3(M/M_\odot) \simeq 3 \times 10^5$cm for $M \simeq M_\odot$. Thus we can see that R should be comparable with R_s and may fall below it in the extreme case, if the fossil-field hypothesis is applied to the generation of the magnetic field in magnetars.

There is another possibility: a microscopic origin due to ferromagnetism or spin polarization of hadron matter [10]. Actually Makishima suggested the hadronic origin of the magnetic field in binary X-ray pulsars or radio pulsars [11], since the accumulation of the observational data shows a peak with a narrow width. Sometimes high-density nuclear matter is expected to show a resemblance with ^{3}He system, by the scaling argument[12], where quantum effects are essential and it shows the ^3P-type superfluidity in some physical conditions. Correspondingly neutron matter has been shown to exhibit the $^3P_2$ superfluidity around the nuclear density [12]. It is also well-known that ^{3}He is very close to a ferromagnetic state under the high pressure due to the large magnetic susceptibility [13, 14]. Microscopic calculations of nuclear matter have been repeatedly performed to find out a possibility of spontaneous spin polarization inside pulsars, but the negative results have been reported so far [15]. The magnetic susceptibility of nuclear matter increases monotonously with density, so that the ferromagnetic phase can not be expected in nuclear matter at any density. At high density we can also expect quark matter and we may be tempted to ask the possibility in quark matter instead. Nowadays there have been actively studied about the possible evidences of hadron-quark transition in astrophysical phenomena as well as relativistic heavy-ion collisions [4, 16, 17]. If such phase transition occurs inside compact stars, it may affect thermal and magnetic properties as well as equation of state (EOS). Here we ask a microscopic origin of the magnetic field by considering uniform quark matter inside compact stars [18].

We consider the possibility of spontaneous spin polarization by the use of QCD. First, the evaluation of the total energy for spin-polarized quark matter has been done by using the one-gluon-exchange (OGE) interaction at zero temperature [18, 19, 20]. Subsequently, the magnetic susceptibility has been studied within the Landau Fermi-liquid theory [13, 21], taking into account the screening effect for gluon propagation [22, 23, 24]. Finally we present a magnetic phase diagram on the density-temperature plane.

We also intend to address another interesting magnetic aspect of quark matter, which is characterized by the spatial modulation of the magnetic moment [25]. It is an inhomogeneous phase accompanying the chiral transition. In the standard scenario, the quark condensate, which is the order parameter of spontaneous symmetry breaking (SSB) of chiral symmetry, decreases due to the Pauli principle as density is increased. It is eventually vanished at some density-temperature point, which means the restoration of chiral symmetry. However, this may not be a unique scenario. Recall the FFLO state in the superconductivity; it has a spatially non-uniform order parameter and is considered to appear in the vicinity of the critical point, when two Fermi spheres with different spins have different sizes [26]. Recent studies have shown that there possibly appear various kinds of inhomogeneous phases in the vicinity of the chiral-symmetry restoration [27, 28, 29]. Among them we consider a special one called dual-chiral-density-wave (DCDW) state in detail, because it exhibits many interesting theoretical features and is expected to bear various implications in the light of compact-star phenomena.

Nowadays there have been done many theoretical studies about the deconfinement transition in high-density nuclear matter by using the MIT bag model [30] or other effective models of QCD to find the EOS including quark degrees of freedom. Unfortunately the lattice simulation is not possible there, because the numerical calculation suffers from so called the sign problem [31]. Thus there are ambiguities about the properties and the critical density of the deconfinement transition. Here we only assume the presence of quark matter without recource to EOS and the details of the deconfinement transition.

2. Spontaneous Spin Polarization in Quark Matter

A simple idea about spontaneous spin polarization owes to Bloch, who first discussed the emergence of ferromagnetism in electron gas at low density [32, 33]. Consider the electron gas in the positively charged background to compensate the electromagnetic charge of electrons. Then the Coulomb interaction is classically vanished, but the Fock exchange interaction between electrons with the same spin gives an attractive effect due to the Pauli principle; in the non-relativistic approximation it reads

$$E_{\rm ex} = -V\frac{9}{4}e^2\left(\frac{2}{9\pi}\right)^{1/3} n^{4/3}\left[(1+p)^{4/3}+(1-p)^{4/3}\right], \tag{15}$$

where n is the total number density and p is the polarization parameter defined by $p = (n_{s=+1} - n_{s=-1})/n$ with the number densities, n_s, of electrons with the spin $s = \pm 1$. This spin dependence may be easily understood by observing that the repulsive Coulomb interaction is effectively avoided for electrons with the same spin, since they cannot approach due to the Pauli principle. On the other hand the kinetic energy is easily evaluated as

$$E_{\rm kin} = V\frac{\hbar^2}{2m}\frac{6}{5}\pi\left(\frac{9\pi}{2}\right)^{1/3} n^{5/3}\left[(1+p)^{5/3}+(1-p)^{5/3}\right], \tag{16}$$

for non-relativistic electrons. Note that the density dependence of each term is peculiar. Searching the minimum of the total energy $\varepsilon_{\rm total}(p) = (E_{\rm kin} + E_{\rm ex})/V$ with respect to p by fixing the total number density $n_{+1} + n_{-1} = n$, we can observe that electron gas is completely polarized for $n^{1/3} < n_c^{1/3} \equiv \alpha/(1+2^{-1/3})$ with $\alpha \equiv 5/(6\pi^2)e^2 m/\hbar^2(9\pi/2)^{1/3}$. The phase transition is of weakly first order in this case.

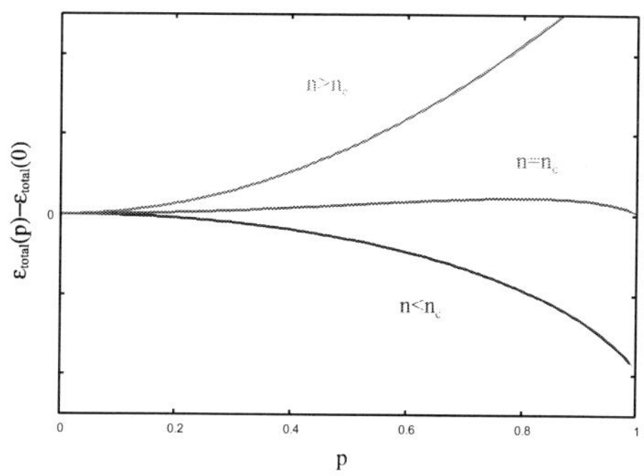

Figure 3. Energy density (arbitrary scale) as the function of the polarization parameter $p = (n_{+1} - n_{-1})/n_q$.

This is the result within the Hartree-Fock approximation, but it has been also shown by the quantum Monte Carlo simulation that the electron gas is in the ferromagnetic phase at very low density [34]. Recently it has been experimentally observed [35]. A lesson we learned here is that we need no spin dependent interaction in the original Lagrangian to see spontaneous spin polarization, but a symmetry principle plays an important role in this context.

When we consider quark matter within QCD, we notice that the situation seems to be similar to the electron gas; quark matter is color neutral as a whole, so that the exchange term of the one-gluon-exchange (OGE) interaction gives a leading-order contribution to the total energy. In ref. we have calculated the interaction energy of relativistic quarks in a perturbative way, and demonstrated the spontaneous spin polarization around nuclear density, $n_q \simeq \rho_0 \simeq 0.16{\rm fm}^{-3}$.

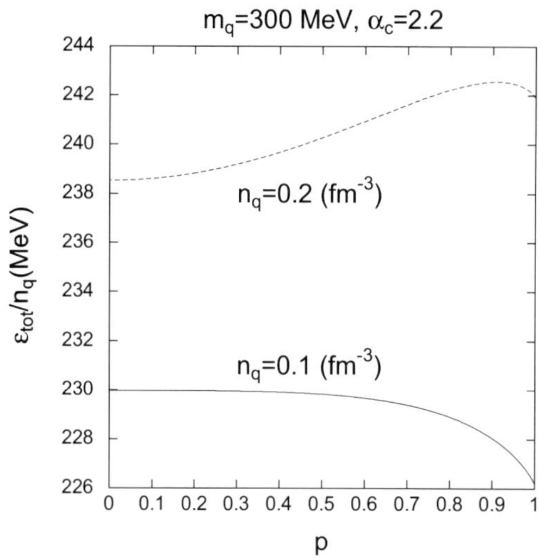

Figure 4. Energy per particle for quark matter with the OGE interaction as a function of the polarization parameter, p.

Assuming that spin-polarized quark matter with density n_q inside compact stars, we can roughly estimate the magnetic field at the surface (see Eq. 4)),

$$
\begin{aligned}
B_{\max} &= \frac{8\pi}{3}\left(\frac{r_q}{R}\right)^3 \mu_q n_q \\
&\simeq 10^{15}\mathrm{G}\left(\frac{r_q}{R}\right)^3\left(\frac{\mu_q}{\mu_N}\right)\left(\frac{n_q}{0.1\mathrm{fm}^{-3}}\right)
\end{aligned}
\tag{17}
$$

for the quark core of the radius r_q, where μ_q is the quark magnetic moment and μ_N the nuclear magneton.[1]

2.1. Magnetic Susceptibility within the Landau Fermi-liquid Theory

In the recent papers we have studied the magnetic susceptibility within the Landau Fermi-liquid theory [22, 23], to get more insight about the phase transition and derive a phase diagram in the density-temperature plane.

Applying a tiny and uniform magnetic field $\boldsymbol{B}$ (along the z-axis), we can study the linear response of quark matter by calculating the change of magnetization, $\mathbf{M}$, by the external field. Then the magnetic susceptibility χ_M is defined as $\chi_M = \partial\mathbf{M}/\partial\boldsymbol{B}|_{\boldsymbol{B}=0}$. We can easily see that χ_M measures the spin-spin correlation in the normal quark matter, or the curvature of the free energy at the origin with respect to the order parameter, the magnetization $\mathbf{M}$. The free energy is given as a function of the external magnetic field,

[1] One may wonder about the back-reaction of the generated magnetic field to quarks. The Fermi energy is $O(100)$MeV and the quark mass is rather heavy in this case, so that the Landau orbiting and the magnetic interaction are safely discarded.

$F(\boldsymbol{B})$, from which $\mathbf{M}$ is given by

$$\mathbf{M} = -\frac{\partial F(\boldsymbol{B})}{\partial \boldsymbol{B}}. \tag{18}$$

Defining the Legendre transform of F such that

$$G(\mathbf{M}) = F + \mathbf{M} \cdot \boldsymbol{B}, \tag{19}$$

G satisfies the reciprocity relation ,

$$\frac{\partial G(\mathbf{M})}{\partial \mathbf{M}} = \boldsymbol{B}. \tag{20}$$

Therefore χ_M can be written as

$$\chi_M = \left[\frac{\partial^2 G(\mathbf{M})}{\partial \mathbf{M}^2}\right]^{-1}, \tag{21}$$

which is the inverse of the curvature of G at the origin, assuming $\mathbf{M} = 0$ for $\boldsymbol{B} = 0$. Adding $-\mathbf{M} \cdot \boldsymbol{B}$ to the free energy $G(\mathbf{M})$ we find the fluctuation amplitude at T,

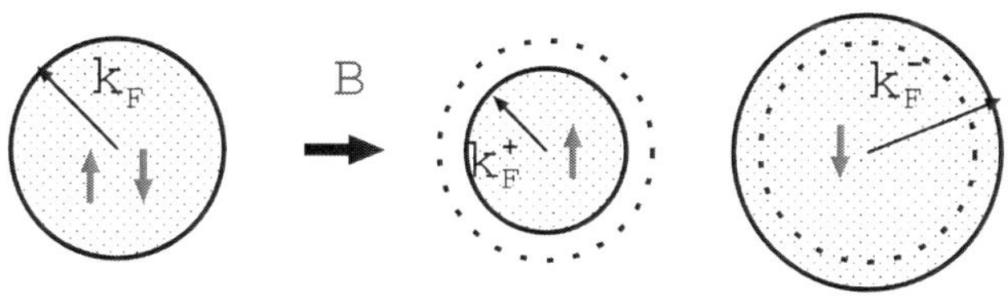

Figure 5. Two Fermi spheres with different spins have different sizes to generate the magnetization after applying an external magnetic field.

$$\langle M^2 \rangle|_{\boldsymbol{B}=0} = T\chi_M, \tag{22}$$

which means the fluctuation amplitude diverges as χ_M [36].

χ_M can be written in terms of the Landau-Migdal parameters derived from the quark-quark interaction,

$$\chi_M = \left(\frac{\bar{g}_D \mu_q}{2}\right)^2 \frac{N(T)}{1 + N(T)\bar{f}^a}, \tag{23}$$

where $\bar{g}_D \equiv \int_{|\mathbf{k}|=k_F} d\Omega_{\mathbf{k}}/4\pi g_D(\mathbf{k})$ is the effective gyromagnetic ratio. $N(T)$ is the effective density of states around the Fermi surface, which is written as

$$N(0) = \frac{N_c k_F^2}{\pi^2 v_F} \tag{24}$$

at $T = 0$, where the Fermi velocity v_F is given by the spin-independent Landau-Migdal parameter f_1^s, $k_F/E_F - (N_c k_F^2/3\pi^2) f_1^s$. $\bar{f}^a$ is the spin-dependent Landau-Migdal parameter. The divergence of χ_M or $1 + N(T)\bar{f}^a = 0$ implies the phase transition to the ferromagnetic phase.

We present here a heuristic argument by the use of the naive OGE interaction in QCD.

$$\mathcal{L}_{\rm QCD} = -\frac{1}{2}G^{\mu\nu}G_{\mu\nu} + \sum_f \bar{q}_f\left(i\gamma^\mu D_\mu - m_f\right)q_f, \tag{25}$$

where

$$\begin{aligned} G_{\mu\nu} &= \partial_\mu A_\nu - \partial_\nu A_\mu - ig[A_\mu, A_\nu] \\ D_\mu q_f &\equiv (\partial_\mu - igA_\mu)q_f \end{aligned} \tag{26}$$

with the gluon field, $A_\mu = \sum_{a=1}^{8} A_\mu^a \lambda^a/2$. Since spin is coupled with motion in relativistic theories, we must define spin polarization in a proper way. A relevant one is to introduce the space-like four vector a^μ with the constraints, $a\cdot k = 0$ and $a^2 = -1$ for a particle with momentum $\boldsymbol{k}$. The standard choice may be

$$\boldsymbol{a} = \boldsymbol{\zeta} + \frac{\boldsymbol{k}(\boldsymbol{\zeta}\cdot\boldsymbol{k})}{m(E_k+m)}, \quad a^0 = \frac{\boldsymbol{k}\cdot\boldsymbol{\zeta}}{m}, \tag{27}$$

where the three vector $\boldsymbol{\zeta}$ specifies the direction of spin in the rest frame of each particle [37]. Then we have the polarization density matrix $\rho(k,\zeta)$,

$$\rho(k,\zeta) = \frac{1}{2m}(\not{k} + m)P(a), \tag{28}$$

with the projection operator, $P(a) = (1 + \gamma_5 \not{a})/2$. Given the quasiparticle interaction $f_{\boldsymbol{k}\zeta a;\boldsymbol{q}\zeta' b}$ between quarks with momentum $\boldsymbol{k}$, spin polarization ζ and color a and $\boldsymbol{q}$, ζ' and b, the color symmetric interaction is generally given as

$$f_{\mathbf{k}\zeta,\mathbf{q}\zeta'} = \frac{1}{N_c^2}\sum_{a,b} f_{\mathbf{k}\zeta a,\mathbf{q}\zeta' b} = \frac{m}{E_k}\frac{m}{E_q}M_{\mathbf{k}\zeta,\mathbf{q}\zeta'}, \tag{29}$$

with the invariant matrix element within the OGE interaction,

$$M_{\mathbf{k}\zeta,\mathbf{q}\zeta'} = -g^2\frac{1}{N_c^2}{\rm tr}\left(\lambda_\alpha/2\lambda_\alpha/2\right)M^{\mu\nu}(k,\zeta;q,\zeta')D_{\mu\nu}(k-q). \tag{30}$$

Here $D_{\mu\nu}$ is the gluon propagator and $M^{\mu\nu}(k,\zeta;q,\zeta')$ the interaction tensor,

$$M^{\mu\nu}(k,\zeta;q,\zeta') = {\rm tr}\left[\gamma^\mu\rho(k,\zeta)\gamma^\nu\rho(q,\zeta')\right], \tag{31}$$

Taking the Feynman gauge for the gluon field, the invariant matrix element can be explicitly written as

$$M_{\boldsymbol{k}\zeta,\boldsymbol{q}\zeta'} = g^2\frac{N_c^2-1}{4N_c^2m^2}\left[2m^2 - k\cdot q - m^2 a\cdot b\right]\frac{1}{(k-q)^2}. \tag{32}$$

In the non-relativistic limit, it is reduced to

$$M_{\boldsymbol{k}\zeta,\boldsymbol{q}\zeta'} \to -g^2\frac{N_c^2-1}{4N_c^2}\frac{1+\boldsymbol{\zeta}\cdot\boldsymbol{\zeta}'}{(\boldsymbol{k}-\boldsymbol{q})^2} < 0. \tag{33}$$

Thus a pair with the parallel spin ($\boldsymbol{\zeta} = \boldsymbol{\zeta}'$) gives an attractive interaction in the nonrelativistic limit. This makes a base of the Bloch mechanism [33]: the Fock exchange interaction gives rise to an attractive effect for the pair of the parallel spin due to the Pauli principle. Then the quasiparticle interaction on the Fermi surface is given as

$$f_{\boldsymbol{k}\zeta,\boldsymbol{q}\zeta'}\Big|_{|\boldsymbol{k}|=|\boldsymbol{q}|=k_F} = g^2\frac{N_c^2-1}{8N_c^2E_F^2}\left[2m^2 - E_F^2 + k_F^2\cos\theta_{\widehat{\boldsymbol{kq}}} - m^2 a\cdot b\right] \times \frac{1}{-k_F^2(1-\cos\theta_{\widehat{\boldsymbol{kq}}})}. \tag{34}$$

We can immediately see that it diverges for the collinear momenta, $\boldsymbol{k} = \boldsymbol{q}$. Substituting the explicit formula of the spin vector in Eq. (27), we can see that it consists of two parts, the spin-independent ($f^s_{\mathbf{k},\mathbf{q}}$) and spin-independent ($f^a_{\mathbf{k},\mathbf{q}}$) terms;

$$f_{\mathbf{k}\zeta,\mathbf{q}\zeta'} = f^s_{\mathbf{k},\mathbf{q}} + \zeta\zeta' f^a_{\mathbf{k},\mathbf{q}}, \tag{35}$$

from which we can derive the Landau-Migdal parameters by the angle-integral over the Fermi surface. For the spin-independent Landau-Migdal parameter, f_1^s,

$$f_1^s = -3\frac{g^2(N_c^2-1)}{8N_c^2E_F^2}\frac{m^2}{2k_F^2}\int_{-1}^{1} d(\cos\theta)\frac{\cos\theta}{1-\cos\theta} \to \infty, \tag{36}$$

and for the spin-dependent one, $\bar{f}^a$,

$$\begin{aligned}\bar{f}^a &\equiv \int\frac{d\Omega_{\boldsymbol{k}}}{4\pi}\int\frac{d\Omega_{\mathbf{q}}}{4\pi} f^a_{\mathbf{k},\mathbf{q}}\Big|_{|\mathbf{k}|=|\mathbf{q}|=k_F} \\ &= -\frac{g^2(N_c^2-1)}{8N_c^2E_F^2}\frac{m(2E_F+m)}{3k_F^2} + f_1^s/3. \end{aligned} \tag{37}$$

Putting Eqs. (36) and (37) in Eq. (23), we have the final expression,

$$(\chi_M/\chi_{\mathrm{Pauli}})^{-1} = 1 - \frac{C_f g^2}{12\pi^2 E_F k_F} m(2E_F+m) \tag{38}$$

with the Casimir operator per color, $C_f = \frac{N_c^2-1}{2N_c}$, where χ_{Pauli} is the susceptibility of non-interacting Fermi gas, $\chi_{\mathrm{Pauli}} = \bar{g}_D^2\mu_q^2 N_c k_F E_F/4\pi^2$. Thus we can observe that infrared divergences cancel each other in the expression of χ_M. It is to be noted that χ_M is proportional to the quark mass. Thus, we may say that heavier quarks favor the spontaneous magnetization. Taking the non-relativistic limit, $m \gg k_F$, and replacing C_f and g by one and e, respectively, we can recover the standard formula for electron gas interacting with the Coulomb potential[33],

$$(\chi/\chi_0)^{-1} = 1 - \frac{e^2 m}{k_F\pi} \tag{39}$$

In Fig. 2.1 we present a result by using the MIT bag model parameters.

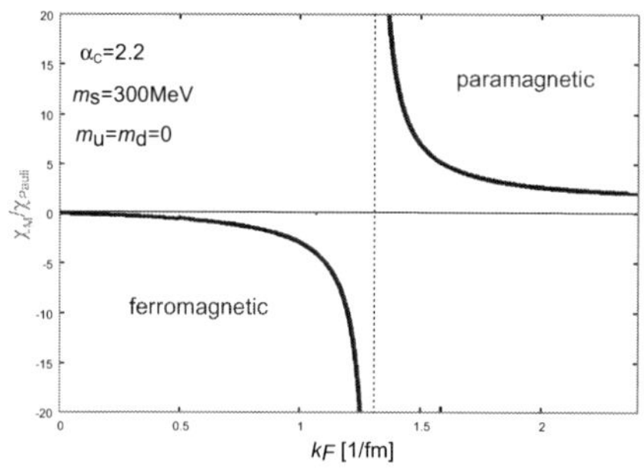

Figure 6. Spin susceptibility χ_M at $T = 0$ within the Landau's Fermi liquid theory. $\chi_{\rm Pauli}$ is the Pauli paramagnetism. It diverges around $k_F \simeq 1.4{\rm fm}^{-3}$, the order of nuclear density, ρ_0, and quark matter is in the ferromagnetic phase at lower density side.

2.2. Magnetic Phase Diagram

We have also studied the screening effect for OGE interaction on χ_M, since it is well known that such many-body effect is important in the electron gas. Actually the screening effect disfavors the spin alignment [38]. Moreover, it is necessary to include it when we consider the gauge interaction like QED/QCD, since the infrared behaviour is singular and it is improved by the screening effect. In QCD it has another interesting effect, depending on the number of flavors: it may favor the ferromagnetic transition for a large number of flavors. In the following let us briefly see the screening effect.

The gluon propagator $D_{\mu\nu}(p)$ is generally written as

$$D_{\mu\nu}(p) = P^t_{\mu\nu} D_t(p) + P^l_{\mu\nu} D_l(p) - \xi \frac{p_\mu p_\nu}{p^4}, \tag{40}$$

where $P^{t(l)}_{\mu\nu}$ is the projection operator onto the transverse (longitudinal) mode,

$$\begin{aligned} P^t_{\mu\nu} &= (1-g_{\mu 0})(1-g_{\nu 0})\left(-g_{\mu\nu} - \frac{p_\mu p_\nu}{|\mathbf{p}|^2}\right) \\ P^l_{\mu\nu} &= -g_{\mu\nu} + \frac{p_\mu p_\nu}{p^2} - P^t_{\mu\nu}. \end{aligned} \tag{41}$$

Correspondingly, $D_{t(l)}(p)$ are the propagator of transverse (longitudinal) gluons, modified by the medium effect, which can be described by the Debye screening [39]. The self-energies for the transverse and longitudinal gluons are calculated by the hard dense loop (HDL) resummation to give

$$\begin{aligned} \Pi_l(p_0, \mathbf{p}) &= \sum_{f=u,d,s} \left(m^2_{D,f} + i \frac{\pi m^2_{D,f}}{2u_{F,f}} \frac{p_0}{|\mathbf{p}|} \right) \\ \Pi_t(p_0, \mathbf{p}) &= -i \sum_{f=u,d,s} \frac{\pi u_{F,f} m^2_{D,f}}{4} \frac{p_0}{|\mathbf{p}|}, \end{aligned} \tag{42}$$

in the limit $p_0/|\mathbf{p}| \to 0$, with the Fermi velocity $u_{F,f} \equiv k_{F,f}/E_{F,f}$ and the Debye mass, $m_{D,f}^2 \equiv g^2\mu_f k_{F,f}/2\pi^2$ for each flavor f [39].[2]

The appearance of the imaginary part in Π_t physically means the damping of the transverse gluons due to the interaction with the surrounding quarks (Landau damping). Thus the longitudinal gluons are statically screened by the Debye mass, while the transverse gluons are dynamically screened due to the Landau damping. Accordingly, the screening effect for the transverse gluons is ineffective at $T = 0$, where soft gluons ($p_0/|\mathbf{p}| \to 0$) contribute. At finite temperature, gluons with $p_0 \sim O(T)$ can contribute due to the diffuseness of the Fermi surface and the transverse gluons are effectively screened, which gives rise to another interesting consequence for χ_M. We have seen a non-Fermi liquid behaviour, which is inherent in the gauge theories: there appears $T^2 \ln T$ term in χ_M, besides the usual T^2 term [22, 23].

Taking into account the screening effect, we can evaluate the magnetic susceptibility. To summarize, we present in Fig. 7 the phase diagram of ferromagnetic quark matter on the temperature-density plane.

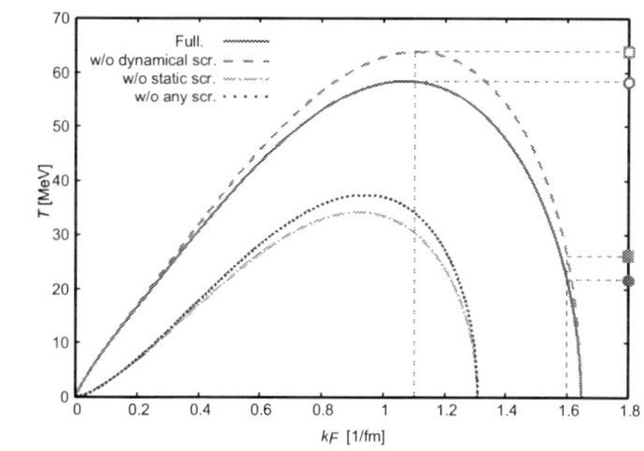

Figure 7. Magnetic phase diagram in the density-temperature plane. The open (filled) circle indicates the Curie temperature at $k_F = 1.1(1.6)$ fm^{-1} while the squares show those without the $T^2 \ln T$ term.

2.3. Spin Wave

Ferromagnetic quark matter is specified by the non-vanishing magnetization M, so that rotation symmetry is violated there, $SO(3) \to O(2)$. Such SSB should accompany the Nambu-Goldstone mode, *spin wave*. Different from the usual treatment of the spin wave in the Heisenberg model [40], we must carefully formulate it since quarks freely move there. For the ferromagnetism of itinerant electrons, Herring took an intuitive but correct approach called the *spiral* approach [33, 41]. He introduced the collective variables as variants from the spin direction. In ref.[42] we have proposed a formulation within the path integral of the spin coherent-state. Since the spin wave or its quantized magnon plays important roles for the thermal evolution as well as the magnetic evolution of compact stars, more elaborate studies are needed about its properties including the dispersion relation: specific heat and thermal conductivity may be affected by the magnons and a new cooling process may open

[2] The Debye mass is given as $e^2\mu^2 u_F/\pi^2$ for electron gas in QED.

through the absorption and emission of magnons in the interaction vertices. Moreover, the magnon-exchange interaction may work between quarks in the ferromagnetic phase, which gives, e.g., an exotic superconductivity [43].

3. Chiral Transition and Inhomogeneous Phases

Here we consider another type of magnetism in quark matter. Recently there are many works about the inhomogeneous phases accompanying the chiral transition [27, 28, 29]. According to the usual discussion about the chiral transition, the $q\bar{q}$ condensate is decreased as density or temperature is increased, and eventually vanished at some density [16]. Since the dynamical mass is proportional to the condensate, it also decreases.

We, hereafter, consider two-flavor (u, d) symmetric quark matter for simplicity. Chiral symmetry is then represented by $SU(2)_L \times SU(2)_R$ algebra, under which the quark field ψ is transformed s.t.

$$\begin{aligned} \psi \to \psi' &= \psi + i\alpha^a(\tau^a/2)\psi \\ \psi \to \psi' &= \psi + i\beta^a(\tau^a/2)\gamma_5\psi, \end{aligned} \tag{43}$$

for small parameters, $\alpha^a, \beta^a \ll 1$. The first one is the usual isospin transformation, while the second one, called chiral transformation, mixes the different parity states. Many studies have shown that Lagrangian should be chirally invariant for massless quarks, but it is spontaneously broken in the vacuum, which is specified by the non-vanishing $q\bar{q}$ condensate, $\langle\bar{\psi}\psi\rangle \neq 0$. As important consequences, the quark mass can be dynamically generated by the spontaneous symmetry breaking (SSB) and pions emerge as a collective mode of quarks and antiquarks. At finite density the Fermi sea prevents the formation of the condensate due to the Pauli principle, and chiral symmetry should be eventually restored at some density.

Recent studies have suggested that the homogeneous ground state becomes unstable for producing the inhomogeneous phases in prior to the chiral transition. Here we consider one of such instabilities in detail by using the effective model of QCD, Nambu-Jona Lasinio (NJL) model [25, 44]. In the paper [25] we have demonstrated formation of a density wave in quark matter in the chiral limit: where not only scalar density but also pseudoscalar density takes a non-vanishing value,

$$\begin{aligned} \langle\bar{\psi}\psi\rangle &= \Delta\cos(\boldsymbol{q}\cdot\mathbf{r}) \\ \langle\bar{\psi}i\gamma_5\tau_3\psi\rangle &= \Delta\sin(\boldsymbol{q}\cdot\mathbf{r}). \end{aligned} \tag{44}$$

Note that we can utilize the pseudoscalar degree of freedom in quark matter, different from the vacuum, which must be an eigenstate of parity. Consequently the above configuration means a violation of parity. It should be interesting here to see a similarity with the FFLO state in superconductivity [26]: a combination of both densities can be written as $\langle\bar{\psi}\psi\rangle + i\langle\bar{\psi}i\gamma_5\tau_3\psi\rangle = \Delta\exp(i\boldsymbol{q}\cdot\mathbf{r})$, which corresponds to the complex energy-gap parameter due to the condensation of the Cooper pairs [45]. In the FFLO state, the gap function, $\Delta(\mathbf{r}) \equiv \langle\psi_\downarrow(\mathbf{r})\psi_\uparrow(\mathbf{r})\rangle$, modulates in space with the wave number characterized by the center-of-mass momentum of the Cooper pairs. The similarity indicates a general mapping between superconductivity and magnetism [46]. It might be also interesting to see the phase transition in terms of the susceptibility. Similarly to the magnetic susceptibility in the previous section, we can introduce the susceptibility or the density correlation function in scalar or pseudoscalar channel, $\chi_{\rm s(ps)}(\omega, q)$ [25]. In the ferromagnetic case, $\chi_M(\omega = 0, q = 0)$ diverges at the critical point, but $\chi_{\rm s(ps)}(\omega = 0, q \neq 0)$ in this case. The finite value of q is characteristic in the inhomogeneous phase.

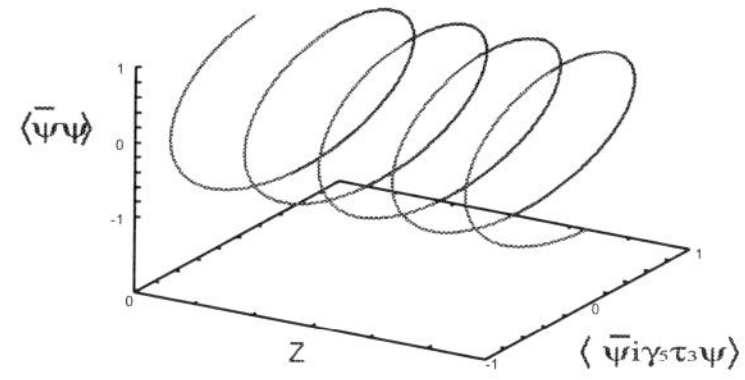

Figure 8. Dual chiral density wave in the chiral space.

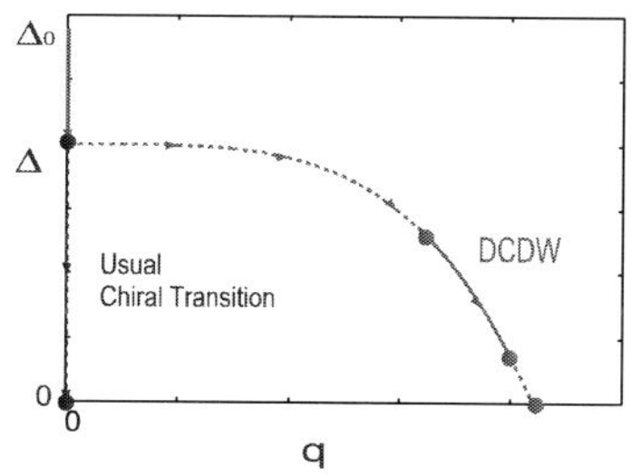

Figure 9. Possible chiral restoration paths on the $\Delta - q$ plane. Dotted lines show the first order phase transitions.

Consider the 2-flavor Nambu-Jona Lasinio model as an effective model of QCD at low density,

$$\mathcal{L}_{\rm NJL} = \bar{\psi}(i\partial\!\!\!/ - m_c)\psi + G\left[(\bar{\psi}\psi)^2 + (\bar{\psi}i\gamma_5\boldsymbol{\tau}\psi)^2\right], \tag{45}$$

where G is the coupling constant and m_c the current mass.[3] Putting Eq. (44) in the Lagrangian under the mean-field approximation, we have

$$\mathcal{L}_{\rm MF} = \bar{\psi}\left[i\partial\!\!\!/ - M^*\left(\cos(\boldsymbol{q}\cdot\boldsymbol{r}) + i\gamma_5\tau_3\sin(\boldsymbol{q}\cdot\boldsymbol{r})\right)\right] - \frac{M^{*2}}{4G}, \tag{46}$$

where the effective mass M^* is defined by way of $M^* = -2G\Delta$. Introducing a new spinor by using a local chiral transformation,

$$\psi(\boldsymbol{r}) = e^{-i\tau_3\gamma_5\boldsymbol{q}\cdot\boldsymbol{r}/2}\psi_w(\boldsymbol{r}), \tag{47}$$

$\mathcal{L}_{\rm MF}$ is recast into

$$\mathcal{L}_{\rm MF} = \bar{\psi}_w\left[i\partial\!\!\!/ - M^* - \tau_3\gamma_5\boldsymbol{\gamma}\cdot\boldsymbol{q}/2\right]\psi_w - \frac{M^{*2}}{4G}. \tag{48}$$

This is a kind of the Weinberg transformation [47], which makes the Lagrangian in the presence of inhomogeneous DCDW to the one with the homogeneous axial-vector field.

[3] We, hereafter, consider the chiral limit where $m_c = 0$ in this subsection.

Table 1. Separation of amplitude and phase degrees of freedom by way of the Weinberg transformation.

non-uniform		uniform
$\langle\bar{\psi}\psi\rangle = \Delta\cos\boldsymbol{q}\cdot\boldsymbol{r}$	$\rightarrow$	$\langle\bar{\psi}_w\psi_w\rangle = \Delta$
$\langle\bar{\psi}i\gamma_5\tau_3\psi\rangle = \Delta\sin\boldsymbol{q}\cdot\boldsymbol{r}$	$\rightarrow$	$\langle\bar{\psi}_w i\gamma_5\tau_3\psi_w\rangle = 0$
—	$\rightarrow$	$\gamma_5\tau_3\boldsymbol{\gamma q}/2$

Then the quark wave function is still given by the plane-wave solution $\psi_k(\boldsymbol{r})$ with the energy eigenvalue,

$$E^{\pm}(\boldsymbol{p}) = \sqrt{\boldsymbol{p}^2 + M^{*2} + \boldsymbol{q}^2/4 \pm \sqrt{(\boldsymbol{p}\cdot\boldsymbol{q})^2 + M^{*2}\boldsymbol{q}^2}}. \tag{49}$$

Accordingly the Fermi seas are deformed in oblate and prolate shapes, depending on the spin degree of freedom. In Fig.9.10 we show energy spectra for example.

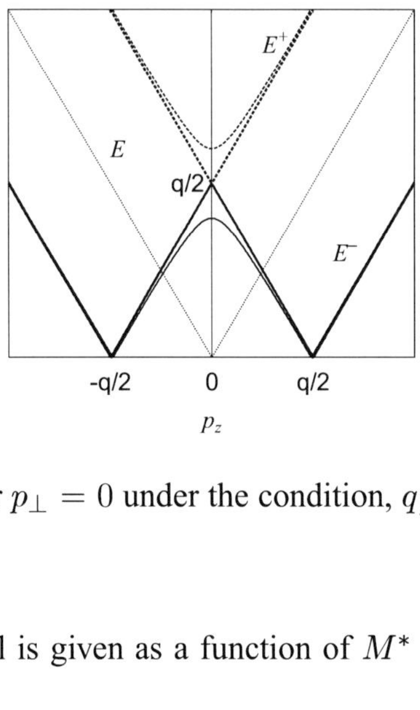

Figure 10. Energy spectra for $p_\perp = 0$ under the condition, $q/2 > M^*$. E is the one in the absence of DCDW.

Thermodynamic potential is given as a function of M^* and q at finite temperature T and chemical potential μ:

$$\begin{aligned}\Omega(q, M^*) &= -T\log Z/V \\ &= -N_f N_c \int \frac{d^3p}{(2\pi)^3} \sum_{s=\pm} \left\{ T\ln\left[e^{-\beta(E^s-\mu)}+1\right]\left[e^{-\beta(E^s+\mu)}+1\right] + E^s \right\} \\ &+ \frac{M^{*2}}{4G}.\end{aligned} \tag{50}$$

The last term in the curly bracket denotes the vacuum contribution, which has an essential role in the chiral transition. Since single-particle spectra are deformed in momentum space, the simple momentum cut-off is irrelevant and sometimes leads to unphysical results. We employ the proper-time regularization with cutoff Λ to evaluate the vacuum contribution.

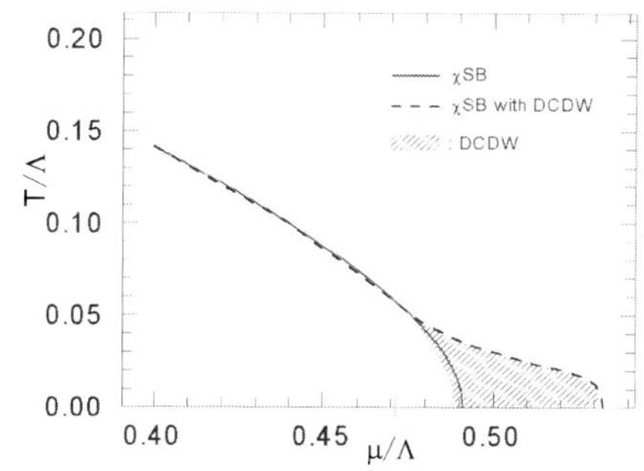

Figure 11. Phase diagram of DCDW on the density-temperature plane. DCDW emerges in the vicinity of the chiral transition.

3.1. "Nesting" Mechanism

We discuss here the physical mechanism for the emergence of DCDW. First we consider the one dimensional case. Assume the presence of the spin density wave (SDW) $\exp(\pm iqz)$ coupled with the non-relativistic electrons like

$$H_{\rm int} = -V\left(e^{-iqz}u^\dagger\sigma_+ u + e^{iqz}u^\dagger\sigma_- u\right). \tag{51}$$

Then eigenvalues are

$$E_k = \frac{1}{2}(\epsilon_k + \epsilon_{k+q}) \pm \left[\frac{1}{2}(\epsilon_k - \epsilon_{k+q})^2 + 4V^2\right]^{1/2}, \tag{52}$$

with the free particle energy, $\epsilon_k = k^2/2m$ (Fig. 12). In the case of $q = 2k_F$, there is produced a gap at the Fermi surface of the free electrons. Thereby the total energy is always decreased by the interaction, *independent* of the strength of the coupling. This is called a nesting effect of the Fermi surface [48, 49]. We can also see this by calculating the correlation function; it is given by the Lindhard function $L(\omega = 0, q)$, which exhibits a logarithmic divergence at $q = 2k_F$.

We can see the similar mechanism should be also responsible to the emergence of DCDW, but in somewhat different manner. By putting $\boldsymbol{p}_\perp = 0$, we can consider quasi-one dimensional case. The energy spectra exhibit different features, depending on the conditions, $q/2 > M^*$ and $q/2 < M^*$. It is not obvious which condition holds in the DCDW phase, but the numerical results shows that the former condition always holds. Then we can understand the mechanism of the DCDW formation by two steps. First looking at Fig. 12, we can see that the levels of massless quarks cross each other at $p_z = 0$. After switching on the mass term the level crossing is finally disentangled to give $E^\pm$. So, if we set $q = 2\mu$, there is always the energy gain, irrespective of the dynamics. Actually it has been shown that the chiral spiral, which is an analog of DCDW in the NJL$_2$ model, exactly gives the

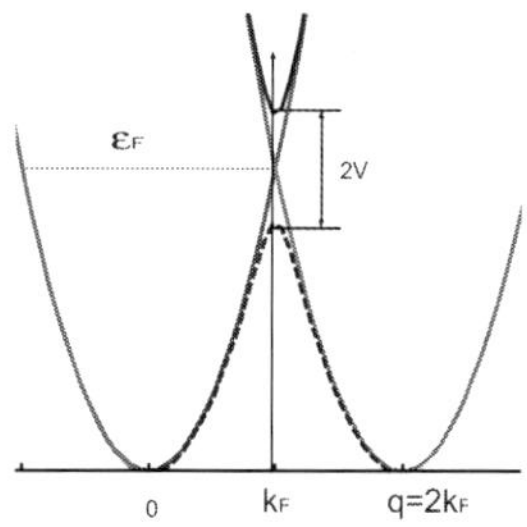

Figure 12. Energy spectra of one dimensional electron gas in the presence of SDW.

magic relation $q = 2\mu$, and is the lowest-energy state among the inhomogeneous states, irrespective of density or temperature [27]. Note that this mechanism is robust only in the 1+1 dimensional case (complete nesting) and the situation becomes subtle, depending on the dimensionality [49]. It has been shown that the results for DCDW are reminiscent of the nesting effect [25]. However, recent study has suggested the real kink crystal (RKC) has lower energy than that of DCDW in the chiral limit [28]. Thus more studies are needed to clarify the conditions for appearing DCDW in the QCD phase diagram.

3.2. Some Implications

The presence of DCDW has some implications on compact star phenomena. Let's consider, for example, the β-decay of quarks in the DCDW state,

$$d(p_1) \to u(p_2) + e^-(p_3) + \bar{\nu}_e(p_4). \tag{53}$$

From Eq. (47) the DCDW state can be represented as a chirally rotated state,

$$|\psi_w\rangle = \hat{U}(\mathbf{q})|\psi\rangle, \tag{54}$$

with

$$\hat{U}(q) = \exp(i\int A_3^0 \mathbf{q}\cdot\mathbf{r} d^3x), \tag{55}$$

the quark current $h^\mu_{1+i2} \equiv \bar{\psi}\gamma^\mu(1-\gamma_5)\tau_+\psi$ is transformed as

$$\tilde{h}^\mu_{1+i2} \equiv \hat{U}(q)^\dagger h^\mu_{1+i2}\hat{U}(q) = \exp(i\mathbf{q}\cdot\mathbf{r})h^\mu_{1+i2}, \tag{56}$$

Thus we can see that DCDW supply an extra momentum to modify the momentum conservation in the β-decay process. It is easily shown that the neutrino emission due to the quark β-decay is suppressed by the energy-momentum conservation in normal quark matter: the momenta of all the particles should be collinear for reaction (53) for free massless quarks at low temperature. So DCDW catalyzes the reaction (53), which may give rise to a fast cooling of compact stars [50].

The symmetry breaking pattern in the DCDW phase is as follows: the original translational symmetry along z-axis (the generator is $\hat{p}_z$) and $U(1)$ subgroup of chiral symmetry (the generator is Q_5^3) is broken at the same time, but it is still invariant under a combination of these operations,

$$T_{\hat{p}_z} \times U_{Q_5^3}(1) \to U_{\hat{p}_z + Q_5^3}(1). \tag{57}$$

Actually we can see that the single-particle wave function is the eigenfunction of the operator,

$$\exp(i\hat{p}_z a)\exp(-i\gamma_5\tau_3\theta/2)\psi_k(\boldsymbol{r}) = \exp(ik_z a)\psi_k(\boldsymbol{r}) \tag{58}$$

with $\theta = qz$. The collective modes ("phasons") have then hybrid properties of phonons and "pions". We can derive the dispersion relation of the collective modes by way of the Ginzburg-Landau theory. If we consider the phase fluctuations, $u(\mathbf{r}, t)$, the free energy should be written as

$$F = F_0(\Delta, q) + \delta F(u(\mathbf{r}, t)), \tag{59}$$

with

$$\delta F = \frac{1}{2}\int \left[A(\nabla_z u)^2 + B(\nabla_\perp^2 u)^2 + C\left(\frac{\partial u}{\partial t}\right)^2 \right] d^3r, \tag{60}$$

like liquid crystal (smectics) [51], where the coefficients A, B, C are the function of Δ, q[48, 49]. The Euler-Poisson equation for u then reads,

$$C\frac{\partial^2 u}{\partial t^2} - A\frac{\partial^2 u}{\partial z^2} + B(\nabla_\perp^2)^2 u = 0. \tag{61}$$

The plane-wave solution, $u = u_0 e^{i(\omega t - \boldsymbol{k}\cdot\mathbf{r})}$, is obtained with the anisotropic dispersion relation,

$$\omega^2 = (Ak_z^2 + Bk_\perp^4)/C. \tag{62}$$

Note that the dispersion relation exhibits a hybrid nature of type I and II Nambu-Goldstone bosons [52]: $\omega \propto k_z$ for the longitudinal excitation with $k_\perp = 0$, while $\omega \propto k_\perp^2$ for the transverse excitation with $k_z = 0$.

3.3. Magnetic Aspect of DCDW

Here we reveal the anomolous magnetic aspect of DCDW [25]. Since explicit form of the spinor can be analytically obtained within the mean-field approximation, one can evaluate various expectation values in the DCDW phase.

By the use of the spinor ψ_w one can evaluate the expectation values of bilinear form of gamma matrices $\mathcal{O}$ in a simpler way,

$$\langle\psi^\dagger(\mathbf{r})\mathcal{O}\psi(\mathbf{r})\rangle = \int \frac{d^3p}{(2\pi)^3}\langle\psi_w^\dagger(\boldsymbol{p})e^{i\tau_3\gamma_5\boldsymbol{q}\cdot\boldsymbol{r}/2}\mathcal{O}e^{-i\tau_3\gamma_5\boldsymbol{q}\cdot\boldsymbol{r}/2}\psi_w(\boldsymbol{p})\rangle. \tag{63}$$

Spin expectation value, for instance, $\mathcal{O} = \gamma_0\gamma_5\gamma_3/2 \equiv \Sigma_z$, vanishes because it is proportional to the stationary condition with respect to wave number q,

$$\int \frac{d^3p}{(2\pi)^3}\langle\psi_w^\dagger(\boldsymbol{p})\Sigma_z\psi_w(\boldsymbol{p})\rangle \propto \left.\frac{\partial\Omega}{\partial q}\right|_{\rm DCDW} = 0. \tag{64}$$

On the other hand, there is formed a spatial modulation of magnetic moment along with DCDW. The Gordon decomposition of the gauge coupling term provides the magnetic interaction in presence of external gauge field: $\frac{Q}{2M^*}\bar{\psi}\sigma_{\mu\nu}\psi F^{\mu\nu}$. Only z component of the magnetic moment remains finite,

$$\langle\bar{\psi}(\mathbf{r})\sigma_{12}\psi(\mathbf{r})\rangle = \int\frac{d^3p}{(2\pi)^3}\langle\psi_w^\dagger(\boldsymbol{p})\gamma_0\sigma_{12}\psi_w(\boldsymbol{p})\rangle\cos(\boldsymbol{q}\cdot\boldsymbol{r}),$$
$$\langle\bar{\psi}(\mathbf{r})\sigma_{23}\psi(\mathbf{r})\rangle = \langle\bar{\psi}(\mathbf{r})\sigma_{31}\psi(\mathbf{r})\rangle = 0, \tag{65}$$

where

$$\langle\psi_w^\dagger(\boldsymbol{p})\gamma_0\sigma_{12}\psi_w(\boldsymbol{p})\rangle = \frac{2M^*}{\sqrt{M^{*2}+p_z^2}}\left[n_+(\boldsymbol{p}) - n_-(\boldsymbol{p})\right] \tag{66}$$

with $n_\pm$ being the Fermi-Dirac distribution function with energy spectra $\epsilon_\pm$. Therefore, the magnetic expectation value reflects the anisotropy in momentum distribution of the DCDW phase, and depends on finiteness of the effective mass M^*.

To see a possible consequence of the spatially modulated magnetic order, we evaluate its local magnetic-flux density with effective charge $Q = \left(\frac{2}{3} - \frac{1}{3}\right)e$,

$$\Phi = \frac{Q}{2M^*}\frac{\langle\bar{\psi}(\mathbf{r})\sigma_{12}\psi(\mathbf{r})\rangle}{\langle\psi^\dagger(\mathbf{r})\psi(\mathbf{r})\rangle}3\rho_B, \tag{67}$$

where we have evaluated magnetic expectation value par quark, and multiplied by quark density. For densities normalized by the normal nuclear density $\rho_B/\rho \sim 3-4$, the local flux Φ is estimated to be $O(10^{16})$ G, which well exceeds the critical magnetic field $B_{\rm QED}$.

It might be interesting to recall that the stable Hartree-Fock solution for the case of the short-range potential is always either the normal paramagnetic state or the uniform ferromagnetic state [33]. We have considered the zero-range interaction between quarks within the NJL model, but spin density wave emerges in our case.

3.4. Deformed DCDW

We must take into account the symmetry breaking effect or the finite current mass for the extension of the above framework [53]. It is important for a realistic discussion of the phase transition at moderate densities. Moreover, it may become important to elucidate the appearance of various inhomogeneous phases. The real kink crystal (RKC) can be easily generalized to include massive quarks, and has been studied on the temperature-density plane [28]. Their results show an interesting change of the domain of the RKC as the current mass is changed; it looks to shrink for larger mass.

Recently we have shown that the symmetry breaking effect can be taken into account by a variational approach, without spoiling the original features [10]. Using the ansatz for the condensates,

$$\begin{aligned}\langle\bar{\psi}\psi\rangle &= \Delta\cos\theta(\boldsymbol{r})\\ \langle\bar{\psi}i\gamma_5\tau_3\psi\rangle &= \Delta\sin\theta(\boldsymbol{r}),\end{aligned} \tag{68}$$

represented by the spatially dependent chiral angle $\theta(\boldsymbol{r})$, one must solve the Hartree equation self-consistently in the presence of current-mass term, but it should be a hard task.

Instead, we apply approximate methods, like a perturbative method [54]. Here we take a variational method, which may give better results than the perturbative one. Anyway the symmetry breaking effects should be small in the light of the success of the chiral symmetry approach to various phenomena. So we take into account only the deformation of the chiral angle, leaving the quark wave function unchanged. Then we can see that θ must satisfy the sine-Gordon equation,

$$f_\pi^{*2}\frac{d^2\theta(z)}{dz^2} + m_c\Delta\sin\theta(z) = 0 \tag{69}$$

at the leading order of derivative, where f_π^* is the pion decay constant in medium within the NJL model,

$$f_\pi^{*2} = -\frac{N_cN_fM^{*2}}{8\pi^2}\mathrm{Ei}(-M^{*2}/\Lambda^2), \tag{70}$$

with the exponential integral $\mathrm{Ei}(-x)$. It should be interesting to observe the appearance of the sine-Gordon equation. The relevant solution is then written as

$$\theta = \pi + 2\mathrm{am}\left(m_\pi^* z/k, k\right), \tag{71}$$

in terms of the Jacobian elliptic function, where we introduced the effective pion mass by the relation, $f_\pi^{*2}m_\pi^{*2} = -m_c\Delta$, and $\mathrm{am}(\xi, k)$ is the amplitude with modulus k. In the limit, $k \to 0$ with keeping m_π^*/k =const.($\equiv q/2$), $\theta \to \pi + qz$ so that we have the DCDW solution again in the chiral limit. On the other hand, $\theta \to 4\tan^{-1}\left[e^{m_\pi^* z}\right]$, in the opposite limit, $k \to 1$, which is the well-known kink solution. Thus our variational ansatz may be regarded as an embedding the sine-Gordon kink crystal in $1+1$ dimension into $1+3$ dimensional quark matter; actually we can easily see that $f_\pi^{*2} \to f_\pi^2$ (f_π: pion decay constant) and Eq. (69) is reduced to the standard sine-Gordon equation in $1+1$ dimension,

$$\frac{d^2\theta(z)}{dz^2} - m_\pi^2\sin\theta(z) = 0, \tag{72}$$

by way of the Gell-Mann-Oakes-Renner relation , $f_\pi^2 m_\pi^2 = -m_c\langle\mathrm{vac}|\bar{q}q|\mathrm{vac}\rangle$ [55].

4. Summary and Concluding Remarks

It should be important to study EOS at high-density region from various viewpoints. Here we tried to extract information about the properties of hadron or quark matter and EOS, confronting the magnetic phenomena of pulsars. We have seen two kinds of the magnetic properties in quark matter: one is spontaneous spin polarization (SSP) and the other is the spin density wave. The former is similar to the itinerant electrons in QED, while the latter is related to the chiral symmetry of QCD. These discussions may be in a rather primitive level and must be verified by observations or more elaborate theoretical studies toward more realistic description. Comprehensive study of these magnetic properties is also needed to study their competition. One of the promising ways may be a unified description of these magnetic properties within the NJL model. We have discussed the possibility of SSP within the MIT bag model and discarded any nonperturbative effect, while the NJL model involves a part of nonperturbative QCD and should give more realistic description at moderate densities. Actually ref. [56] discussed the possible SSP within the NJL model, and indicated an important role of the dynamical mass.

Energetics or the mechanism of the giant flares observed in some magnetars may give a direct hint about the origin of the magnetic field [4]. Thermal evolution of compact stars should be important for us to get the information about the EOS and properties of high-density matter. Actually the magnetic properties is not directly related to EOS but may manifest through the thermal activities of compact stars.

There have been done many works about color superconductivity (CSC) in high-density matter [57]. It also has implications on compact star phenomena. So it should be interesting to study the interplay of CSC and magnetic properties in quark matter. In ref.[58] we have discussed a coexistence of CSC and ferromagnetic order.

More elaborate studies are necessary for inhomogeneous phases: relations among the various phases should be figured out as well as their properties. For example, there seem many resemblances between pion condensation in hadron matter [59] and the DCDW state. One may find a kind of hadron-quark continuity across the deconfinement transition. They may also appear during relativistic heavy-ion collisions. So it is interesting to consider how we can observe them in this context. Theoretically it is interesting and important to figure out the existence and properties of the Lifshitz point in the QCD phase diagram, which is the multicritical point for homogeneous phases and inhomogeneous one [60]. It has been shown that the chiral end point is the Lifshitz point for the real kink crystal [28]. The critical behavior of the DCDW phase is still unclear and must be explored in detail. The relation to the chiral end point or the Lifshitz point must be also clarified in the QCD phase diagram.

References

[1] S. L. Shapiro and S. A. Teukolsky, *Black Holes, White Dwarfs, and Neutron Stars* (John Wiley & Sons Inc, 1983).

[2] J.D. Jackson, *Classical Electrodynamics* (John Wiley & Sons Inc, 1975).

[3] P.M. Woods and C. Thompson, *Compact stellar X-ray sources* (Cambridge Univ, Press, 2006), p.547.

[4] W. Becker ed., *Neutron Stars and Pulsars* (Springer, 2009).

[5] A.K. Harding and D. Lai, *Rep. Prog. Phys.* 69 (2006) 2631.

[6] S. Mereghetti, *A&A Rev.* 15 (2008) 225.

[7] P. Haensel, A.Y. Pothekhin and D.G. Yakovlev, *Neutron Stars 1: Equation of State and Structure* (Springer, 2007).

[8] F.P. Gavriil et al., Science 319 (2008) 1802.
M.A. Livingstone, V.M. Kaspi and F.P. Gavriil, arXiv:1007.2829.

[9] G. Chanmugam, *Ann. Rev. Astron. Astrophys.* 30 (1992) 143.

[10] T. Maruyama, E. Nakano and T. Tatsumi, *Horizons in World Physics, Vol. 276*, (Nova Science Pub., NY, 2011), p.209.

[11] K. Makishima, *Prog. Theor. Phys. Suppl.* 151 (2003) 54.

[12] R. Tamagaki, *Prog. Theor. Phys. Suppl.* 112 (1993) 1.

[13] G. Baym and C.J. Pethick, *Landau Fermi-Liquid Theory* (WILEY-VCH, 2004).

[14] V. Goudon et al., *J. Phys.: Conf. Ser.* 150 (2009) 032024.

[15] For recent papers, I. Bombaci et al., *Phys. Lett.* B632 (2006) 638.
G.H. Bordbar and M. Bigdeli, *Phys. Rev.* C76 (2007) 035803.

[16] T. Schaefer, arXiv:hep-ph/0509068.
P. Braun-Munzinger and J. Wambach, *Rev. Mod. Phys.* 81 (2009) 1031.
K. Fukushima and T. Hatsuda, arXiv:1005.4814.

[17] I. Sagart et al., *Phys. Rev. Lett.* 102 (2009) 081101.
H. Sotani, N. Yasutake, T. Maruyama and T. Tatsumi, *Phys. Rev.* D 83 (2011) 024014.

[18] T. Tatsumi, *Phys. Lett.* B489 (2000) 280.

[19] A. Niegawa, *Prog. Theor. Phys.* 113 (2005) 581.

[20] K. Pal, S, Biswas, A.K. Dutt-Mazumder, *Phys. Rev.* C79 (2009)015205; C80 (2009) 024903.

[21] J.W. Negele and H. Orland, *Quantum Many-Particle Systems* (Addison-Wesley Pub., 1988).
A.A. Abrikosov, L.P. Gorkov and I.E. Dzaloshinski, *Methods of Quantum Field Theory in Statistical Physics* (Prentice-Hall. Inc., 1963).
A.B. Migdal, *Theory of finite Fermi systems* (Intersci. Pub., 1967).

[22] T. Tatsumi and K. Sato, *Phys. Lett.* B663 (2008) 322.

[23] T. Tatsumi and K. Sato, *Phys. Lett.* B672 (2009) 132.
K. Sato and T. Tatsumi, *Nucl. Phys.* A 826 (2009) 74.

[24] K. Pal, A.K. Dutt-Mazumder, *Phys. Rev.* C80 (2010) 054911; C81 (2010) 054906.

[25] T. Tatsumi and E. Nakano, hep-ph/0408294.
E. Nakano and T. Tatsumi, *Phys. Rev.* D71 (2005) 114006.

[26] P. Fulde and R.A. Ferrell, *Phys. Rev.* 135(1964) A550.
A.I. Larkin and Y.N. Ovchinikov, *JETP* 20 (1965) 762.

[27] G. Basar and G.V. Dunne, *Phys. Rev. Lett.* 100 (2008) 200404; *Phys. Rev.* D78 (2008) 065022.
G. Basar, V.Dunne and M. Thies, *Phys. Rev.* D79 (2009) 105012.

[28] D. Nickel, *Phys. Rev. Lett.* 103 (2009) 072301; Phys. Rev. D80 (2009) 074025.

[29] B. Bringoltz, *Phys. Rev.* D 79 (2009) 125006.

[30] T.A. DeGrand, R.L. Jaffe, K. Johnson and J. Kiskis, *Phys. Rev.* D12 (1975) 2060.

[31] For reviews, O. Philipsen, *Eur. Phys. J. ST* 152 (2007) 29.
A. Bazavov et al., *Phys. Rev.* D80 (2009) 014504.

[32] F. Bloch, *Z. Phys.* 57 (1929) 545.

[33] C. Herring, *Exchange Interactions among Itinerant Electrons: Magnetism IV* (Academic press, New York, 1966).
K. Yoshida, *Theory of magnetism* (Springer, Berlin, 1998).

[34] C. Ceperly, *Nature* 397 (1999) 386.

[35] D.P. Young et al., *Nature* 397 (1999) 412.

[36] P.W. Anderson, *Basic notions of condensed matter physics* (Addison-Wesley Pub., 1984).

[37] V.B. Berestetsii, E.M. Lifshitz and L.P. Pitaevsii,
Relativistic Quantum Theory (Pergamon Press, 1971).

[38] K.A. Brueckner and K. Sawada, *Phys. Rev.* 112 (1957) 328.
B.S. Shastray, *Phys. Rev. Lett.* 38 (1977) 449.

[39] J.I.Kapusta, *Finite-Temperature field theory* (Cambridge Univ. Press, 1989).
M. LeBellac, *Thermal Field Theory* (Cambridge Univ. Press, 1996).

[40] A. Altland and B. Simons, *Condensed Matter Field Theory* (Cambridge Univ. Press, 2010).

[41] C. Herring, Phys. Rev. 85 (1952) 1003.

[42] T. Tatsumi, *Exotic states of nuclear matter*, *Proc. of Int. Sympo. EXOCT07* (World Scientific, 2008).

[43] N. Karchev, *Phys. Rev.* B67 (2003) 054416.

[44] M. Kutschera, W. Broniowski and A. Kotlorz, *Nucl. Phys.* A516 (1990) 566.
M. Sadzikowski and W. Broniowski, *Phys. Lett.* B488 (2000) 63.

[45] J.R. Schrieffer, *Theory of Superconductivity* (Westview Press, 1999).

[46] A.J. Schofield, *Physics* 2 (2009) 93.

[47] S. Weinberg, *Quantum Theory of Fields II* (Cambridge U. Press, 1996).

[48] A.W. Overhauser, *Phys. Rev. Lett.* 4 (1960) 462; *Phys. Rev.* 128 (1962) 1437.

[49] G. Grüner, *Density Waves in Solids* (Addison-Wesley Longman, 1994).

[50] T. Muto and T. Tatsumi, in preparation.

[51] P.G. de Gennes and J. Prost, *The physics of liquid crystals* (Oxford Univ. Press, 1974).

[52] H.B. Nielsen and S. Chadha, *Nucl. Phys.* B105 (1976) 445.

[53] C. Boehmer, M. Thies and K. Urichs, *Phys. Rev.* D75 (2007) 105017.
C. Boehmer, U. Fritsch, S. Kraus and M. Thies, *Phys. Rev.* D78 (2008) 065043.

[54] S. Maedan, *Prog. Theor. Phys.* 123 (2010) 285.
T.L. Partyka, arXiv:1005.2667.

[55] V. Schön and M. Thies, *Phys. Rev.* D62 (2000) 096002.

[56] S. Maedan, *Prog. Theor. Phys.* 118 (2007) 729.

[57] D. Bailin and A. Love, *Phys. Rep.* 107 (1984) 325.
For reviews, K. Rajagopal and F. Wilczek, hep-ph/0011333.
M.G. Alford, K. Rajagopal, T. Schaefer, A. Schmitt. *Rev. Mod. Phys.* 80 (2008) 1455.

[58] E. Nakano, T. Maruyama and T. Tatsumi, *Phys. Rev.* D68 (2003) 105001.
T. Tatsumi, E. Nakano and K. Nawa, *Dark Matter* (Nova Science Pub., New York, 2006), p.39.

[59] T. Takatsuka et al., *Prog. Theor. Phys.* 59 (1978) 1933.
A. Akmal and V.R. Pandharipande, *Phys. Rev.* C79 (1997) 2261.

[60] R.M. Hornreich, M. Luban and S. Shtrickman, *Phys. Rev. Lett.* 35 (1975) 1678.
A. Michelson, *Phys. Rev.* B16 (1977) 577; 585.

In: Neutron Stars
Editors: Hiroshi Uechi et al, pp. 259-287
ISBN: 978-1-62081-123-8

Chapter 10

GRAVITATIONAL COLLAPSE OF MASSIVE STARS AND EOS

Shoichi Yamada*
Department of Physics, School of Science & Engineering,
Waseda University

Abstract

The importance of equations of state (EOS's) will be elucidated by their applications to gravitational collapse of massive stars, the common event that occurs at the end of their lives. Future detections of neutrinos (and gravitational waves) from nearby events will provide us with the opportunity to probe the properties of hadronic matter that may not be accessible by terrestrial experiments, particularly in the event of black hole formations. In this chapter, thermodynamical states in the collapsing stellar cores and EOS's that have been employed frequently to describe them in the literature will be reviewed for sub- and supra-nuclear densities separately. The roles these EOS's play in the dynamics of gravitational collapse and some observational consequences that their uncertainties may have will be presented from exemplary numerical simulations.

1. Introduction

Stellar evolutions are dictated mainly by their masses [1]. Although there are still some uncertainties, stars that have more than about 10 solar masses ($M_\odot$) on the main sequence will end their lives with the gravitational collapse of central cores and formation of compact objects such as neutron stars and black holes [2]. During this catastrophic event, the density of stellar core increases by more than 4 orders of magnitude and temperature becomes nearly hundred times higher [3, 4, 5]. It is hence no surprise that thermodynamical states in the core at the beginning of the collapse are quite different from those at the end.

Equations of state (EOS's) of the stellar matter provide all the information on them. The core matter consists of hadrons as well as electrons and photons. The latter two are rather simple and can be treated as ideal Fermi and Bose gases. On the other hand, the hadronic contributions are the source of complexities. At low densities, for example, some

*E-mail address: shoichi@waseda.jp

nucleons form various bound states or nuclei [6]. The mixtures of nucleons and nuclei may be regarded as mixed states in the liquid-gas phase transition. It is well known that nuclei become highly non-spherical in a certain density regime and are often referred to as "nuclear pastas" [7, 8]. At high densities, hadronic matter becomes uniform, strongly interacting with each other, and gives a dominant contribution to the matter pressure. Possible emergences of hyperons and meson condensations have been extensively studied over the years [9, 10]. Deconfinement of quarks is certainly another possibility [11]. It is hence clear that EOS has to deal with quite a lot.

As if this is not demanding enough, EOS is expected to give more. As explained more in detail in the following sections, not only thermodynamical variables such as pressure and sound velocity but energy transport by neutrinos are also critically important for the dynamics of massive star cores. Since emissions and absorptions as well as other interactions of neutrinos depend on the thermodynamical states of matter, EOS is indispensable in evaluating the interaction rates and computing the neutrino transport in the core.

From the astrophysical point of view, EOS is an important input in the study of the core collapse and formation of compact objects. In fact, since the dynamics of core collapse is highly nonlinear and complex, numerical simulations are almost mandatory for its quantitative understanding. In so doing, all the thermodynamical variables have to be provided by EOS. It is almost obvious, however, that no theory can deal with all the ingredients exactly and it is also true that there are some density/temperature regimes, which are unaccessible to current experiments. Any EOS, therefore, includes some unavoidable uncertainties particularly at high densities. It is then a natural idea to employ astrophysical objects or phenomena as an alternative probe into the properties of dense matter: we look for observable consequences that these uncertainties may have and future observations can find out. In this sense astrophysics is not only an end user but also a potential contributor to hadron physics. In this chapter we demonstrate this possibility in the following sections.

1.1. Gravitational Collapse of Massive Stars

In order to expedite the understanding of the following sections for those who are not familiar with astrophysics, we briefly review at this point how the gravitational collapse of massive star cores proceeds. The core just prior to collapse consists of iron group nuclei, electrons and photons. Just like white dwarfs, the compact stellar objects formed at the end of intermediate mass stars, the degenerate electrons are the main contributor to the pressure that sustains the hydrostatic equilibrium of the core. As known well for spherically symmetric configurations, the stability of hydrostatic equilibrium against radial perturbations is determined by the adiabatic index, $\Gamma \equiv (\partial \ln p / \partial \ln \rho)_s$, where p, ρ and s are the pressure, density and specific entropy of matter, respectively [12]. The critical value is $4/3$, at which the increase of gravitational attraction by radial compressions equals to that of pressure repulsion. If the adiabatic index becomes smaller than this critical value, the gravity prevails and the collapse occurs. EOS is hence a key player here. Note that highly relativistic, degenerate electrons have the adiabatic index of $4/3$ and the hydrostatic configuration supported mainly by them is marginally stable.

Fig. 1 displays schematically the regions in the $\rho - T$ diagram, in which matter has an adiabatic index smaller than the critical value, together with the causes. As quasi-static

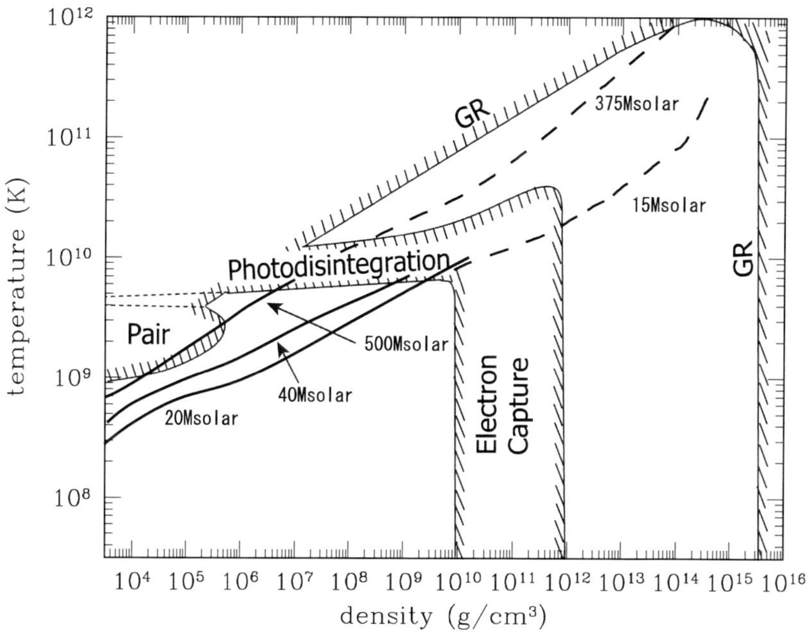

Figure 1. The schematic drawing of unstable regions in the ρ-T diagram. The adiabatic indices in the hatched regions are smaller than the critical value for the reasons displayed in the figure (refer to the body for details) and spherically symmetric configurations in hydrostatic equilibrium are unstable against radial perturbations. Solid lines are rough trajectories of the quasi-static evolutions of stars with indicated masses. Dashed lines show the paths of gravitational collapses of 15 and $375M_\odot$ stars [15, 16].

stellar evolutions proceed, the central density and temperature increase as shown schematically in the figure for representative cases. In the case of the low-mass end of massive stars ($\sim 10M_\odot$), for example, the trajectory hits an instability region induced by electron captures. As mentioned above, the hydrostatic configuration at this density and temperature is supported by electrons. The electron captures on free protons and nuclei reduce the number density of electrons, which then leads to the depletion of the electron pressure. For more massive stars, the trajectories in the $\rho - T$ diagram are shifted upwards. They then encounter the instability region produced by photo-dissociations of nuclei. Although the thermal pressure is sub-dominant, it is not negligible at high temperatures ($\sim 10^{10}$K). Since the dissociations are endothermic reactions, they reduce the thermal pressure and cause the gravitational collapse. For even more massive stars ($\gtrsim 150M_\odot$), the instability is induced by electron-positron pair creations, which are also endothermic [13, 14]. For hypothetical supermassive stars ($\gtrsim 10^3M_\odot$), a small general relativistic correction increases the gravitational pull slightly and otherwise marginally stable configurations supported mainly by photons become unstable [17]. The collapses of these very massive objects begin before the iron core is formed, and total disruptions by explosive oxygen burnings, which are referred to as pair instability supernovae, or black hole formations are supposed to occur. These cases will not be mentioned further in this chapter.

Once the gravitational collapse sets in and the density and temperature are increased, both the electron captures and photo-dissociations of nuclei are enhanced, since the increase of the chemical potential of degenerate electrons is faster than that of non-degenerate neutrons and the depletion of electrons are energetically favored; the high temperatures, on the other hand, favor light nuclei, which increase entropy per baryon. Thus the positive feedback accelerates the collapse. The electron capture or neutronization can be regarded as a relaxation to β-equilibrium, in which the electron capture balances with its inverse reaction. The process is irreversible and produces entropy. Electron-type neutrinos are copiously emitted in the process and extract energy from the matter, which decreases entropy. It turns out that these competing effects almost cancel each other and entropy is increased slightly [18]. In the early phase of core collapse ($\rho \lesssim 10^{11}$g/cm^3), the neutrinos have nearly no interaction with matter and escape the star freely. As the density increases, however, even the weak interactions become appreciable and neutrinos start to be trapped in the core. At $\rho \sim 10^{12}$g/cm^3, the inverse reactions by the trapped neutrinos become frequent enough and the β-equilibrium is established.

The main source of the opacity for neutrinos is the coherent scattering on nuclei. If the sizes of nuclei are smaller than the wave length of neutrinos, the neutrinos scattered by nucleons inside the nuclei interfere positively with each other, making the cross section proportional not to the number of nucleons but to its square [19]. Once the neutrinos start to populate inside the core, they are degenerate and their chemical potentials increase just like electrons. Since the rate of coherent scattering is proportional to the neutrino energy squared, the matter becomes even more opaque. The absorption of neutrinos on neutrons, the inverse process of the electron capture, is also enhanced and slows down the neutronization of the matter. Remaining electrons in turn make nuclei become bigger, since the Coulomb energy is reduced. This too increases the coherent scattering rates. As a result of these positive feedbacks, neutrinos are trapped even more strongly as the density increases [20]. After the β-equilibrium is reached at $\rho \sim 10^{12}$g/cm^3, there are no interaction remaining to produce entropy and the collapse proceeds adiabatically thereafter.

The iron cores of massive stars have a rather low entropy per baryon, $\sim 1k_B$, owing to the neutrino cooling after carbon burning, where k_B is the Boltzmann constant. As mentioned above, since the gravitational collapse proceeds with a nearly constant entropy per baryon, the changes of density and temperature at the center are approximately given by an adiabat with $s \sim 1k_B$. Some representative adiabats are shown in Fig. 2 for the Shen's [21] and Lattimer & Swesty's [22] EOS's, which are currently the standard EOS's employed most frequently in core-collapse simulations. One important consequence of the low entropies in the stellar cores is immediately discernible: the adiabats run through the mixed phase in the liquid-gas phase transition. In this description, nucleons are either contained in nuclei as a dense liquid or a dilute gas dripped out of nuclei. As a result, nuclei exist at densities of relevance, 10^{10}g/cm$^3 \lesssim \rho \lesssim 10^{14}$g/cm^3, and the uniform liquid phase or uniform nuclear matter is formed by coalescence of these nuclei around the nuclear saturation density [6].

When the density exceeds the nuclear saturation density, the pressure of nuclear matter rises quickly owing to repulsive interactions among dense nucleons and overwhelms the pressure of electrons, which has been dominant so far. The adiabatic index ($\Gamma \gtrsim 2$) becomes considerably larger than the critical value ($\Gamma_c = 4/3$) and the core recovers stability against collapse and its contraction starts to decelerate. It is important to note that the collapsing

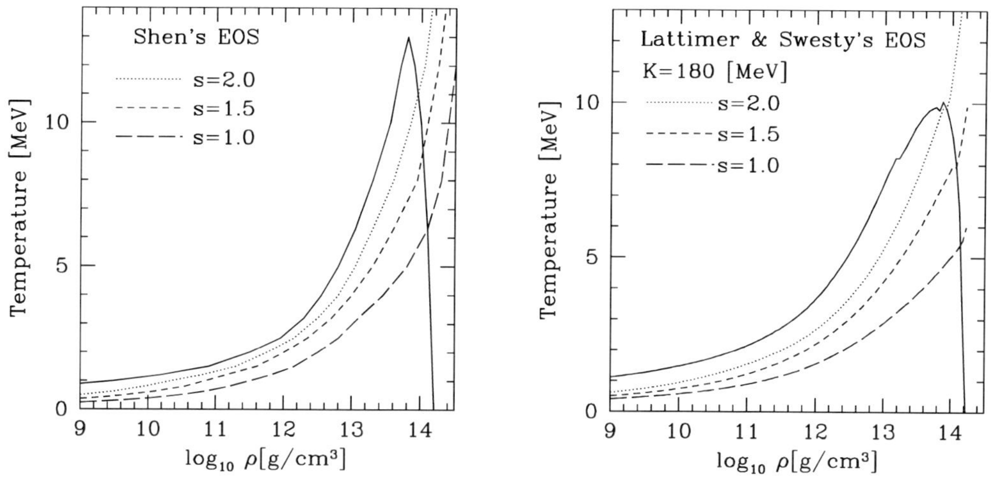

Figure 2. The phase diagrams and some representative adiabats in the ρ-T plane for the Shen's [21] (left) and Lattimer & Swesty's [22] (right) EOS's. The proton fraction is $Y_p = 0.3$. The Solid lines represent the phase boundary, under which nuclei exist.

core is divided into two parts: the inner core that shrinks subsonically and homologously ($v \propto r$ with v and r being the infall velocity and radius, respectively) and the outer core that falls supersonically [23]. The stiffening of the matter leads to a collective bounce of the inner core. Since the outer core is causally disconnected from the inner core, the former collides with the latter at their boundary, producing a shock wave there. The inner core is quickly settled in a hydrostatic equilibrium whereas the infall of the outer core is reversed by the out-going shock wave. As it propagates outwards, the shock wave is weakened by photo-dissociations of nuclei in the outer core, which are endothermic as mentioned earlier, as well as through neutrino cooling after the shock reaches the region that is transparent for neutrinos. As a result, the shock wave ceases to go any farther and becomes a so-called standing shock. Matter in the unshocked outer core and the envelope continues to accrete through the standing shock onto a nascent protoneutron star.

As mentioned shortly, how far the shock wave can reach is important for what happens next and is determined by the initial location and strength of the shock wave. Since it is generated at the boundary between the inner and outer cores and the inner core mass is proportional to the electron fraction squared ($\propto Y_e^2$) as is the Chandrasekhar mass or the maximum mass of white dwarfs, the electron captures that occur up to the neutrino trapping are the key to the initial location of the shock wave. As explained in the next section, since the electron capture rates are dependent on the abundance of nuclei, EOS at sub-nuclear densities plays an important role here. EOS at supra-nuclear densities is also crucial for the initial strength of the shock wave. In producing and pushing outwards the shock wave, the inner core works as a spring and it is intuitively understandable that the softer EOS tends to produce the stronger shock wave. According to numerical simulations, the maximum central density is about twice the nuclear saturation density and the electron fraction is $Y_e \sim 0.3$ at the bounce. The stiffness of this neutron-rich matter at this density is very important information that EOS should provide.

From the stalled shock wave an explosion has to occur somehow. Nobody knows exactly how at present, though. The most promising scenario is that the standing shock wave is revived by neutrino heating [4, 5]. If the radius of the stalled shock wave is sufficiently large ($r \gtrsim 100$km), the matter just behind the shock wave is cool enough to absorb neutrinos diffusing out from deeper inside. If the shock radius is smaller than the so-called gain radius, the heating is more than compensated for by the cooling via local neutrino emissions. As mentioned above, EOS sets the condition for the neutrino-heating and hence is one of the key players in this mechanism. If the luminosity and/or energy of neutrinos are high enough, the shock wave is indeed revived. The problem is that the mechanism seems to work successfully only for less massive stars ($M \lesssim 15\mathrm{M}_\odot$) and that the explosion, if obtained, appears underenergetic ($E_{\mathrm{exp}} \sim 10^{50}$ergs) [24].

For more massive stars ($M \gtrsim 20\mathrm{M}_\odot$), different mechanisms may operate to produce explosions. It may be also possible, however, that some massive stars fail to explode and produce black holes rather than neutron stars with no optical event like the supernova. The stellar evolution theory suffers from uncertainties in mass loss for these massive stars and the theoretical predictions of the progenitor structure differ substantially from modeler to modeler [25, 26, 27]. The fates of very massive stars may not be a monotonic function in fact. Some authors argue, however, for the existence of channels to black hole based on the results of progenitor surveys [28] and light curve analyses [29] (but see also [30]).

If the channels really exist, it is an intriguing question how the event can be observed and what information we can extract from them. As demonstrated later in section 3, the black hole formation is not a quiet death of massive stars. To the quite contrary, it is as bright as or even brighter than supernovae in terms of neutrino emissions and their luminosity and average energy carry unique information both on the black hole formation and on EOS at supra-nuclear densities. The important point is that very high densities that cannot be reached in supernovae can be probed in the event. In fact, the protoneutron star continues to accrete matter until the maximum mass is reached and the second collapse ensues, leading to the black hole formation. Hence we can probe the density regime that only the maximum-mass configuration can reach. It should be stressed that this maximum mass of the protoneutron star is set by EOS. It is also noted that the compact objects just prior to the second collapse may not be composed of nucleons alone and some exotic phases might emerge.

2. EOS at Sub-nuclear Densities

In this section, we look into EOS at sub-nuclear densities more in detail, which is important particularly in the early phase of core collapse, that is, from the onset of collapse up to bounce. EOS influences the dynamics of collapse in a couple of ways. Among other things, the stability of the core against collapse is determined by the adiabatic index, $\Gamma = (\partial \ln p / \partial \ln \rho)|_s$. The neutron-richness of the matter at core bounce is determined by the electron capture and matter opacity for neutrinos, both of which depend on the abundance of various nuclei. Since the mass of inner core affects the dynamics after core bounce and is determined mainly by the electron fraction, EOS at sub-nuclear densities is crucially important for later evolutions. Incidentally, the mass of inner core is one of the main factors to determine the maximum density at core bounce, which is in turn one of the critical elements that affect gravitational wave signals produced at bounce if the core rotates. It is also stressed that EOS at sub-nuclear densities is intimately related with EOS at supra-nuclear densities, as will be evident later.

2.1. Nuclear Statistical Equilibrium

The constituent particles in this density regime are nucleons, electrons and photons. As mentioned earlier, the low entropies ($\sim 1k_B$) in the core imply the existence of nuclei. Hence the nucleons reside both inside the nuclei and in the vapor surrounding them. The temperatures in the core, on the other hand, are high enough ($T \gtrsim 10^{10}$K) to ensure that chemical equilibrium is established for all reactions but weak interactions. The β-equilibrium is delayed until neutrinos are sufficiently trapped at $\rho \sim 10^{12}$g/cm^3. The photons can be treated as ideal Bose gas as usual. Since the electrons have large Fermi energies ($\gtrsim$ 10MeV), they are also treated as ideal Fermi gas despite they interact electromagnetically with other charged particles. The nucleons and nuclei are just classical Boltzmann gases at low densities and the abundance of various nuclei is determined by the so-called nuclear statistical equilibrium (NSE) that is established by nuclear reactions such as

$$(Z, N) + \gamma \longleftrightarrow (Z-1, N) + p, \tag{10.1}$$
$$(Z, N) + \gamma \longleftrightarrow (Z, N-1) + n, \tag{10.2}$$

where nuclei are specified by their numbers of proton, Z, and neutron, N. Then the chemical potentials of nucleons, μ_p and μ_n, and nuclei, μ_i, are related with one another via

$$\mu_i = Z_i\mu_p + N_i\mu_n, \tag{10.3}$$

where the subscript i specifies the nucleus, which has Z_i protons and N_i neutrons. The abundance of nuclei can be written down by the chemical potentials of proton and neutron as

$$n_i = g_i \left(\frac{M_i k_B T}{2\pi\hbar^2}\right)^{3/2} \exp\left(\frac{Z_i\mu_p + N_i\mu_n - M_i c^2}{k_B T}\right), \tag{10.4}$$

where M_i and g_i are the mass and statistical weight of nucleus i, respectively, the latter of which accounts for the internal degrees of freedom of the nucleus. The two chemical potentials of nucleons are finally determined as a function of the baryonic number density, n_B , temperature, T, and the proton fraction, Y_p, by the total numbers of proton and neutron both inside and outside nuclei:

$$n_p + \sum_i Z_i n_i = Y_p n_B, \tag{10.5}$$
$$n_n + \sum_i N_i n_i = (1 - Y_p) n_B. \tag{10.6}$$

The abundance of various nuclei obtained for experimental nuclear mass data is shown in the upper panels of Fig. 3 for different densities on the adiabat of $s = 1k_B$. It is clear that at the lower density, tightly bound iron group nuclei are abundant. At the higher density, on the other hand, lighter nuclei become abundant. This is in sharp contrast to what we

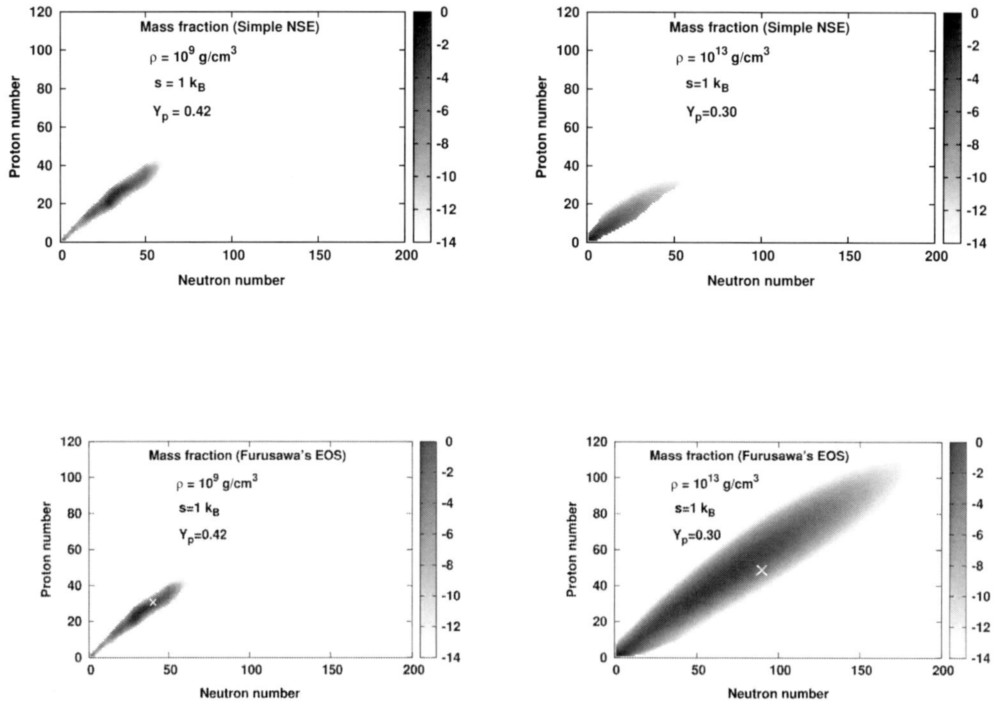

Figure 3. The abundances of nuclei at two different densities on the adiabat with $s = 1k_B$ for the ordinary NSE EOS employing experimental nuclear masses (upper panels) and the Furusawa's EOS [6] (lower panels). The $\times$ symbols indicate the representative heavy nuclei given by the Shen's EOS [21].

commonly see in core-collapse simulations: the heavier nuclei become populated, which is demonstrated in the lower right panel(see subsection 2.3 for more details). Although a smooth transition to uniform nuclear matter should take place at some density, the simple NSE EOS just obtained does not behave that way as demonstrated in Fig. 4 for the Shen's EOS [21], one of the standard EOS's for the core-collapse simulations. This should not be surprising, however, since uniform nuclear matter cannot be expressed by the collection of ordinary nuclei included in the NSE EOS. This is most simply understood from the fact that the binding energy per baryon of iron, $\sim$ 9MeV, is different from the one of uniform nuclear matter, $\sim$ 16MeV.

The problem with the simple NSE EOS is obvious: nuclei in massive stars' core are not isolated but immersed in electrons and surrounded by other nuclei and nucleons, and hence the nuclear mass obtained in laboratory should be modified [31]. The existence of electrons lowers the Coulomb energy in nuclei and tends to favor bigger nuclei. Nuclei form a Coulomb lattice at high densities, where the so-called plasma parameter , $\Gamma_p = (\overline{Z}\mathrm{e})^2/(d\,k_B T)$, expressed with the average atomic number, $\overline{Z}$, and inter-particle distance, d, exceeds a critical value, ~ 171 [32]. This also affects the Coulomb energy of system. The nucleons dripped out of nuclei, on the other hand, exert pressure on nuclei and modify their

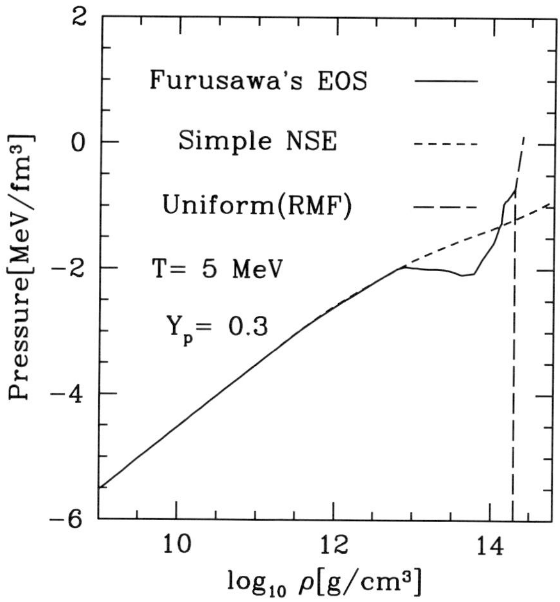

Figure 4. Comparison of the simple NSE EOS at sub-nuclear densities with the Shen's EOS [21] at supra-nuclear densities. The temperature and proton fraction are $T = 5$MeV and $Y_p = 0.3$, respectively. The Furusawa's multi-species EOS [6] is also displayed.

structures at high densities. The surface structure is also changed. In fact, as the density increases and the transition to uniform nuclear matter is approached, the surface should disappear. It is also known that spherical nuclei do not give the minimum free energy and nuclear pastas emerge near the nuclear saturation density [7, 8]. It should be noted further that the vapor nucleons outside nuclei are not free but interacting with each other and should be treated in the same way as uniform nuclear matter is. (This is also the case for the bulk energy of nuclei.) At this point, the connection of the EOS at sub-nuclear densities with the one at supra-nuclear densities manifests itself. The former cannot and should not be constructed independently of the latter.

Ideally speaking, the structures of nuclei in the proper environment should be computed microscopically. This is a challenging task indeed. The microscopic treatment of finite nuclei itself is not easy, but we are now required to take into account the effects of environment, which is in turn determined by the abundance of nuclei and hence affected by the nuclear structures. This means that the problem is actually to calculate both the structures and abundance of nuclei simultaneously and consistently. To accomplish this for a large number of nuclei is all but impossible. In most of EOS's used for core collapse simulations, the single nucleus approximation (SNA) has been adopted, in which a collection of nuclei is represented by a single, presumably the most abundant nuclear species. Then the structure of this representative nucleus is solved with the vapor nucleons, other nuclei and electrons being taken into account. In so doing, various sophistications of approximations have been employed in the literature. We will look at them next, taking the standard EOS's for core-collapse simulations as an example.

2.2. Single Nucleus Approximation and Current Standard EOS's

The simplest approach adopted in the EOS by Lattimer & Swesty [22] is the employment of the compressible liquid drop model. In this approximation, the single representative nucleus is described by a uniform spherical drop, whose bulk energy is expressed by the free energy for uniform nuclear matter. The Coulomb and surface energies are added to determine the size of nucleus. Since the nucleus is usually neutron-rich, the symmetry energy has to be taken into account both in the bulk and surface energies. The nucleon vapor surrounding the nucleus is assumed to be uniform and treated in the same way as uniform nuclear matter, thus the interactions between nucleons being automatically included. In order to evaluate the Coulomb energy with the existence of other nuclei being taken into consideration, the so-called Wigner-Seitz cell (W-S cell) is introduced. The mechanical and chemical equilibrium between the nuclei and the nucleon vapor is obtained as a result of minimization of the free energy with respect to the parameters included: the number densities of protons and neutrons both inside and outside the nucleus, the size of W-S cell and the radius of nuclei. In the Lattimer & Swesty's EOS, the bubble phase (one of the nuclear pastas) is incorporated by modifying the Coulomb and surface energies. Moreover, alpha particles are treated separately from other nuclei as a representative of light nuclei. It is noted that the smooth transition from EOS at sub-nuclear densities to one at supra-nuclear densities is guaranteed in this approach by the fact that the bulk energy of nucleus and the free energy of vapor nucleons are treated in the same manner as uniform nuclear matter.

In the above approximation, the density distribution of nucleons in the W-S cell is not solved but assumed. In the Thomas-Fermi approximation, on the other hand, it is solved [31]. In this approximation, the free energy of nucleus is a functional of the density distribution of nucleons. Since it includes the derivatives, the surface energy is included automatically. The free energy is assumed to be given point-wisely by the same function of density as for uniform nuclear matter. The periodic boundary condition is imposed at the boundary of W-S cell, which essentially corresponds to taking account of the existence of other nuclei. The density profiles in and the size of W-S cell are variational parameters to minimize the free energy. In the Shen's EOS [21], which is one of the current standard EOS's for core-collapse simulations and employs the Thomas-Fermi approximation for nuclei at sub-nuclear densities, the authors take a step further to make the minimization process easier: the density profiles of neutron and proton are parametrized, and not the function but those small number of parameters are varied to obtain the free energy minimum. In this EOS, the pasta phase is not taken into account. As a representative of light nuclei, alpha particles are treated separately from other nuclei just as in the Lattimer & Swesty's EOS.

EOS's that adopt more sophisticated treatments of nuclear structures are, unfortunately, not available for the public use in core-collapse simulations at present. The EOS developed by Wolff & Hillebrandt [33] and still employed by the Munich group is based on the Hartree-Fock approximation with the Skyrme interactions [34]. As discussed later, both the electron capture rates and the abundance of nuclei are affected by shell structures of nuclei. Hence quantum mechanical treatments such as the Hartree-Fock approximation are certainly more desirable.

Although the baryonic contribution to the matter pressure is minor during the collapsing

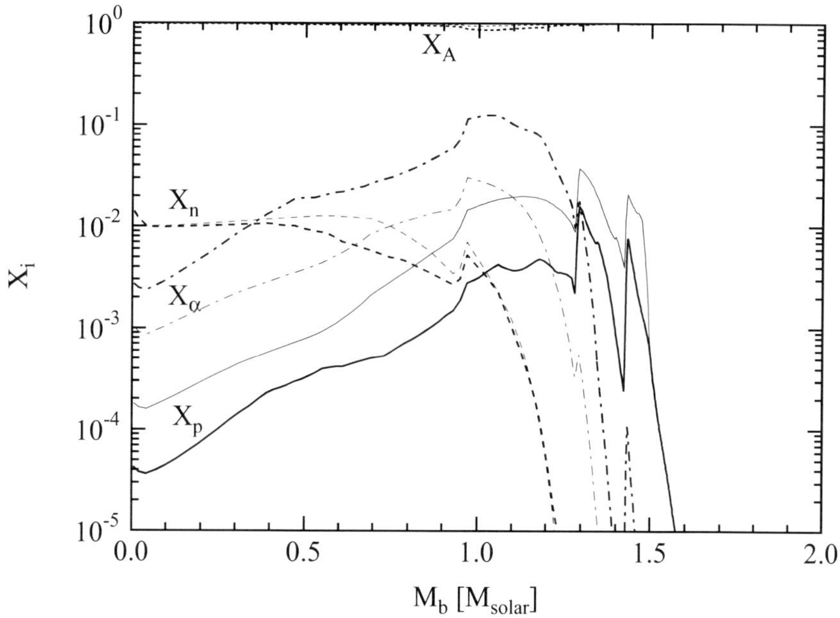

Figure 5. The abundances of nucleons, alpha particles and representative heavy nuclei in the collapsing core for the Lattimer & Swesty's [22] (thin lines) and Shen's [21] EOS's (thick lines). The progenitor is a $15M_\odot$ star. The density is $\rho = 10^{11}$g/cm^3. Solid, dashed, dotted and dot-dashed lines show the mass fractions of protons, neutrons, nuclei and alpha particles, respectively [15].

phase, the baryonic EOS is still important, since it provides the information on the abundance of nuclei, which is needed for the evaluation of the rates of weak interactions with nuclei. As mentioned already, the electron captures on nuclei are the dominant factor for the destabilization of stellar core, reducing the adiabatic index, and the neutrino trapping is controlled by the coherent scattering on nuclei, the rate of which is proportional to the average mass number squared of nuclei. In fig. 5, the abundances of nucleons, alpha particles and representative heavy nuclei are shown for the Lattimer & Swesty's and Shen's EOS's. These are taken from the 1D numerical simulation by Sumiyoshi et al. [15] of the collapse of $15M_\odot$ progenitor. The density is $\rho = 10^{11}$g/cm^3, at which matter are opaque and neutrinos are trapped in the core.

The difference is evident although it is not very large. The mass fraction of free protons in the Shen's EOS, for example, is smaller than that in the Lattimer & Swesty's EOS by a factor of ~ 5. This is caused by a larger symmetry energy in the former. The smaller free proton fraction reduces the electron captures on free protons. Note that the electron capture on nuclei with $N \geq 40$ is suppressed in these simulations by the Bruenn's prescription [18]. This point will be revisited shortly. Liebendörfer et al. [35] pointed out that there is a negative feedback operating in the deleptonization of the core: if the electron capture rates are small, higher free-proton fractions will reduce the electron fraction more strongly and it turns out that the resultant electron fractions are not significantly different as we see below.

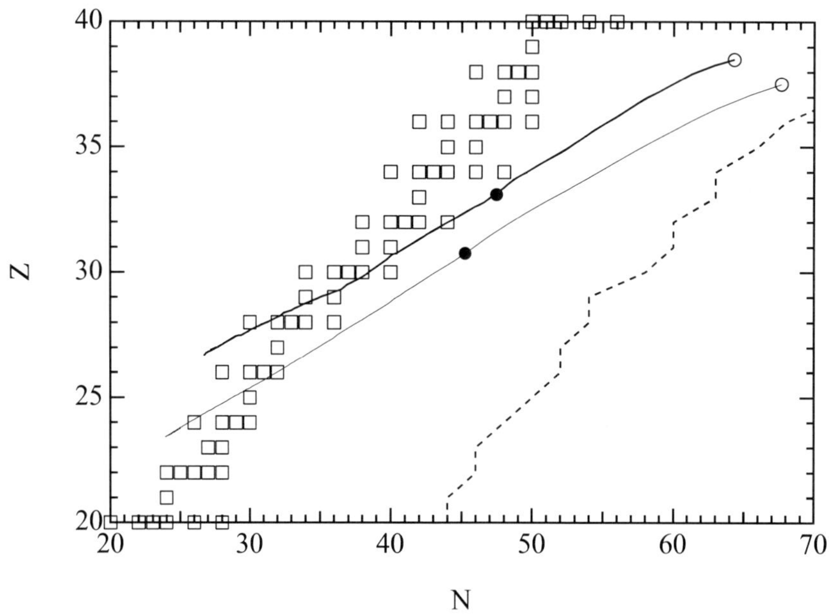

Figure 6. The time evolutions of the representative nuclei for the Lattimer & Swesty's [22] (thin line) and Shen's [21] (thick line) EOS's during the collapsing phase in the nuclear chart. Stable nuclei and the neutron drip line are shown by the open square symbols and dashed line, respectively. The nuclear species at the central densities of $\rho_c = 10^{11}$ and 10^{12}g/cm^3 are marked by the solid and open circles, respectively [15].

The nuclear species appearing in the central core during collapse are shown in the nuclear chart (Fig. 6). The nuclei in the Shen's EOS are systematically less neutron-rich than those in the Lattimer & Swesty's EOS, the latter having several more neutrons for the same proton numbers. This is again due to the effect of symmetry energy, which puts nuclei closer to the stability line in the former EOS. Note that the mass number reaches ~ 80 and ~ 100 when the central density becomes $\rho = 10^{11}$g/cm^3 and 10^{12}g/cm^3, respectively.

The profiles of lepton fractions at bounce are shown in Fig. 7. The central electron fraction in the Shen's EOS is $Y_e = 0.31$, which is slightly higher than $Y_e = 0.29$ in the Lattimer & Swesty's EOS. The central lepton fractions including neutrinos for the two EOS's are rather close to each other, being $Y_L = 0.36$ and 0.35, respectively. The difference of lepton fraction results in a different size of the inner core. The larger lepton fraction in the Shen's EOS leads to a larger inner core of $0.61 M_\odot$, whereas it is $0.55 M_\odot$ in the Lattimer & Swesty's EOS. The central density reaches 3.4×10^{14} and 4.4×10^{14}g/cm^3 in the Shen's and Lattimer & Swesty's EOS's, respectively. The difference in the maximum density reflects mainly those in the stiffness of matter at supra-nuclear densities in the two EOS's, which will be discussed in the next section. The initial shock energy, which is roughly estimated by the gravitational binding energy of the inner core at bounce, turns out to be not drastically different because the lower density in the Shen's EOS is compensated for by the larger mass.

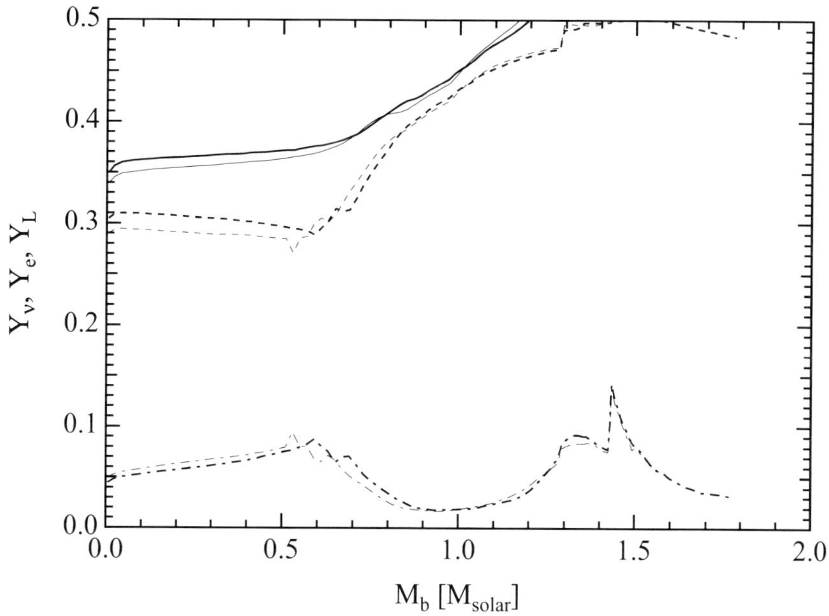

Figure 7. The lepton fractions at bounce as a function of baryon mass coordinate for the Lattimer & Swesty's [22] (thin lines) and Shen's [21] (thick lines) EOS's. The solid, dashed and dot-dashed lines show the lepton, electron and neutrino fractions, respectively. The progenitor is a $15M_\odot$ star [15].

In the past few years, we have seen a substantial improvement in the theoretical prediction of the rates of electron captures on heavy nuclei [5]. In the conventional simulations such as shown above the simple prescription based on the independent particle approximation has been adopted, in which the electron captures on nuclei with $N \geq 40$ are completely suppressed by the final state blocking [18]. Langanke and his collaborators have demonstrated, based on the shell model and random phase approximation, however, that the approximation is too crude [36]. In fact, the strength distributions of Gamow-Teller transitions were found to be systematically different from those predicted in the independent particle approximation, the fact which leads to the reduction of electron capture rates at relatively low densities, where nuclei with $A \lesssim 65$ are dominant. At higher densities, on the other hand, it was found that finite temperatures and residual interactions between orbits reduce the Fermi-blocking for the final states and the electron captures on nuclei are enhanced as a result, leading to lower values of Y_e when the neutrino trapping is completed. Another important point is that the greatest contributor to the electron captures is neither the most abundant nucleus nor the one with the highest capture rate but the one with the largest product of abundance and capture rate. This was demonstrated by Hix et al. [37], who combined the updated electron capture rates for various nuclei with the abundance obtained in the simple NSE EOS. They found that the inner core mass is decreased by $\sim 0.1M_\odot$, which is not a small amount because the shock wave consumes $\sim 10^{51}$erg as it passes through every $0.1M_\odot$ of the outer core.

It should be stressed that the electron fraction is determined not only by the capture rate of each nucleus but also by the abundance of nuclei as well as by the matter opacity for neutrinos and that the latter two are provided by EOS. Hence it should be now easy to understand that one of the possible extensions of EOS at sub-nuclear densities will be to accommodate both the abundance and structural change of various nuclei. In the rest of this section is devoted to this issue.

2.3. Multi-nuclear-species EOS

Hempel et al. [38] was the first to take this step (see also [39]). They modified the ordinary NSE to allow for the fact that nuclei are embedded in the electron sea and surrounded by other nuclei and nucleons. They first adopted experimental mass data [40] of nuclei with $Z \lesssim 100$. They then made a correction to the Coulomb energy of nuclei by employing the W-S cell as usual. Moreover, the nucleons residing in the vapor outside nuclei were described by the relativistic mean field theory (see Section 3) in the same way as uniform nuclear matter at supra-nuclear density. Hence the interactions among nucleons are included and the smooth transition from sub-nuclear to supra-nuclear densities is ensured. On the other hand, possible modifications for the bulk and surface energies in the mass of nuclei as well as the nuclear pasta phase were ignored in their treatment. Furusawa et al. [6] took these issues into consideration, albeit crudely. Their formulation was based on the conventional mass formula and the experimental nuclear mass data were accommodated in this framework. The prescription made it possible to include nuclei with the atomic number $\gtrsim 100$, for which there are no experimental data available but are abundant in the collapsing core. The so-called bubble phase was explicitly accounted for in the evaluation of the Coulomb and surface energies and other pastas were approximated by an interpolation of the ordinary and bubble phases. The nucleons dripped out of nuclei were described by RMF just like uniform nuclear matter at supra-nuclear densities. The bulk energy of nuclei is modified from the one extracted from the experimental data so that it should approach the free energy per nucleon of uniform nuclear matter as the density rises. The surface energy was reduced according to the nucleon-density contrast between inside and outside nuclei. These measures, although phenomenological, guarantee a more smooth transition from sub-nuclear to supra-nuclear densities.

The lower panels of Fig. 3 show the abundances of nuclei at two densities on the adiabat with $s = 1k_B$ in the nuclear chart. For comparison, the representative nuclei obtained by the Shen's EOS for the same thermodynamic conditions are marked by $\times$ symbols. It is evident that the most abundant nuclei obtained in the multi-species calculations do not agree with the representative nuclei. This is mainly due to the fact that the shell effect is not included in the Shen's EOS. It is also clear from the lower right panel of the figure that nuclei with $Z \gtrsim 100$ are abundant at this high density. This is in sharp contrast to the simple NSE EOS given in the lower left panel, where lighter nuclei become abundant.

In Fig. 8, the average mass number squared, $\overline{A^2}$, is compared with the square of average mass number, $\overline{A}^2$. As mentioned earlier, the coherent scattering rate is proportional to the former. In SNA employed in the currently standard EOS's, however, the former is replaced by the latter with $\overline{A}$ being the mass number of the representative nucleus. The validity of the approximation should be examined. It is found from the figure that the two values are

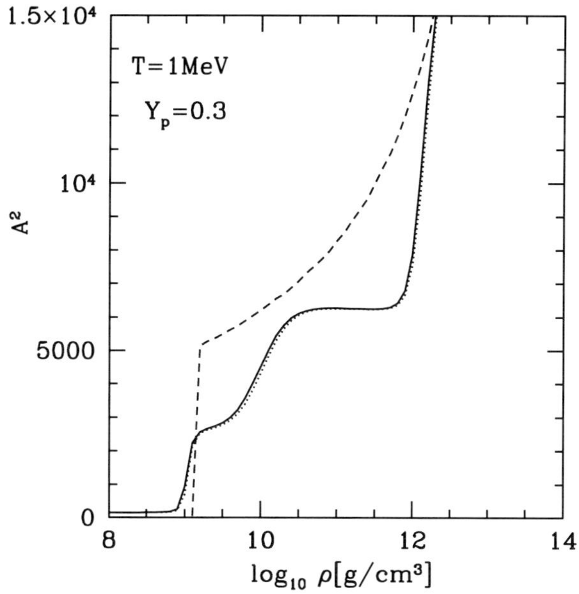

Figure 8. The comparison of the average mass number squared (solid line), $\overline{A^2}$, the square of average mass number (dotted line), $\overline{A}^2$ [6]. For comparison, the mass number squared for the representative nuclei in the Shen's EOS [21] is also shown by the dashed line.

almost coincident with each other. This is again due to the shell effect and nuclei tend to be populated in the vicinity of the neutron magic numbers. In this sense, the use of $\overline{A}^2$ in the evaluation of the coherent scattering rate is justified. It is noted, however, that the average mass number, $\overline{A}$, is systematically shifted from that of the representative nucleus in the EOS with SNA. What ramification this may have in core-collapse simulations is remaining to be investigated.

The difference in thermodynamical quantities such as pressure and entropy is rather minor in spite of the fact that the abundance of nuclei is considerably different as mentioned just now. This is partially because the multi-species EOS by Furusawa et al. [6] is designed so that it would make a smooth transition to the Shen's EOS at the point where the latter predicts the emergence of uniform nuclear matter; at low densities, on the other hand, it just reproduces the ordinary NSE, which the Shen's EOS represents approximately as well. The small remaining deviation is caused by SNA and the neglect of the pasta phase in the Shen's EOS. The difference in the abundance of light nuclei with $Z \leq 5$ is the source of deviation in the free energy and entropy per baryon between the Furusawa's and Shen's EOS's. In fact, only alpha particles are taken into account as the light nuclei in the Shen's EOS. The Furusawa's EOS predicts substantial populations of other light nuclei such as deuteron. These extra degrees of freedom produce the difference found in the free energy and entropy per baryon.

Although the extension of EOS at sub-nuclear densities described above is an important step forward, further sophistication is undoubtedly necessary. Among other things, structures of nuclei should be solved with the influences of other nuclei, nucleons in vapor and electrons being properly taken into account. This will not be an easy task, since the environment of nuclei, that is, the abundance of nuclei and number densities of nucleons and electrons, is itself dependent on the structures of nuclei. Hence the environment and nuclear structures must be determined consistently and simultaneously. As discussed above, since the abundance of nuclei is sensitive to their shell structures, quantum treatments are desirable. Moreover, an appropriate handling of the pasta phase will be an important ingredient to ensure the smooth transition to uniform nuclear matter.

3. EOS at Supra-nuclear Densities

The EOS's at sub-nuclear densities that have been discussed so far and are meant particularly for the collapsing phase before core bounce are also important in the sense that they set the physical condition for the later evolution. As seen in the previous section, the difficulty in the calculation of them mainly lies in the treatment of inhomogeneous matter, that is, nuclei and they depend mainly on the properties of nuclear matter around the nuclear saturation density, which are, from nuclear physics point of view, relatively well known experimentally at least for almost symmetric matter. On the other hand, as described in Introduction, matter near the core center has supra-nuclear densities after core bounce. If the shock wave produced by bounce penetrates the core somehow, the so-called protoneutron star , which is hot (a few tens MeV) and proton-rich ($Y_p \sim 0.3$), is left behind. As it cools down to the ordinary neutron star, emitting neutrinos copiously, the density rises up to $2 \sim 3$ times the nuclear saturation density. If the shock fails to escape the core, matter continues to accrete onto the protoneutron star and a black hole is formed eventually. In this case, even higher densities (and temperatures) will be realized before matter is swallowed by the black hole.

In this late phase of core collapse, the dynamics as well as cooling is influenced by the properties of hadronic matter at supra-nuclear densities, where, as a matter of fact, hadronic pressures dominate over leptonic ones. EOS in this density regime is required to deal with strong interactions between the constituent particles, and both the description of the interactions and the statistical mechanics calculation of free energy are major challenges although matter is homogeneous. It is also interesting from hadron physics point of view that constituent particles may not be nucleons alone at these high densities. The possibilities of various exotic phases have been investigated by many authors over the years ([10] and references therein). To mention some representative ones, mesons like pions and kaons may appear as a Bose-Einstein condensate; strangeness may emerge as hyperons: quarks may be deconfined with substantial population of strange quarks. EOS at supra-nuclear densities is required to address these issues.

From the point of view of numerical simulations of massive star collapse, the pressure exerted by the hadronic matter is the main concern. The black hole formation, for example, is usually triggered by the collapse of the protoneutron star when its mass exceeds the maximum mass, which is sensitive to the stiffness of hadronic matter at very high densities. The vertical border line near the highest density in Fig. 1 corresponds to this situation. An interesting thing to note here is that gravity is so strong that the Newtonian description is not allowed and general relativity plays a critical role in setting the maximum mass. In fact,

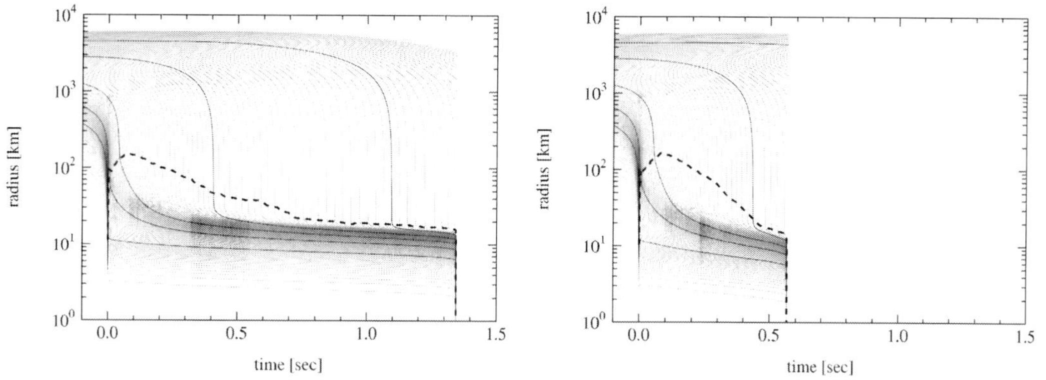

Figure 9. Radial trajectories of mass elements in the core of a $40M_\odot$ star as a function of time after bounce for the Shen's [21] (left) and Lattimer & Swesty's [22] (right) EOS's. The locations of shock wave are shown by thick dashed lines [41]

the maximum mass corresponds to the onset of instability to radial perturbations. In the Newtonian dynamics, as discussed in Introduction, the critical adiabatic index is $4/3$. This is no longer true of general relativistic configurations. As a matter of fact, the gravitational pull is stronger in general relativity and larger pressure is needed to sustain a hydrostatic equilibrium. As a result, the critical adiabatic index becomes larger than $4/3$. Roughly speaking, the collapse of the protoneutron star is caused not by softening of EOS but by strengthened gravity.

Regardless of the cause, the maximum mass exists for the protoneutron star and the important thing is that it is different from the one for the cold neutron star in neutrino-less β-equilibrium owing mainly to the finite temperature (a few tens of MeV) of the protoneutron star. In Fig. 9, we show, as an example, numerical simulations of the second collapse of protoneutron star [41]. The lines in the figure give the trajectories of representative mass elements in the massive star core. The left and right panels correspond to the results obtained with the Shen's [21] and Lattimer & Swesty's [22] EOS's, respectively. Their vertical turns near the end of simulations represent the second collapse to the black hole. The gravitational (baryonic) masses of the protoneutron stars just prior to this collapse are $2.4\,(2.7)$ and $2.0\,(2.1)M_\odot$ for the Shen's and Lattimer & Swesty's EOS's, respectively. The maximum gravitational masses of cold neutron stars, on the other hand, are 2.2 and $1.8M_\odot$ for the same EOS's. This is the finite temperature effect. Note also that the matter stiffness depends on the proton fraction. The example is a clear demonstration that it is almost mandatory for EOS at supra-nuclear densities meant for the use in core collapse simulations to take a proper account of the finite temperatures and the proton fractions that are neither 0.5 (symmetric matter) nor 0 (pure neutron matter).

This is one of the main reasons why we do not have many EOS's available for core-collapse simulations at present compared with the choices for the cold neutron star. In fact, as mentioned several times so far, only two EOS's are adopted these days, that is, the Shen's EOS [21] and the Lattimer & Swesty's EOS [22]. The former is based on the so-called relativistic mean field theory (RMF), in which nuclear interactions are described by the exchange of mesons. The Lagrangian employed in the Shen's EOS is the following:

$$\begin{aligned}\mathcal{L}_{RMF} &= \bar{\psi}\left[i\gamma_\mu\partial^\mu - M - g_\sigma\sigma - g_\omega\gamma_\mu\omega^\mu - g_\rho\gamma_\mu\tau_a\rho^{a\mu}\right]\psi \\ &+\frac{1}{2}\partial_\mu\sigma\partial^\mu\sigma - \frac{1}{2}m_\sigma^2\sigma^2 - \frac{1}{3}g_2\sigma^3 - \frac{1}{4}g_3\sigma^4 \\ &-\frac{1}{4}W_{\mu\nu}W^{\mu\nu} + \frac{1}{2}m_\omega^2\omega_\mu\omega^\mu + \frac{1}{4}c_3\left(\omega_\mu\omega^\mu\right)^2 \\ &-\frac{1}{4}R^a_{\mu\nu}R^{a\mu\nu} + \frac{1}{2}m_\rho^2\rho^a_\mu\rho^{a\mu}. \end{aligned} \tag{10.7}$$

In the above expression, the notation is the same as in [21]: ψ, σ, ω and ρ denote nucleons (proton and neutron), scalar-isoscalar meson, vector-isoscalar meson and vector-isovector meson, respectively, and $W_{\mu\nu} = \partial^\mu\omega^\nu - \partial^\nu\omega^\mu$ and $R^a_{\mu\nu} = \partial^\mu\rho^{a\nu} - \partial^\nu\rho^{a\mu} + g_\rho\epsilon^{abc}\rho^{b\mu}\rho^{c\nu}$. The nucleon-meson interactions are given by the Yukawa couplings and the isoscalar mesons (σ and ω) interact with themselves, which are expressed as the cubic and quartic terms. These self-interactions are tuned to reproduce more fundamental Dirac-Brückner-Hartree-Fock theory [42]. In the mean field theory, the mesons are assumed to be classical and replaced by their ensemble averages whereas the Dirac equation for nucleons is quantized and the free energy is evaluated based on the energy spectrum of nucleons obtained this way. M is the mass of nucleons and assumed to be 938MeV. Note that the effective mass of nucleon is obtained consistently from the expectation value of the σ meson in this theory. The masses of mesons, m_σ, m_ω, m_ρ, and the coupling constants, g_σ, g_ω, g_ρ, g_2, g_3, c_3, are determined so that not only the saturation of uniform nuclear matter but also the properties of finite nuclei could be best reproduced [43].

The Lattimer & Swesty's EOS [22] employs a free energy per baryon with the Skyrme-type parametrization :

$$f(n, Y_p, T) = E(n, Y_p, T)/n - Ts(n, Y_p, T), \tag{10.8}$$

where the internal energy density and the entropy per baryon are given respectively as

$$E(n, Y_p, T) = \sum_t \frac{\hbar^2\tau_t}{2m_t^*} + \left[a + 4b\,Y_p(1 - Y_p)\right]n^2 + c\,n^{1+\delta} - Y_p n\,\Delta, \tag{10.9}$$

$$s(n, Y_p, T) = \sum_t \left(\frac{5\,\hbar^2\tau_t}{6\,m_t^*\,T} - n_t\,\eta_t\right)/\,n. \tag{10.10}$$

In these expressions, the subscript t stands for the isospin of nucleon, τ_t is the kinetic energy density, m_t^* is the effective mass, which is fixed to the nucleon mass in the actual implementation, and Δ is the neutron-proton mass difference. The degeneracy parameter, η_t, is given by the chemical potential of nucleon, μ_t, and the potential, $V_t \equiv \delta E/\delta n_t$, as $\eta_t = (\mu_t - V_t)/k_B T$. The other parameters in this expression, a, b, c, δ, are determined by the properties of zero-temperature symmetric nuclear matter at its saturation density, namely the binding energy, bulk symmetry energy and bulk incompressibility parameter. The free energy is hence constructed to reproduce the thermodynamic properties of bulk nuclear matter accurately for any Skyrme-type interaction. Reflecting uncertainties in these properties themselves, however, the authors provide three sets of EOS with different incompressibilities, $K = 180, 220, 375$MeV.

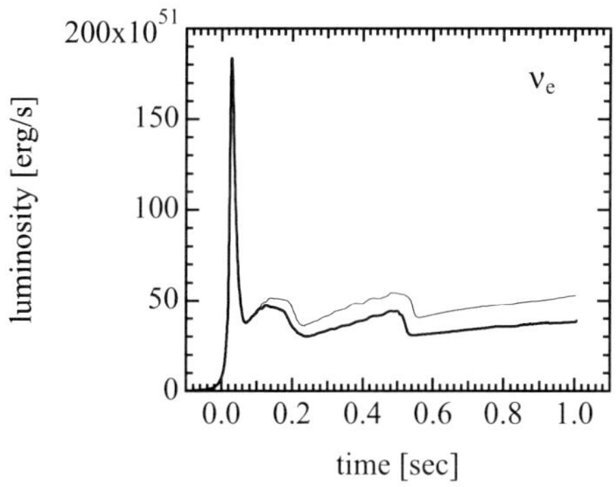

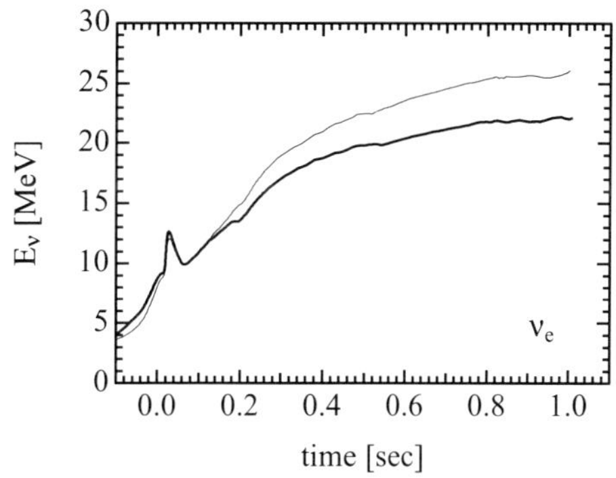

Figure 10. The luminosity (left) and average energy (right) of electron-type neutrino as a function of time for the Shen's [21] (thick line) and Lattimer & Swesty's [22] (thin line) EOS's. The progenitor is a $15M_\odot$ star [15].

The two sets of supernova EOS are different from each other, reflecting the characteristics of their nuclear theoretical frameworks and the inputs of nuclear data used to constrain the EOS. The RMF EOS tends to be stiff and the density dependence of symmetry energy is strong (see, e.g., Sumiyoshi et al. 1995b): the incompressibility and symmetry energy of the Shen's EOS are 281 and 36.9MeV, respectively. The Lattimer & Swesty's EOS with the incompressibility $K = 180\text{MeV}$ among three choices has been frequently used in core-collapse simulations so far. The symmetry energy of this EOS is 29.3MeV. The maximum gravitational masses of cold neutron stars are 1.8 and $2.2M_\odot$ for the Lattimer & Swesty's and Shen's EOS's, respectively. Recent discovery of $1.97M_\odot$ neutron star suggests that the incompressibility $K = 180\text{MeV}$ in the Lattimer & Swesty's EOS is too small. Although it is unavailable publicly and not explained in detail above, the Wolff & Hillebrandt's EOS [33], which is based on the Hartree-Fock approximation with a Skyrme force for the nucleon-nucleon interaction with parameters given by Köhler [34], is close to but somewhat stiffer than the Shen's EOS and has the incompressibility of $K = 263\text{MeV}$ and the symmetry energy of 32.9MeV.

We now turn to the influences that the differences in EOS have on core collapse simulations. Shown in Fig. 10 are the luminosity and average energy of electron-type neutrinos as a function of time, which are obtained in the spherically symmetric simulation of a $15M_\odot$ progenitor with the Shen's and Lattimer & Swesty's EOS's [15]. The differences between the two EOS's are not very large particularly in the early phase ($t \lesssim 300\text{ms}$) but increase as the time passes. Since the Lattimer & Swesty's EOS with $K = 180\text{MeV}$ adopted in this computation is softer than the Shen's EOS, the protoneutron star is more compact for the former EOS. As a result, the luminosity and average energy of neutrinos are higher, since larger amount of gravitational energy is liberated and matter temperature in the protoneutron star is higher owing to the stronger compression. The average shock radius is found to be smaller for the Lattimer & Swesty's EOS. It is noted that this spherical model failed to produce an explosion. In fact, it is now a common recognition that spherically symmetric models are unlikely to give an energy sufficient for shock revival.

As mentioned earlier, it is a current consensus of the supernova society that the core dynamics is intrinsically non-spherical after core bounce. This is due to hydrodynamical instabilities inherent to the post-shock flow. The so-called gain region, where the neutrino heating dominates over the cooling, has a negative entropy gradient and is hence unstable to convection. It has been recognized over the last few years that the stagnated shock wave is also unstable to non-spherical deformation, which is referred to as the standing accretion shock instability or SASI. The negative gradient of the lepton fraction is another driving mechanism of convection in the protoneutron star. These instabilities tend to enhance the neutrino heating and influences the shock position and its eventual revival. Although the features observed above for the spherically symmetric model are expected to be robust qualitatively, multi-dimensional simulations should be consulted for more quantitative comparisons. Marek & Janka [24], for example, made a comparison of the Wolff & Hillebrandt's and Lattimer & Swesty's EOS with $K = 180$MeV in 2-dimensional, axisymmetric simulations for a couple of progenitor models. Unfortunately, the result for the Shen's EOS has not been published yet. As already mentioned, however, the Wolff & Hillebrandt's EOS is rather similar to the Shen's EOS. It was found also in this study that the neutrino luminosity is higher for the softer Lattimer & Swesty's EOS. Accordingly the neutrino heating rate is higher. Interestingly, the average shock radius is almost the same because of the greater heating and more vigorous non-spherical motions in the Lattimer & Swesty's EOS. Note that a shock revival was obtained for the Lattimer & Swesty's EOS in these simulations although the explosion energy is still very small $\lesssim 10^{50}$erg at the end of simulations. Whether this is the final answer to the long-standing supernova puzzle is remaining to be investigated further. It is evident, however, that EOS has a critically important influence on the success of shock revival.

The failed model of spherically symmetric supernova presented earlier in this section demonstrated that the difference in EOS becomes more apparent in the later phase (see Fig. 10). It is true that the results cannot be taken at face value, since we expect that the shock will revive at some point and the accretion will not continue until such late times. But a natural question arises: what about a black hole formation, in which the accretion onto the protoneutron star continues up until the second collapse occurs? In such an event, the differences in dynamics originated from the uncertainties in EOS may manifest themselves ever more clearly as the time passes. The problem, however, is that we know little of the progenitors. One might think naively that very massive stars will produce black holes anyway. This is not true unfortunately. Mass losses of massive stars are an increasing function of stellar mass in general and the core mass just prior to collapse will not be a monotonic function of stellar mass [25, 26, 27]. In one of the recent supernova surveys researchers claim that there is no compelling evidence that black holes are produced by some mass-range of massive stars [30]. There are also other observations that suggest an explosion and a subsequent formation of neutron star instead of black hole from very massive stars of $\gtrsim 40 - 50\mathrm{M}_\odot$ [44]. Bearing these caveats in mind, we will discuss in the following the putative black hole formation from very massive stars, since there is a possibility that such an event will provide us with invaluable information on EOS at very high densities ($\rho \gtrsim 3\rho_s$), which would not be probed otherwise.

In Fig. 9 the trajectories of mass elements in the core of non-rotating $40\mathrm{M}_\odot$ star are shown for the Shen's EOS and Lattimer & Swesty's EOS with $K = 180$MeV. We see no hint of outward motions. Although the failure of explosion may be due to unknown intrica-

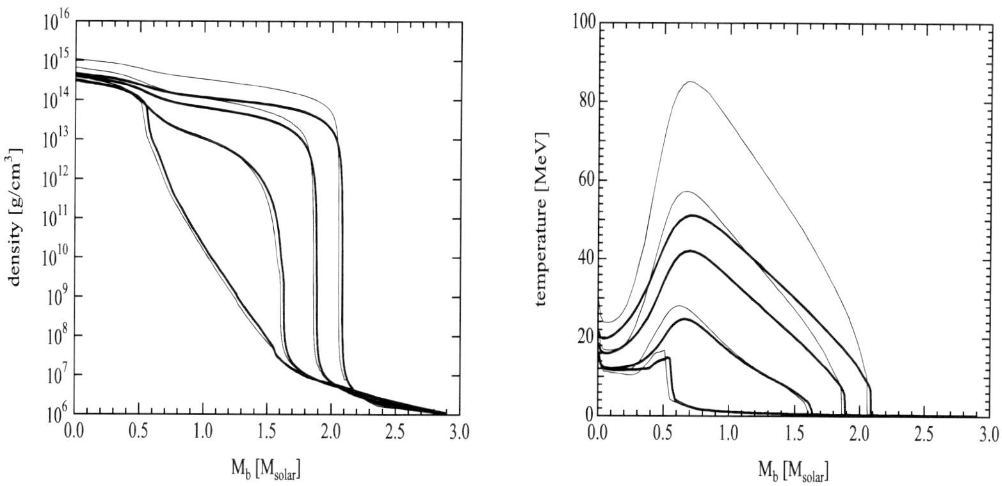

Figure 11. The density (left) and temperature (right) profiles in the protoneutron stars for the Shen's [21] (thick lines) and Lattimer & Swesty's [22] (thick lines) EOS's. The progenitor is a $40M_\odot$ star [41].

cies as in supernova simulations, we assume here that it is real: such a massive star indeed fails to revive the stalled shock wave and does not expel a stellar envelope and that the ever lasting accretion leads eventually to a second collapse of protoneutron star and a subsequent black hole formation as in the figure. The shock waves shown by the dashed lines start to recede at $t_b \sim 100$ms, where t_b denotes the time after core bounce, and hover just above the protoneutron star. The inner trajectories bend downward abruptly at $t_b \sim 1.3$s and ~ 0.6s for the Shen's and Lattimer & Swesty's EOS's, respectively, which marks the second collapse that is immediately followed by the formation of an event horizon or a black hole. The remarkable thing is that the difference in the times of black hole formation is quite large, ~ 0.7s. The second collapse or the black hole formation occurs when the protoneutron star acquires a mass that exceeds the maximum value, which is crucially determined by EOS. Since the accretion rate dictates the time evolution of protoneutron star mass, the time lag for the black hole formation reflects the difference in the maximum mass for the two EOS's. In fact, as mentioned earlier, the conserved baryonic mass of the protoneutron star just prior to the second collapse is $2.7M_\odot$ for the stiffer Shen's EOS whereas it is $2.1M_\odot$ for the softer Lattimer & Swesty's EOS.

The differences of the two EOS's are also reflected in the profiles of density and temperature in the protoneutron star. As shown in the left panel of Fig. 11, the central density is consistently higher for the Lattimer & Swesty's EOS, which in turn leads to higher temperatures for the same EOS. The central density is $\sim 10^{15}$g/cm^3 at the time of the second collapse and reaches $\sim 10^{16}$g/cm^3 at the time of black hole formation. The temperature increases beyond 100MeV during the recollapse and reaches ~ 200MeV at the peak position, which is off center, for the Lattimer & Swesty's EOS as shown in the right panel of Fig. 11. Although the corresponding values for the Shen's EOS are a bit lower, it is obvious that the density and temperature regimes realized in the protoneutron star just prior to the black hole formation are those that would not be accessible otherwise.

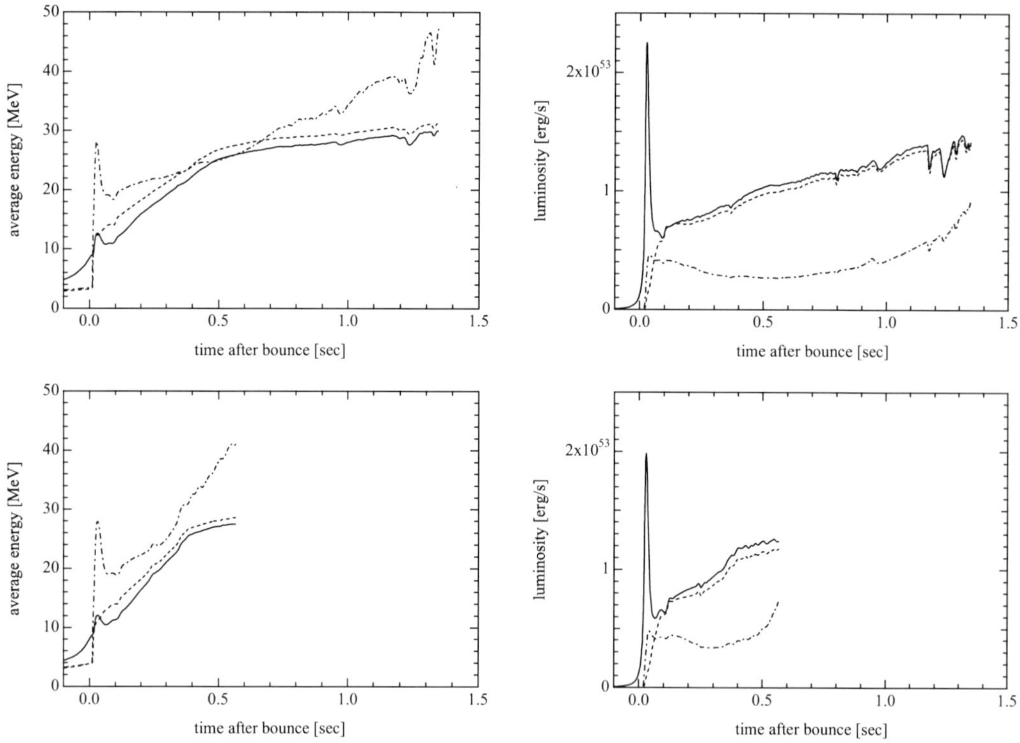

Figure 12. The average energies (left panels) and luminosities (right panels) of ν_e (solid lines), $\overline{\nu}_e$ (dashed lines) and $\nu_{\mu/\tau}$ (dot-dashed lines) for the Shen's[21] (upper panels) and Lattimer & Swesty's [22] (lower panels) EOS's. The progenitor is a $40 M_\odot$ star [41].

If the event really occurs, we cannot expect an optical brightening unlike ordinary supernovae and only neutrinos will be carriers of information on what is going on deep inside the star. It is hence important to investigate the neutrino signals from the black-hole-forming collapse and explore the influence of EOS on them. In Fig. 12 the average energies and luminosities of emitted neutrinos are shown as a function of time for the two EOS's. The durations of neutrino emission directly reflect the times up to the black hole formation discussed above and differ significantly between the two EOS's. It is apparent that the neutrino energies increase monotonically (except for the initial peaks just after bounce) because of the rapid temperature increases. This behavior is different from what one expects for ordinary supernovae and can be used to identify the black hole formation. The luminosities after the neutronization burst also increase toward the black hole formation. This is mainly due to the enhancement by matter accretion especially for ν_e and $\overline{\nu}_e$. Since the $\mu-$ and τ neutrinos and anti-neutrinos are produced and emitted deeper inside and less affected by the hyper accretion pertinent to the event. As a result, their luminosities are much smaller than those of ν_e and $\overline{\nu}_e$, which is another interesting feature of this phenomenon.

Since the duration of neutrino emissions probes the matter stiffness at very high densities, EOS's that give a similar stiffness cannot be distinguished by this information alone. In fact, the Lattimer & Swesty's EOS with $K = 220$MeV gives a second-collapse time of ~ 0.7s that is in between those for the Lattimer & Swesty's EOS and for the Shen's

EOS [45]; as shown later, an RMF EOS incorporating hyperons [46] also gives a similar time of second collapse [47]. A possible variety of progenitors is another concern; the combination of a soft EOS and a light progenitor may give a second-collapse time that is similar to the one given by the opposite combination. In fact, O'Connor et al. [48] demonstrated that there is indeed such degeneracy. Therefore, more detailed analyses are required to distinguish these EOS's. For example, employing the time evolutions of neutrino luminosity, Nakazato et al. [45] performed a Kolmogolov-Smirnov test to demonstrate that the hyperon EOS can be distinguished from the Lattimer & Swesty's EOS for a galactic event. In their analysis, the neutrino oscillations are fully taken into account. This kind of attempts should be further pursued for a larger collection of EOS's. It is also important to quantify the effects of multi-dimensionality of dynamics, such as rotation, hydrodynamical instabilities and magnetism, on the neutrino signals.

It should be stressed that EOS's that incorporate nucleons alone as hadronic degrees of freedom will be inappropriate at the density and temperature regime that we have just observed. Indeed, at such high densities, where the inter-particle distance is comparable to or even smaller than the size of nucleons, $\sim$ 1fm, it is very unlikely that they maintain their identity. It is also interesting that the maximum temperature achieved in the protoneutron star prior to the black hole formation, which is much higher than $\sim$ 100MeV, may be sufficient for quark deconfinement [11]. Hence it makes sense to attempt to access the possible effects of new degrees of freedom on the dynamics of the black hole formation as well as the neutrino signals that we will observe from such an event.

Fig. 13 displays the results of such an attempt by Nakazato et al. [49], in which the emergence of deconfined quark matter is taken into account. The EOS for the quark phase is based on the so-called MIT bag model, in which the up, down and strange quarks are approximated as ideal Fermi gases in a vacuum that has an elevated energy represented by the bag constant, which is chosen to be 250MeV/fm^3. The EOS for the hadronic phase is the Shen's EOS and the two phases are connected by the Gibbs condition for the first order phase transition. As shown in the right panel, the quarks start to populate at $\rho \sim 10^{15}\text{g/cm}^3$. The second collapse soon follows because of the softening of the EOS by the phase transition as shown in the left panel of the figure. The time that elapses from bounce to the black hole formation falls in between those for the two nucleonic EOS's discussed so far. This depends on the choice of the bag constant, however, and the second collapse would have occurred much earlier if a smaller bag constant had been adopted. In fact, Sagert et al. [50] recently found a shock revival induced by the second collapse in their simulations of spherically symmetric collapse of 10.2 and $13\text{M}_\odot$ stars. The assumed bag constants were small, $B^{1/4} = 162$ and 165MeV with $\alpha_s = 0$ and $B^{1/4} = 155$MeV with $\alpha_s = 0.3$, where α_s is the coupling constant in QCD. As a result, quarks started to populate just after core bounce in their models. The second collapse that occurred a few hundreds millisecond after the first bounce was again bounced, since the mass of "protohybrid star" had not reached the maximum mass yet. The shock wave was produced, propagated outwards and eventually merged with the stalled first shock, pushing further ahead together. Although this is an interesting suggestion, the results cannot be taken at face value, since the parameter they chose gives the maximum mass of quark star that is much smaller than $2\text{M}_\odot$.

It is another possibility that strangeness emerges in bound states such as hyperons. The consequence of larger hadronic degrees of freedom is again a softening of EOS, which will

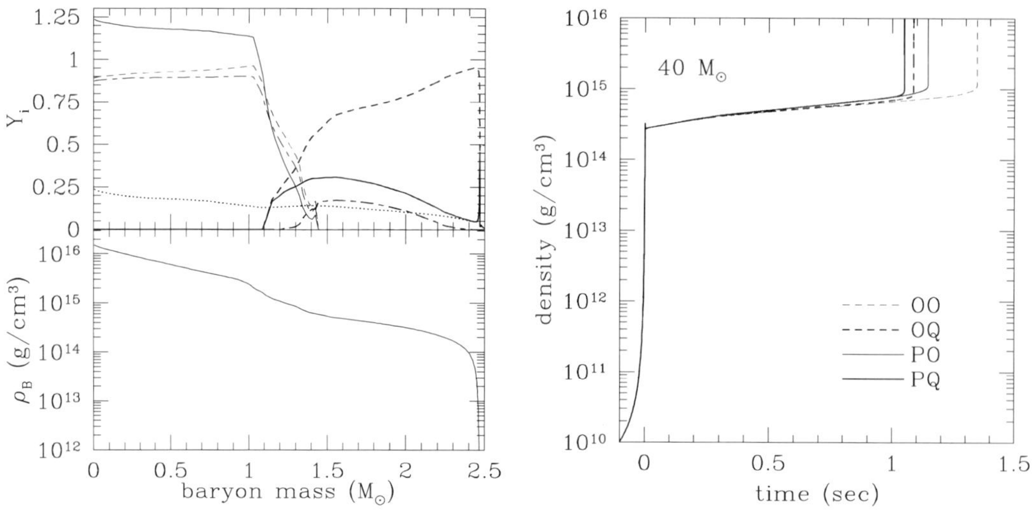

Figure 13. The abundance of constituent particles at the moment of black hole formation (left) and the trajectories of mass elements in the core-collapse of $40M_\odot$ star [49]. In the upper-left panel, the thick solid and dashed lines show the abundance of proton and neutron, respectively, whereas the dotted and dash-dotted lines denote those of electron and thermal pion, respectively. The thin solid, dashed and dash-dotted lines represent up, down and strange quarks, respectively. In the right panel, 00, 0Q, P0, PQ stand for the cases with only nucleon, with the phase transition to quark matter, with thermal pion and with the phase transition and pion, respectively.

then produce an earlier second collapse in the black-hole-forming collapse. Sumiyoshi et al. [47] investigated such an event in detail, employing a hyperonic EOS that Ishizuka et al. [46] constructed. It is an extension of the Shen's EOS to include hyperons. The hadronic interactions are described by mesons, which are also extended to the scalar and vector mesons, ζ and ϕ, with hidden strangeness ($\bar{s}s$). Although the most up-to-dated knowledge on the coupling constants of hyperon-related interactions is incorporated in this EOS, it is still unsatisfactory. In fact, the maximum mass of cold "hyperon star" is $1.8M_\odot$ in this EOS, which is too small as mentioned above. Hence the results shown here cannot be taken at face value, either. They are still useful, however, to infer the effects of hyperonic degrees of freedom on the black hole formation.

The dynamics is almost identical to the one for the Shen's EOS until hyperons appear. As expected, the second collapse occurs much earlier than for the Shen's EOS and the time of black hole formation is rather close to that for the Lattimer & Swesty's EOS with $K = 220$MeV as shown in Fig. 14. The degeneracy makes it difficult to distinguish them from the duration of neutrino emissions alone. The evolutions of luminosities and average energies of neutrinos may help us in this respect, though. In fact, Nakazato et al. [45] performed Kolmogolov-Smirnov tests to see if the neutrino light curves for the Lattimer & Swesty's and hyperonic EOS's can be distinguished and concluded that they are indeed distinguishable for a Galactic event. This sort of analysis should be extended further to other degenerate cases.

Last but not least, Meson condensations may be the true trigger of black hole forma-

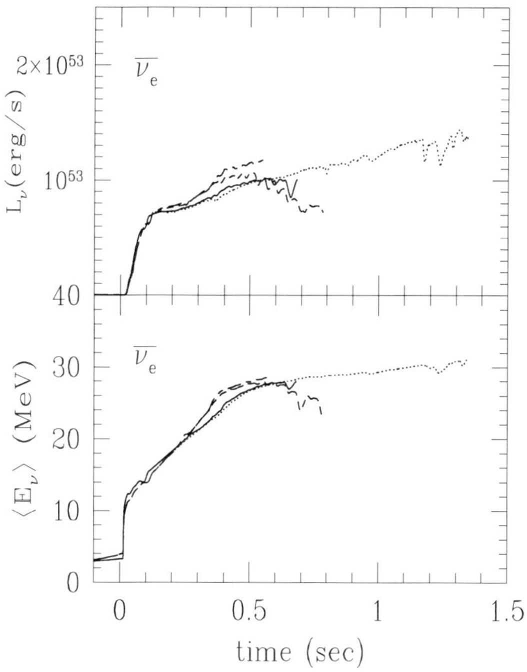

Figure 14. The luminosities (upper panel) and average energies (lower panel) of $\overline{\nu}_e$ in the core collapse of $40M_\odot$ star for four different EOS's, the Lattimer & Swesty's EOS with $K = 180$MeV (dash-dotted lines) and 220MeV [22] (dashed lines), the Shen's EOS [21] (dotted lines) and the hyperon EOS [46] (solid lines) [45].

tion. The possibilities of pion and/or Kaon condensations at high densities have been investigated over the years ([10] and references therein). Their possible effects on the protoneutron star cooling attracted much interest in the interpretation of the neutrino signals from SN1987A [51]. Although their consequences in the black-hole-forming collapse have not been studied yet because of no publicly available EOS, they are certainly interesting and important and warrant further investigation.

4. Conclusion

In this chapter we have elucidated the important roles that EOS plays at both sub- and supra-nuclear densities in the gravitational collapse of massive stars and the subsequent formation of compact stars. At sub-nuclear densities, nucleons are mostly assembled to nuclei and their abundances crucially affect the dynamics of collapse through weak interactions with neutrinos. The electron captures on nuclei as well as the coherent scatterings of neutrinos with nuclei set the electron fractions in the stellar core, which then determine the initial location and strength of the shock wave produced by core bounce. The main effort in constructing EOS at this density regime is devoted to the treatment of the spacial inhomogeneity or nuclei. In so doing, the effects of surrounding particles on the structure of nuclei have to be taken into account. The conventional approach is the single nucleus approximation, in which only a single representative heavy nucleus is considered except for alpha particles. By comparing one of the standard SNA EOS's with the newly developed NSE-based multi-species EOS, we have demonstrated in this chapter that the mass and atomic numbers of the most abundant nucleus may be systematically deviated from those for the representative nucleus predicted by the SNA EOS owing to the shell effect. What implication this may have for the dynamics of core-collapse is the issue to be studied next. Even in the multi-species EOS, the estimation of nuclear masses is phenomenological and certainly needs improvement. Ideally speaking, the treatment of the nuclear structure should be consistent with the calculations of the weak interaction rates. It should be also noted that EOS at sub-nuclear densities is intimately related with EOS at supra-nuclear densities.

At supra-nuclear densities, both the description of strong interactions between constituent particles and the many body calculation of the free energy are the major challenge. In the recent supernova simulations, the maximum density is at most twice the nuclear saturation density. Since the properties of nearly symmetric uniform nuclear matter are rather well constrained around the nuclear saturation density by laboratory experiments, the difference in dynamics that is induced by the uncertainties in EOS is not very remarkable. In general, the softer EOS is, the higher the density and temperatures become and, as a result, the higher luminosities and energies of neutrinos are realized, which are advantageous for successful supernova explosions in the neutrino-heating scenario, supposedly the most promising one at present. Although the supernova mechanism is still elusive, it is no doubt that EOS is one of the key ingredients in supernova theory. The EOS's that have been introduced in this chapter and most frequently used in the core collapse simulations are rather phenomenological. Although the non-vanishing temperatures and proton fractions different from 0 and 0.5 are the main obstacles, EOS's based on more fundamental descriptions of nuclear interactions and more sophisticated calculations of free energy are highly welcomed.

Since the differences in EOS manifest themselves more clearly at higher densities, it is no surprise that the black-hole-forming collapses, if they really occur, will be a better probe of the properties of dense matter. In fact, the recent numerical simulations have demonstrated that the duration of neutrino emissions alone provide us with invaluable information on the stiffness of matter at densities much higher than the nuclear saturation density and that there is a chance to resolve the degeneracy of the durations for EOS's with similar stiffnesses by more detailed analyses. The maximum density and temperature achieved in the stellar core prior to the black hole formation indicate clearly the necessity to consider the quark deconfinement and/or the emergences of hyperons and/or meson condensations. So far publicly available EOS's for these exotica are very limited and are phenomenological if they exist. More experimental data and better theoretical treatments are definitely needed.

As already mentioned repeatedly, the observationally estimated maximum mass of neutron star has been increased to $\sim 2M_{\odot}$ by the detection of Shapiro delay for the binary of radio pulsar PSR J1614-2230 with the white dwarf binary J1045-4509 [52]. Not suffering from uncertainties inherent to astrophysical modeling, the result seems highly reliable. And stringent as well. In fact, some of the soft EOS's that have been used commonly in core collapse simulations so far and are cited in this chapter do not meet the requirement on the maximum mass, and possible modifications have been already discussed in the literature. This is probably a good opportunity for us all to give a renewed consideration to EOS.

Acknowledgment

The author thanks K. Sumiyoshi, H. Suzuki, K. Nakazato, S. Furusawa, A. Ohnishi and C. Ishizuka for their collaborations on his core collapse investigations. He gratefully acknowledges the contributions of K. Sumiyoshi, K. Nakazato and S. Furusawa to the preparation of figures. K. Sumiyoshi and S. Furusawa kindly read the manuscript carefully and gave useful comments, which helped the author correct errors and improve descriptions. The author is solely responsible for any remaining mistakes. Some of the author's works presented in this chapter were partially supported by Grant-in-Aid for Scientific Research of the Ministry of Education, Culture, Sports, Science and Technology (17540267, 19104006, 21540281)

References

[1] D. D. Clayton, *Principles of Stellar Evolution and Nucleosynthesis*, Univ. Chicago Press, 1983.

[2] A. Heger, C. L. Fryer, S. E. Woosley, N. Langer and D. H. Hartmann, *Astrophys. J.* 591 (2003) 288-300.

[3] H. A. Bethe, *Rev. Mod. Phys.* 62 (1990) 801-866.

[4] K. Kotake, K. Sato and K. Takahashi, *Rep. Prog. Phys.* 69 (2006) 971-1143.

[5] H. -T. Janka, K. Langanke, A. Marek, G. Martinez-Pinedo and B. Müller, *Phys. Rep.* 442 (2007) 38-74.

[6] S. Furusawa, S. Yamada, K. Sumiyoshi and H. Suzuki, submitted to *Astrophys. J.* (2011), arXiv:1103.6129v1 [astro-ph.HE].

[7] D. G. Ravenhall, C. J. Pethick and J. R. Wilson, *Phys. Rev. Lett.* 50 (1983) 2066-2069.

[8] K. Oyamatsu, M. Hashimoto and M. Yamada, *Prog. Theor. Phys.* 72 (1984) 373-375.

[9] N. K. Glendenning, *Compact Stars: Nuclear Physics, Particle Physics and General Relativity*, Springer-Verlag, 2000.

[10] J. M. Lattimer and M. Prakash, *Science* 304 (2004) 536-542.

[11] F. Weber, R. Negreiros and P. Rosenfield, in *Neutron Stars and Pulsars*, Astrophysics and Space Science Library 357 (2009) 213-245.

[12] S. L. Shapiro and S. A. Teukolsky, *Black Holes, White Dwarfs, and Neutron Stars*, John Wiley & Sons Inc, 1983.

[13] S. E. Woosley, A. Heger and T. A. Weaver, *Rev. Mod. Phys. A* 74 (2002) 1015-1071.

[14] H. Umeda and K. Nomoto, *Astrophys. J.* 565 (2002) 385-404.

[15] K. Sumiyoshi, S. Yamada, H. Suzuki, H. Shen, S. Chiba and H. Toki, *Astrophys. J.* 629 (2005) 922-932.

[16] K. Nakazato, K. Sumiyoshi and S. Yamada, *Astrophys. J.* 645 (2006) 519-533.

[17] H. -Th. Janka, in the proceedings of ESO Astrophysics Symposia, *Lighthouses of the Universe: The Most Luminous Celestial Objects and Their Use for Cosmology*, 2002, Eds. Gilfanov, Sunyeav and Churazov, pp357-368.

[18] S. W. Bruenn, *Astrophys. J. Suppl.* 58 (1985) 771-841.

[19] A. Burrows, S. Reddy and T. A. Thompson, *Nucl. Phys.* A777 (2006) 356-394.

[20] K. Sato, Prog. Theor. Phys. 54 (1975) 1325-1338.

[21] H. Shen, H. Toki, K. Oyamatsu and K. Sumiyoshi, *Nucl. Phys.* A637 (1998) 435-450. See also arXiv:1105.1666v1 [astro-ph.HE] for recent update.

[22] J. M. Lattimer and F. D. Swesty, *Nucl. Phys.* A535 (1991) 331-376.

[23] A. Yahil, Astrophys. J. 265 (1983) 1047-1055.

[24] A. Marek and H. -Th. Janka, *astrophys. J.* 694 (2009) 664-696.

[25] R. Hirschi, G. Meynet and A. Maeder, *Astron. Astrophys.* 433 (2005) 1013-1022.

[26] M. Limongi, A. Chieffi, *Astrophys. J.* 647 (2006) 483-500.

[27] S. E. Woosley and A. Heger, *Astrophys. J.* 637 (2006) 914-921.

[28] S. J. Smartt, *Ann. Rev. Astron. Astrophys.* 47 (2009) 63-106.

[29] K. Nomoto, S. Wanajo, Y. Kamiya, N. Tominaga and H. Umeda, in the proceedings of IAU Symposium No. 254, *The Galaxy Disk in Cosmological Context*, 2009, Eds. Andersen, Bland-Hawthorn and Nordstrom, pp355-367.

[30] N. Smith, W. Li, A. V. Filippenko and R. Chornock, *Mon. Not. Roy. Astron. Soc.* 412 (2011) 1522-1538.

[31] D. Vautherin, in *Supernovae*, Les Houches, Session LIV, 1994, Eds. Bludman, Mochkovitch and Zinn-Justin, pp346-391.

[32] W. L. Slattery, G. D. Doolen, H. E. DeWitte, *Phys. Rev. A* 21 (1980) 2087-2095.

[33] W. Hillebrandt and R. G. Wolff, in *Nucleosynthesis: Challenges and New Developments*, 1985, Eds. Arnett & Truran, pp131-150.

[34] H. S. Köhler, *Nucl. Phys.* A258 (1976) 301-316.

[35] M. Liebendörfer, O. E. B. Messer, A. Mezzacappa, W. R. Hix, F. -K. Thielemann and K. Langanke, in the proceedins of *11th Workshop on Nuclear Astrophysics*, 2002, Eds. Hillebrandt & Müller, pp126-131.

[36] K. Langanke and G. Martinez-Pinedo, *Rev. Mod. Phys.* 75 (2003) 819-862.

[37] W. R. Hix, O. E. B. Messer, A. Mezzacappa, M. Liebendörfer, J. Sampaio, K. Langanke, D. J. Dean and G. Martinez-Pinedo, *Phys. Rev. Lett.* 91 (2003) 201102-1-4.

[38] M. Hempel and J. Schaffner-Bielich, *Nucl. Phys.* A837 (2010) 210-254.

[39] S. I. Blinnikov, I. V. Panov, M. A. Rudzsky and K. Sumiyoshi, (2009) arXiv:0904.3849v1 [astro-ph.HE].

[40] G. Audi, A. H. Wapstra and C. Thibault, *Nucl. Phys.* A729 (2003) 337-676.

[41] K. Sumiyoshi, S. Yamada and H. Suzuki, *Astrophys. J.* 667 (2007) 382-394.

[42] R. Brockmann and R. Machleidt, *Phys. Rev. C* 42 (1990) 1965-1980.

[43] Y. Sugahara and H. Toki, *Nucl. Phys.* A579 (1994) 557-572.

[44] D. F. Figer, F. Najarro, T. R. Geballe, R. D. Blum and R. P. Kudritzki, *Astrophys. J. Lett.* 620 (2005) L49-L52.

[45] K. Nakazato, K. Sumiyoshi, H. Suzuki and S. Yamada, *Phys. Rev. D* 81 (2010) 083009-1-6.

[46] C. Ishizuka, A. Ohnishi, K. Tsubakihara, K. Sumiyoshi and S. Yamada, *J. Phys. G* 35 (2008) 085201-1-19.

[47] K. Sumiyoshi, K. Nakazato, C. Ishizuka, A. Ohnishi, S. Yamada and H. Suzuki, *Nucl. Phys.* A835 (2010) 295-302.

[48] E. O'Connor and C. D. Ott, *Astrophys. J.* 730 (2011) article id. 70.

[49] K. Nakazato, K. Sumiyoshi and S. Yamada, Astrophys. J. 721 (2010) 1284-1294.

[50] I. Sagert, T. Fischer, M. Hempel, G. Pagliara, J. Schaffner-Bielich, A. Mezzacappa, F. -K. Thielemann and M. Liebendorfer, Phys. Rev. Lett. 102 081101-1-4.

[51] J. A. Pons, J. A. Miralles, M. Prakash and J. M. Lattimer, *Astrophys. J.* 553 (2001) 382-393.

[52] P. B. Demorest, T. Pennucci, S. M. Ransom, M. S. E. Roberts and J. W. T. Hessels, *Nature* 467 (2010) 1081-1083.

INDEX

A

B

C

D

E

F

G

H

I

O

P

Q

R

S

T

U

V

W

Y